Hands-On
Azure Digital Twins

A practical guide to building distributed IoT solutions

Alexander Meijers

Packt>

BIRMINGHAM—MUMBAI

Hands-On Azure Digital Twins

Copyright © 2022 Packt Publishing

Group Product Manager: Rahul Nair

Publishing Product Manager: Meeta Rajani

Senior Editor: Athikho Sapuni Rishana

Content Development Editor: Sayali Pingale

Technical Editor: Rajat Sharma

Copy Editor: Safis Editing

Associate Project Manager: Neil Dmello

Proofreader: Safis Editing

Indexer: Pratik Shirodkar

Production Designer: Vijay Kamble

Marketing Coordinator: Nimisha Dua

First published: January 2022

Production reference: 1200122

Published by Packt Publishing Ltd.
Livery Place
35 Livery Street
Birmingham
B3 2PB, UK.

978-1-80107-138-3

www.packt.com

To my lovely girlfriend and four children, and my global team at Avanade, for their understanding and support during my time writing this book. Their continuous support has helped me to grow and become what I am today, allowing me to share content with others in my own way!

– Alexander Meijers

Contributors

About the author

Alexander Meijers is a professional who inspires, motivates, and supports others and helps them to innovate. His goal is to help organizations achieve more by creating, improving, and working smarter, with the aim of shortening business processes and improving the environment for employees.

As global XR technology lead and Microsoft Windows MVP for mixed reality, working for Avanade, he understands business issues and translates them into logical solutions using technology. Additionally, he supports companies in applying emerging experiences during their digital transition journey.

He works with technologies such as virtual, augmented, and mixed reality, in combination with cloud services including mixed reality services, Azure Digital Twins, and IoT, from the Microsoft Azure platform, Office 365, Power Platform, and Dynamics 365.

His primary focus is manufacturing, utilities, and the engineering and construction sector. However, he certainly does not stay away from other sectors.

He engages in speaking, writing books, blogging, and is an organizer of local and global events such as the Mixed Reality User Group in the Netherlands, and Global XR Talks and the Global XR Conference, both part of the GlobalXR.Community.

In the last few months, he has designed and implemented a solution. Data is collected from IoT and Microsoft Dynamics 365 Field Service. An Azure Digital Twins service is built up dynamically by using Azure Functions, Azure Service Bus, and Logic Apps. A 3D visual is generated by using Microsoft HoloLens 2 as an augmentation device.

Since January 2018, he has been a Windows Development MVP for the Mixed Reality category.

About the reviewer

Sjoukje Zaal is a Microsoft chief technical officer at Capgemini, a Microsoft regional director, and a Microsoft Azure MVP with over 20 years of experience providing architecture, development, consultancy, and design expertise. She mainly focuses on cloud, security, productivity, and IoT. She loves to share her knowledge and is active in the Microsoft community as a cofounder of the user groups Tech Daily Chronicle, Global XR Community, and the Mixed Reality User Group. She is also a board member of Azure Thursdays and Global Azure. Sjoukje is an international speaker, is involved in organizing many events, and has written several books and blogs. She is also part of the MVP Diversity and Inclusion Advisory Board.

Table of Contents

Preface

Section 1: Azure Digital Twin Essentials

1

About Digital Twins

Understanding the concept of a Digital Twin	**4**	Open modeling language	15
		Tools	16
Digital replica	4	Costs	17
Entities	5		
Relationships	5	**Understanding the components of Azure Digital Twins architecture**	**18**
Realities	5		
		Managing Azure Digital Twins	19
Exploring the Digital Twin environment	**6**	Azure Functions	19
		Azure IoT Hub	20
Looking at real-world applications	**9**	Azure IoT Hub Device Provisioning Service	20
Smart building	10	Azure Logic Apps	21
Education	11	Azure Storage	21
Simulation	12	Azure Analytics	21
Historical data	12	Azure Service Bus	22
Insight and control	13		
		Exploring the Azure Digital Twins REST API	**22**
Azure Digital Twins	**14**		
The Azure Digital Twins service	15	**Summary**	**23**
Important terminology	15		

2
Requirements and Installation

Technical requirements	26	Node.js	40
Azure Digital Twins service	26	Azure Digital Twins Explorer	41
Azure account	26	Download	41
Resource group	26	Installation	43
Azure Digital Twins service	29	Compile and run	44
Configuring access control	33		
		Creating your first Digital Twin	48
Microsoft Visual Studio	35	Uploading models	49
The Windows Azure CLI with Windows PowerShell	39	Summary	54

Section 2: Getting Started with Azure Digital Twins

3
Digital Twin Definition Model

Technical requirements	58	Complex schemas	68
Digital Twins Definition Language	58	Field	69
		Object	69
JSON-LD	59	Map	70
Versioning	59	Array	71
		Enum	72
Digital Twin interface	59		
Interface content	62	Geospatial schemas	73
Properties	62	Semantic types	74
Telemetry	63	Validating a model	74
Relationships	64	Summary	77
Component	66	Questions	78
		Further reading	78
Schemas	67		
Primitive schemas	67		

4
Understanding Models

Technical requirements	80	Getting a model	97
Designing models	80	Getting models	98
Modeling a smart building	80		
Designing models	80	**Managing Digital Twins**	**99**
Modeling recommendations	81	Creating a Digital Twin	99
		Updating a Digital Twin	101
Creating an application	82	Deleting a Digital Twin	103
Managing models	88	Getting a Digital Twin	104
Creating a model	89	**Summary**	**105**
Creating multiple models	92	**Questions**	**105**
Inheriting from a model	94	**Further reading**	**106**
Deleting a model	96		

5
Model Elements

Technical requirements	107	**Components**	122
Primitive properties	108	**Summary**	127
Complex properties	113	**Questions**	127
Telemetry	116		

6
Creating Relationships between Azure Digital Twin Models

Technical requirements	130	**Relationship properties**	143
Understanding relationships	130	**Updating relationship**	
Creating relationships	132	**properties**	145
Getting relationships	136	**Summary**	147
Getting a single relationship	136	**Questions**	147
Getting a list of relationships	137	**Further reading**	147
Deleting relationships	141		

7
Querying Digital Twins

Technical requirements	150	Querying using code	166
Setting up a demo graph	150	Querying asynchronous	
Basic querying	154	calls using code	170
Querying by model	160	Summary	172
Querying relationships	162	Questions	173
Filtering results	164	Further reading	173

8
Building Models Using Ontologies

Technical requirements	176	Uploading ontology models	183
Understanding ontologies	176	Summary	192
Modeling strategy	177	Questions	192
Using industry-standard		Further reading	193
ontologies	179		

Section 3: Digital Twins Advanced Techniques

9
APIs and SDKs

Technical requirements	198	Using the Azure CLI	
Understanding the		to manage Azure	
developer landscape	198	Digital Twins	216
Understanding the REST API	199	Understanding service limits	217
Control plane	201	Summary	219
Data plane	207	Questions	219
SDKs	212	Further reading	220
Monitoring API metrics	213		

10
Building a Digital Twin Pipeline

Technical requirements	221	Getting sensor messages on Azure Service Bus	232
Understanding application architecture	222	Summary	241
Setting up a demo sensor using Azure IoT Central	224	Questions	242
		Further reading	242

11
Updating the Model

Technical requirements	243	Publishing the Azure function	262
Updating the digital twin	244	Setting the connection string	266
Updating the sensor model	244		
Creating a storage account	245	Creating a digital twin for the sensor	268
Creating an Azure function	249	Viewing the result	269
Setting the connection string	257	Summary	271
Creating an Azure function placeholder	257	Questions	271
Granting the Azure function permissions	261	Further reading	272

12
Event Routing

Technical requirements	274	Subscribing to event messages	288
Data ingress and egress	274	Monitoring event route messages	294
Event notifications	277		
Understanding event routes	280	Summary	297
Creating an event grid topic	282	Questions	297
Creating an endpoint	284	Further reading	298
Creating an event route	286		

13

Setting up Azure Maps

Technical requirements	300	Creating and validating a dataset	319
Understanding Azure Maps	300		
Creating an Azure Maps account	301	Creating and validating a tileset	322
Creating a Creator resource	304	Creating and validating a feature stateset	325
Building a map	306		
Uploading a map	310	Summary	328
Converting a map	316	Questions	328
		Further reading	329

14

Integrating Azure Maps

Technical requirements	332	Subscribing to Event Grid	351
Updating a feature stateset	332	Monitoring updates	353
Setting up an update Azure Function	334	Visualizing the model	356
		Summary	359
Publishing the Azure Function	345	Questions	360
Configuring application settings	349	Further reading	360

15

Monitoring and Troubleshooting

Technical requirements	362	Viewing metrics	370
Setting up a log analytics workspace	362	Using alerts	373
		Summary	374
Setting up diagnostic settings	365	Questions	374
		Further reading	375
Viewing logs	368		

Section 4: Digital Twin Implementations in Real-world Scenarios

16
Facility of the Future

Understanding the scenario	380	The solution architecture	388
Designing the digital		Summary	389
twin solution	382	Questions	389

17
Creating Digital Twins for Smart Building

Understanding the smart		A smart building	
building ecosystem	392	solution design	396
Sensors	392	The smart building	
Analytics	395	architecture	397
User interfaces	395	Summary	399
Automation	396	Questions	399

18
Simulations Using a Digital Twin

Understanding simulation	402	Training	405
Solution design and		Testing	406
architecture	403	Summary	407
Work preparation	403	Questions	408

Assessments

Chapter 3 – Digital Twin		Relationships between	
Definition Model	409	Azure Digital Twin Models	409
Chapter 4 – Understanding		Chapter 7 – Querying	
Models	409	Digital Twins	410
Chapter 5 – Model Elements	409	Chapter 8 – Building Models	
Chapter 6 – Creating		Using Ontologies	410

Chapter 9 – APIs and SDKs   410

Chapter 10 – Building a Digital
Twin Pipeline   410

Chapter 11 – Updating
the model   410

Chapter 12 – Event Routing   410

Chapter 13 – Setting up
Azure Maps   411

Chapter 14 – Integrating
Azure Maps   411

Chapter 15 – Monitoring
and Troubleshooting   411

Chapter 16 – Facility of
the Future   411

Chapter 17 – Creating Digital
Twins for Smart Building   411

Chapter 18 – Simulations
Using a Digital Twin   411

Index

Other Books You May Enjoy

Preface

Being able to create a real-time digital counterpart from reality gives organizations the ability to build simulations, training environments, and other business solutions. Digital twins are a virtual representation of these solutions. Organizations such as Microsoft provide cloud services to support building the foundation of a digital twin. This book will help you understand what digital twins are and how to build these IoT solutions using the Azure Digital Twins service and other related Azure services

Who this book is for

This book is targeted at Azure developers, Azure architects, or anyone else who wants to learn more about how to implement IoT solutions using Azure Digital Twins and additional Azure services.

What this book covers

Chapter 1, About Digital Twins, explores the concept of a digital twin.

Chapter 2, Requirements and Installation, goes over all the requirements, services, and tools to get up and running with the Azure Digital Twins service.

Chapter 3, Digital Twin Definition Model, discusses and describes each of the metamodels as part of the digital twin definition model.

Chapter 4, Understanding Models, covers models and how to manage them by creating, updating, and removing models.

Chapter 5, Model Elements, discusses several model elements, such as properties, telemetry, and components.

Chapter 6, Creating Relationships between Azure Digital Twin Models, explores the concept of relationships between digital twins and how we can create and delete them.

Chapter 7, Querying Digital Twins, explains the query language and executing different types of queries to retrieve data from an Azure Digital Twins instance.

Chapter 8, Building Models Using Ontologies, explains how to use ontologies – predefined sets of models – to provision a digital twin solution more quickly.

Chapter 9, APIs and SDKs, looks at the differences between APIs and SDKs and how to manage and control Azure Digital Twins instances using the REST API.

Chapter 10, Building a Digital Twin Pipeline, offers a guide to get data into a model by building a pipeline using several Azure services.

Chapter 11, Updating the Model, continues extending the pipeline by getting sensor data from demo sensors in Azure Service Bus.

Chapter 12, Event Routing, goes into more detail about how to send data to other services. We will learn about how notifications are triggered and messages are routed to endpoints.

Chapter 13, Setting Up Azure Maps, goes through setting up the Azure Maps service, which allows us to visualize data on top of a map.

Chapter 14, Integrating Azure Maps, looks at integrating the setup of Azure Maps with the Azure Digital Twins instance using several Azure services.

Chapter 15, Monitoring and Troubleshooting, discusses how to leverage an Azure Log Analytics workspace and diagnostic settings to monitor and troubleshoot our Azure Digital Twin instance.

Chapter 16, Facility of the Future, shows with an example how insights can contribute to different processes and roles within an organization by using an Azure digital twin.

Chapter 17, Creating Digital Twins for Smart Building, shows how an Azure digital twin is used with the smart building concept to automate and control a building's ecosystem.

Chapter 18, Simulations Using a Digital Twin, provides a better understanding of what simulation is and how simulation can benefit from an Azure digital twin.

To get the most out of this book

Building digital twin solutions requires you to have basic knowledge of using Microsoft Azure and standard development tools such as Microsoft Visual Studio, and have intermediate experience in building applications with .NET.

Software/hardware covered in the book	OS requirements
Microsoft Visual Studio Community 2019	Windows
Windows Azure CLI 2.30.0	Windows
Windows PowerShell 5.1.19041.1151	
Node.js 14.15.4	Windows
Azure subscription	Microsoft Azure

All examples can be executed using a trial subscription with Microsoft Azure. All code in the book is expected to work with future version releases of the abovementioned software. While the book focuses on using a Windows computer, several of these tools are available on other platforms, too, such as macOS.

When you have finished the book, start building your own concept around a digital twin to apply what you have learned. Start with something small and easy and extend your solution along the way.

Download the color images

We also provide a PDF file that has color images of the screenshots/diagrams used in this book. You can download it here: `https://static.packt-cdn.com/downloads/9781801071383_ColorImages.pdf`

Download the example code files

You can download the example code files for this book from GitHub at `https://github.com/PacktPublishing/Hands-on-Azure-Digital-Twins`. In case there's an update to the code, it will be updated on the existing GitHub repository.

We also have other code bundles from our rich catalog of books and videos available at `https://github.com/PacktPublishing/`. Check them out!

Conventions used

There are a number of text conventions used throughout this book.

`Code in text`: Indicates code words in text, database table names, folder names, filenames, file extensions, pathnames, dummy URLs, user input, and Twitter handles. Here is an example: "Create a new folder called `chapter6` under the `Models` folder of `SmartbuildingConsoleApp`."

A block of code is set as follows:

```
public string RelationshipId(string twinSourceId, string
twinDestinationId)
{
    return string.Format("{0}-{1}", twinSourceId,
twinDestinationId);
}
```

Bold: Indicates a new term, an important word, or words that you see on screen. For example, words in menus or dialog boxes appear in the text like this. Here is an example: "Click on the **+relationship** icon in the **Graph View** area to start creating a relationship."

> **Tips or Important notes**
> Appear like this.

Get in touch

Feedback from our readers is always welcome.

General feedback: If you have questions about any aspect of this book, mention the book title in the subject of your message and email us at customercare@packtpub.com.

Errata: Although we have taken every care to ensure the accuracy of our content, mistakes do happen. If you have found a mistake in this book, we would be grateful if you would report this to us. Please visit www.packtpub.com/support/errata, selecting your book, clicking on the Errata Submission Form link, and entering the details.

Piracy: If you come across any illegal copies of our works in any form on the internet, we would be grateful if you would provide us with the location address or website name. Please contact us at copyright@packt.com with a link to the material.

If you are interested in becoming an author: If there is a topic that you have expertise in and you are interested in either writing or contributing to a book, please visit authors. packtpub.com.

Share Your Thoughts

Once you've read *Hands-on Azure Digital Twins*, we'd love to hear your thoughts! Scan the QR code below to go straight to the Amazon review page for this book and share your feedback.

https://packt.link/r/1801071381

Your review is important to us and the tech community and will help us make sure we're delivering excellent quality content.

Section 1: Azure Digital Twin Essentials

Our first section is all about understanding and learning the concept and architecture of a digital twin. We will be focusing on the Microsoft Azure Digital Twins service and how to set this up. All tools and services required to start building with the Microsoft Azure Digital Twins service will be explained and installed.

This part of the book comprises the following chapters:

- *Chapter 1, About Digital Twins*
- *Chapter 2, Requirements and Installation*

1
About Digital Twins

This chapter will explore the concept of a Digital Twin. A **Digital Twin** is a virtual representation of the real world combined with real-world data. Digital Twins can be used for a variety of scenarios. Digital Twins can be used to visualize insights or to simulate real-life situations by using a virtual representation and real-life sensory data. Learning about Digital Twins allows you to build solutions around these scenarios.

In this chapter, we'll go through several scenarios to understand Digital Twin implementations. We'll look at Microsoft's Azure Digital Twins service and how it allows us to model a Digital Twin. We'll walk through the layout of the service and how it is incorporated into the model of a Digital Twin. Part of that is a global overview of the architecture, which includes the relationship to other Azure services. This is required to create an actual Digital Twin solution. We will finish with an overview of the available SDKs and APIs for using Azure Digital Twins to create your own Digital Twin solutions. The chapter contains a lot of introductions to different services and tools that will appear again in the following chapters.

In this chapter, we'll go through the following topics:

- Understanding the concept of the Digital Twin
- Exploring the Digital Twin environment
- Looking at real-world applications

- Azure Digital Twins
- Understanding the components of Azure Digital Twins architecture
- Exploring Azure Digital Twins APIs

Understanding the concept of a Digital Twin

You have probably heard someone talking about Digital Twins in the last few years. You could even say that it has been a buzzword for some time. But since 2019-2020, it's become more than just a buzzword. Organizations and people have started to understand the benefits of having a Digital Twin. There has even been a large increase in organizations that want to start and implement a Digital Twin.

But what is a Digital Twin? I get that question a lot. And every time it is difficult to come up with an answer that others will understand. And even referring to the definition on Wikipedia will not make it easy to understand. There are a lot of different definitions you can find online in articles and blog posts. To explain what a Digital Twin is requires a definition to start with followed by a more in-depth explanation of the definition itself. I use the following definition:

A Digital Twin is a digital replica of entities and
their relationships in a reality

You may have noticed that this definition contains several terms: **digital replica**, **entities**, and **reality**. It becomes clearer when explaining each of them in more depth.

Digital replica

A digital replica is a way of storing several entities and their relationships in a specific model. Such a model is stored in a location according to your requirements and needs. An example could be a database or service. Each product on the market that is available to create a Digital Twin has its own way of storing the information that describes the model. That means that the digital replica can describe a real-life situation using definitions and parameters. Think of a machine and whether it is turned on or off. The digital replica would describe the machine and its state. But a digital replica could also be about a collection of machines and their relationships. Think of a machine that is creating a product and the machine that is packing the product. The packing machine requires products to pack anything. That relationship is also described in a digital replica.

Entities

Entities can be different things. An **entity** can be anything from a physical living being such as people to a physical non-living thing such as processes, machines, buildings, equipment, rooms, and devices. When we talk about physical, it means being physically part of the reality from which you create a digital replica. Each of these entities has a specific purpose within the model. An entity is described by its characteristics that are relevant to the model and what you try to achieve in your solution. An entity could be a temperature sensor installed in a room. The characteristics are then the location of that sensor, the temperature the sensor is measuring, and the notifications it is raising when the temperature gets too low or too high. The location in this case is the room where the sensor is located. That characteristic is a relationship to another entity called the room. All these characteristics when developing Digital Twins are described by properties and metadata.

Relationships

An important part of a Digital Twin is the way entities are related to each other. These **relationships** are important as they define the context in which the entities are depending on each other and are a part of the reality on which the model is based. A **relationship** itself defines a set of data based on how the **relationship** is defined between the entities. An example is the **relationship** between the temperature sensor and the room where it is installed. This **relationship** defines what the temperature is within the room. Business rules can be used to take certain actions based on entities and their underlying **relationships**. An example would be switching off the lights within the room when there is no movement for a pre-defined time. In that situation, the lights, motion sensor, and room are each an entity with underlying relationships.

Realities

Each entity is part of a **reality**. Normally the **reality** would be a part of the physical world around us, like the example of the temperature sensor in a room. In that case, we have an actual device, room, and building. But imagine a world that represents a theoretical **reality**. This could be a virtual, generated **reality** that acts as the source for the digital replica. An example would be a digital world created in virtual **reality** or even another Digital Twin.

You have just learned about the concept of Digital Twins and its elements. This is important since it will help you to understand how Digital Twins can be applied to different scenarios. In the next section, we will explore the different parts of the environment around a Digital Twin to implement Digital Twin solutions.

Exploring the Digital Twin environment

It is important to understand that we need to do more than just store a model of entities to use a Digital Twin. Using a Digital Twin requires us to bind information to our entities in the model and use some method of visualization to view the model and its outputs.

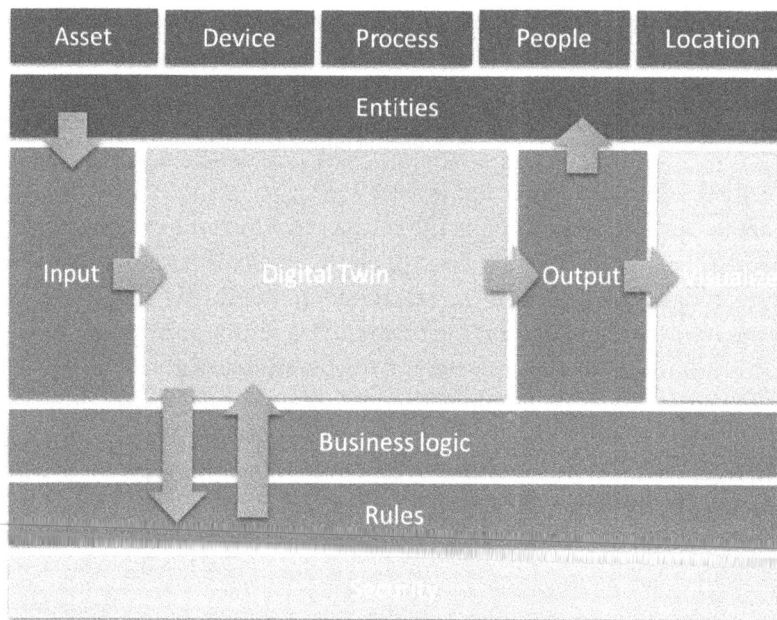

Figure 1.1 – High-level overview of a Digital Twin environment

The model in *Figure 1.1* shows a high-level overview of everything that is in some way used within a Digital Twin environment:

- **Entities** – This part represents the entities from your reality. This is, for example, real-world assets, people, processes, and locations. Data that defines these entities is stored in some way in the Digital Twin.

- **Digital Twin** – This is the digital replica model of the entities in the reality.

- **Input** module – This part of the model provides data from entities into the Digital Twin model. In some situations, this is also used to dynamically generate the model-based structure of the entities. It depends heavily on actual data that flows from the entities being used in the reality.

- **Output** module – The output of the model is in most cases used to visualize the data in some way. But that is not always the case. The output could also be a setting turned on based on business logic and rules that are triggered by the input.

- **Business logic** and **Rules** modules – This is all about building logic and rules around the data in your Digital Twin. The result of this logic can resolve into setting the data of entities in the Digital Twin. You could extend this by connecting to or triggering the actual entity.

- **Visualize** – It is often thought that a Digital Twin is visualized. But that is not always the case. In many situations, the data flows back to the entity itself. But in some situations, a visualization of data could enhance the experience and benefit the business process. Visualization can be reached in many ways. Think of a display at the door of a meeting room displaying availability, an Excel that is filled with output data, or using augmented glasses to create a 3D presentation based on the data from the entities.

- **Security** – Each module needs to have some sort of role-based security. This can influence what data flows in and out of the Digital Twin. It could be used to only view the data that you are allowed to see based on your role in the organization. But it could also be used to view a subset of output data coming from the Digital Twin.

Now let's look at how a Digital Twin is connected to real-world entities.

A Digital Twin needs to be integrated with the physical and non-physical entities that it represents. As shown in the following figure, you will see that a Digital Twin is about being connected:

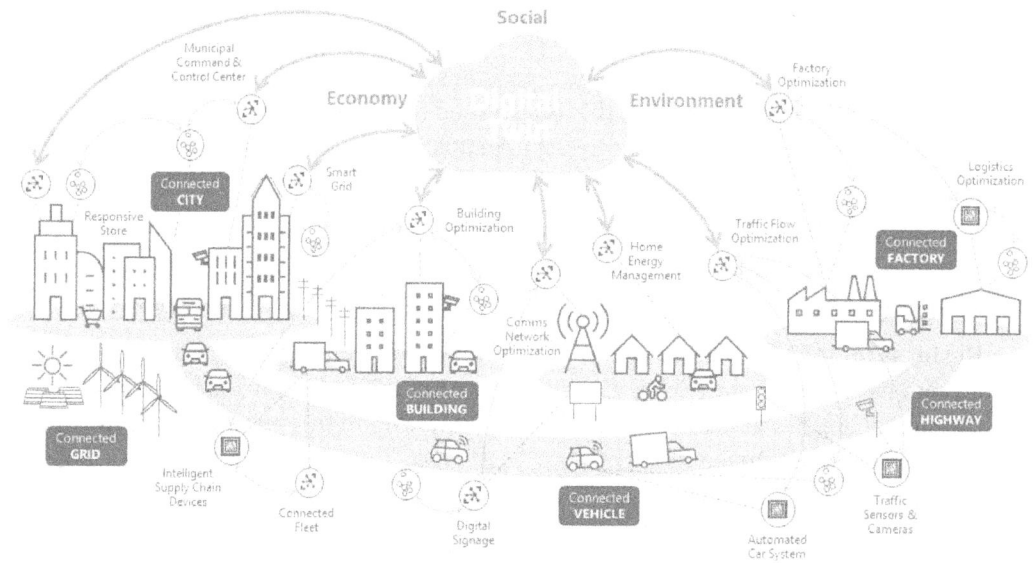

Figure 1.2 – Real-world entities are connected to a Digital Twin

Any device that can generate some form of output information based on sensors, processes, or a manual action can be an input for your Digital Twin. And these same devices, if they have some form of interface, can be controlled using rules based on the received input.

Taking the example of the motion sensor, the entity representing the motion sensor needs to get the values from the actual motion sensor in the room. Then, business logic can define that the lights need to be turned off or on based on the value within that entity. The lights themselves are also an entity in the model and bound to the actual lights within the room. The business logic can then set the value on the entity of the lights.

There are Digital Twins that require no visualization. Those Digital Twins handle and set values that will cause certain physical entities to respond. Like the example with turning on and off the lights. There are situations where a Digital Twin needs to be visualized. Such situations require some sort of presentation of the information from the Digital Twin model. Presentations can be flat out a spreadsheet or a list. But they can also be a visual representation of reality on a desktop screen. And nowadays, with extended reality using augmented glasses, it is even possible to have 3D modeled presentations of your Digital Twin.

You have learned about the different modules and parts that are required when you are going to build a Digital Twins solution. In the next part, we will be looking at several real-world examples.

Looking at real-world applications

In this chapter, we have several scenarios that will help you better understand the idea behind a Digital Twin and what it can contribute to an organization. Since implementing a Digital Twin can be a costly journey to implement, it is important to address the business value for organizations:

- **Gaining insights** – Digital Twins can be used to get better insights into your business operations and allow you to optimize these processes. More insights can be reached by visualizing situations that would normally not be visible or more difficult to understand. Hence, better insights and allowing you to respond differently, more quickly, or in a more streamlined process.

- **Collaboration** – Using Digital Twins to view data in a collaborative way. By mimicking a real situation digitally, you can have multiple users experience and view the same data while not even being at the same location.

- **Education** – Digital Twins give you the ability to create learning environments. These learning environments can be used at schools or at the start of a job to get new employees more quickly up to speed.

- **Simulation** – Simulation by using Digital Twins allows you to create digital replicas of environments that are normally difficult to reach, too dangerous, or just not accessible. It also allows you to experiment with settings to see what the outcome is before changing these settings in the actual environment to prevent downtime or process disturbances.

- **Create experiences** – Experience a process, environment, or other situation by digitizing using a Digital Twin. There is a very clear distinction for people between looking at a flat dashboard and having the situation visualized in three dimensions. People are used to understanding more quickly by looking at something in a three-dimensional way. It can also contribute to investors and management having a clearer overview of what they are managing.

- **Optimize operations and costs** – Use Digital Twins to optimize processes in an organization to optimize operations and reduce costs. Digital Twins can deliver a return on investment in different ways. Better insights reduce time to action, getting new employees up to speed more quickly, or simulate possible optimizations before making them available in the real world.

Smart building

A smart building involves including intelligence in the processes of maintaining a building and its services. In this example, we have the Contoso office building, which has several meeting rooms. These rooms are maintained by facilities. Since this office building is a smart building, several sensors are installed in each room. We have sensors for temperature, humidity, motion, and light. Each of these sensors can be read out through a smart network. Additionally, things such as light can also be controlled by that same smart network.

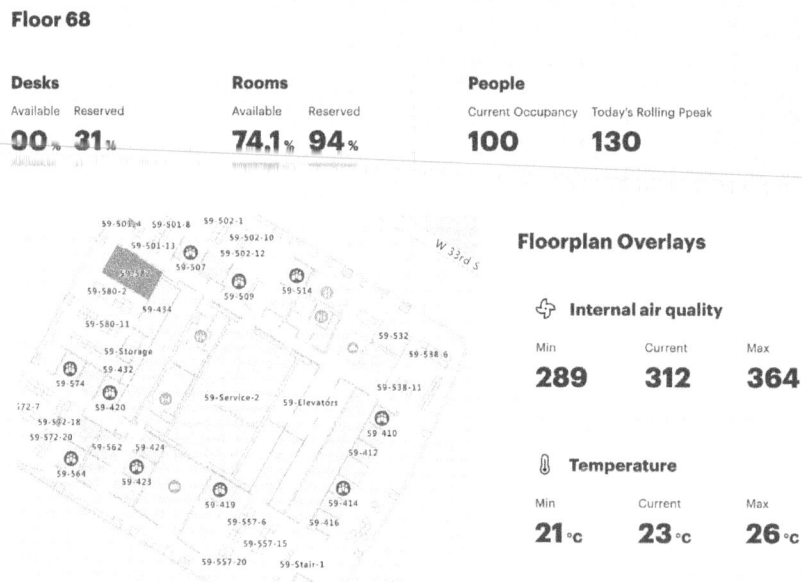

Figure 1.3 – Example of a 2D map interface used for a smart building

The facilities team is responsible for the state of these rooms. One of their responsibilities is, for example, keeping the rooms at a standard temperature. There are several conditions that can influence the temperature in the rooms:

- **Room usage** – The room being used or not for a specific amount of time
- **Group size** – The number of people that are in the room when it is in use
- **Devices** – Devices such as laptops, wall screens, projectors, and the use of light

Another responsibility is to provide a booking system for the meeting rooms. This booking system shows the availability on a small screen at the door and allows employees to book a meeting room through an app or service, such as Outlook.

A Digital Twin can be used to monitor and act based on business logic and rules. It can automatically switch off the lights when there is no movement in the room for a specific period. It can change the temperature based on the conditions mentioned earlier. But it is also used to update the small screen at the door with the latest information and bookings, maintaining availability by freeing up rooms that were booked but do not have someone inside, since we monitor light and movement.

There are many other features that are provided by using a Digital Twin for a smart building.

Education

In this example, we are at the Contoso Water company. Water is coming in from natural sources and goes through a filter system into large reservoirs where it is transported to households. The system needs to be in harmony for optimal performance.

This water company uses large pumps that transport incoming water into the filter system. The configuration of these pumps and their throughput needs to be adjusted over time. It is a delicate task to make changes. With the wrong configurations, the filter system could be overloaded and bring the whole water plant to a stop. The engineers who are allowed to perform these tasks must have many years of experience to perform these tasks without worry.

What happens with a junior engineer who just came from school? The junior engineer needs to learn these tasks too. But you don't want the junior engineer to practice on the actual water pump.

A Digital Twin can help with this. By replicating the important, key parts of the water plant, simulating and visualizing the assets using augmentation or virtual reality, the junior engineer can learn how to execute these tasks. The student engineer can view the digital pump and learn how to control this pump. And by tweaking the values in the model, it is possible to see the influence on the other assets in the plant. For example, by changing the throughput of the water, the student engineer can learn what the effect is on the filter system. This can be achieved by enriching the Digital Twin model using historical data and specifications of the real-world assets such as the water pump and the filter system.

Simulation

Simulation is one of the most powerful features you can create with Digital Twins. Just like the teaching example, you can simulate almost anything in a digital world. Simulation can be used for areas that are normally very dangerous to be in, such as an oil rig or areas that are not easily accessible, such as an operating room in a hospital. But it can also be used to optimize processes, which will in time create more margin.

In this example, we have the Contoso oil plant. Most of the areas in an oil plant are dangerous due to the threat of explosions and fire hazards. Oil plants contain large quantities of assets, each with its own configuration and real-time sensor devices.

For an oil plant, it is important to have the plant running as close to optimal as possible. The more oil produced, the better.

Now assume we use a Digital Twin to create a digital replica of the oil plant. We add the configuration information and real-time IoT data from sensors to the Digital Twin. This digital replica allows us now to change configuration settings for the plant and see what effect this has on oil production. Since we don't know the results upfront, the simulation will help us decide which asset configurations are required to have a positive effect on oil production.

Historical data

An interesting use case is the ability to view historical data using a Digital Twin. The amount of data stored by organizations is phenomenal. But that data is almost never used in a way that the organization could benefit from.

For this example, we will use Contoso Events. Contoso Events is a large organizer that organizes events in big cities. One of their main concerns is handling security and safety. This involves the government, police, health, and fire department. They will need to prove that they have taken enough precautions to manage it. Contoso Events needs to determine what is required to achieve a certain level of security and safety. Part of this is closing roads and defining health points in the city. The organization uses smart cameras during the event to monitor issues and crowd density. They have been collecting this data and data from various access, check, information, and health points.

In this case, the Digital Twin is a digital replica of the city map enriched with the collected data. All data is visualized on a city map. The Digital Twin contains a slider to move through time to view everything that is happening during the event. Based on these insights, their own experiences, and using machine learning, they are able to determine what the best locations are for placing roadblocks, information points, access points, and health points. The information is shared with all the government agencies.

Insight and control

Having better insights into and control over processes can help an organization. It is all about empowering workers at any level, improving and optimizing processes, and getting insights and end-to-end telemetry.

We have a company called Contoso Construction. This company performs construction work on existing facilities worldwide. Their work involves adding and updating new assets. They use large equipment such as trucks and cranes to perform their work. Several workers are onsite performing different tasks to get the job done. Contoso Construction wants to have more insights and control over the work process and the safety of their employees. While they try to prevent incidents, when those incidents happen, they require a command center to monitor and assist from a global level. Each assignment they have gives them asset data. This asset data contains the location of the asset and real-time data from sensors that are connected to the asset. An asset can be anything from a simple machine to a storage location.

By combining all this data into a Digital Twin and connecting it to backend systems containing the assets, Contoso Construction can create a visual representation of the site using augmented glasses, as shown:

Figure 1.4 – Command and control center of a construction site

When an issue occurs, the facility manager can view the issue in real time on a 3D map of the site. The visualization shows which alerts are generated based on sensor data and which cases and tasks are created. It is even possible to assign a task to an engineer. That data flows back into the Digital Twin and updates the connected backend systems.

You have learned about the different scenarios of using a Digital Twin. These examples have shown you how to solve scenarios around gaining insights, education, simulation, and creating experiences with Digital Twins. In the next part, we will explain how Digital Twins reside within Microsoft Azure.

Azure Digital Twins

Microsoft has built a comprehensive cloud platform in the last 10 years called Microsoft Azure. Every year, new services are released on the platform. One of them is Azure Digital Twins. Azure Digital Twins is part of the **Internet of Things (IoT)** platform and is called the next-generation IoT solution that models the real world.

The Azure Digital Twins service

The Azure Digital Twins service is a **platform as a service (PaaS)** that enables you to create digital replicas of an environment. PaaS is a cloud model where a provider delivers hardware and software tools enabling you to build your solutions without worrying about purchasing, installing, and maintaining hardware infrastructure.

It has a robust setup that provides a scalable and secure environment. It is part of IoT-connected solutions within Azure that allow it to connect to assets such as IoT devices as other Azure services and backend systems. The Azure Digital Twins service uses an event system to allow you to build your own business logic and data processing as routing. And it integrates with Azure services such as Azure Storage, Analytics, and Azure Machine Learning to extend the platform with predictiveness. The service can be created through the Azure portal.

Important terminology

Before we move forward, it is important to understand the different terminology used within Azure Digital Twins. Without it, it could become very confusing when explaining how Azure Digital Twins works:

- **Azure Digital Twins instance** – This represents the complete digital replica of reality.
- **Model** – A model is seen as the noun in a description within reality. It is like a class within programming and defines a part of that reality. Examples of a model are *Room*, *Engineer*, *Asset*, *MotionSensor*, and *Device*. Each model is described by properties, relationships, telemetry, components, and commands.
- **Digital Twin** – An instance of a model to represent a certain entity. An example is *CoffeeRoom* based on the *Room* **model**.
- **Twin graph** – A representation of **Digital Twins** and their underlying relationships.

Open modeling language

Models used in Azure Digital Twins are based on the **Digital Twins Definition Language (DTDL)**. DTDL is used as a definition language for Azure Digital Twins and IoT Plug and Play. It is part of the IoT space and helps to describe the model's ability to support provisioning and configuration across the different IoT resources. DTDL uses a variant of **JavaScript Object Notation (JSON)** and is named **JavaScript Object Notation for Linked Data (JSON-LD)**.

> **Important note**
> Azure Digital Twins uses DTDL version 2. While it uses DTDL as its modeling
> language, it does not currently implement DTDL commands.

Tools

There are several tools that are needed when you start working on Azure Digital Twins.
First, you will require an Azure environment. The Azure environment allows us to create
the Azure Digital Twins service and to monitor all the other services that we will be using
throughout the book.

We will be using Azure Digital Twins Explorer to view an Azure Digital Twin that is
created. The tool connects to an instance of the Azure Digital Twins service and allows
you to query models.

In most cases, Azure resources such as Azure Functions, Azure Service Hub, and others
are used to connect to the Azure Digital Twins service. Throughout the book, we will be
using Microsoft Visual Studio to create code and deploy services to build Digital
Twin examples.

In most scenarios, an Azure Digital Twins instance contains digital replicas of devices.
These devices generate IoT data. While normally devices are connected and routed using
Azure IoT Hub, you could also make use of **IoT Central**.

Azure IoT Hub is a managed cloud-hosted service that operates as a message hub for
bi-directional communication between an IoT application and IoT devices.

 IoT Central is a web-based service that allows you to connect a variety of different
devices. But it is also possible to mimic these devices. You will get the available telemetry,
which instead now generates demo IoT data. This allows you easily to test your Azure
Digital Twins instance without the need for actual devices.

Costs

The Azure Digital Twins service does not have any upfront cost or termination fees. You only pay for your consumption. The service has billing at three levels:

Pricing dimension	Description	Measurement	Price calculation
Operations	Each API call made to the Azure Digital Twins API counts as an operation.	An operation is measured at its size part of the returned body. Each 1 KB of size is counted as a single operation.	Per million messages
Messages	Any message sent using event routing to an output destination using Event Grid, Event Hub, or Service Bus counts.	Messages are measured in 1 KB of the payload size.	Per million operations
Querying of units	A query execution against the Azure Digital Twins Query API. (Each query made against the Azure Digital Twins Query API also counts as an operation.)	Each query execution uses CPU, memory, and IOP resources. The return header of the query contains the number of query units consumed for that query.	Per million query units

You have learned about the Azure Digital Twins service and how this represents a Digital Twin within Microsoft Azure. We also talked about the modeling language and the costs of the service. In the next part, you will learn about the different Azure components that act as input and output services when building solutions with the Azure Digital Twins service.

Understanding the components of Azure Digital Twins architecture

A Digital Twin Graph is created based on instances of those models and their relationships. The service provides several ways of managing the models and the Digital Twin Graph. But it depends on other Azure services to create an Azure Digital Twins instance.

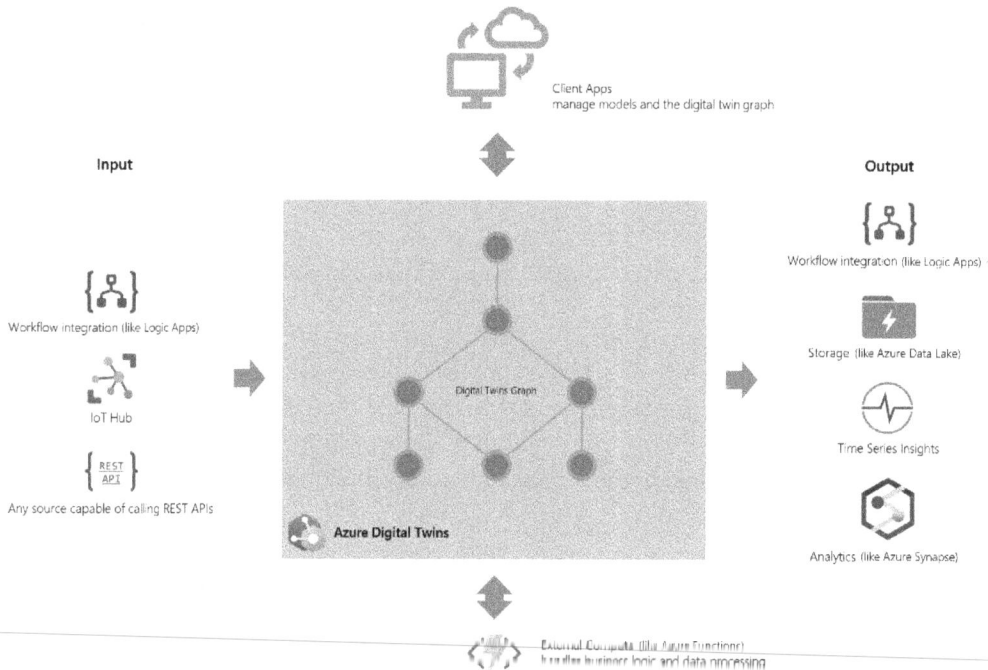

Figure 1.5 – An overview of input and output services that are used with Azure Digital Twins

In *Figure 1.5*, you see some of the Azure services available to interact with Azure Digital Twins. Some of these services can be used for different parts of your Azure Digital Twins instance.

Managing Azure Digital Twins

Managing is all about defining and creating models, creating instances of models, and laying out relationships between those instances to create a Digital Twins Graph. To access the Azure Digital Twins service, you have REST APIs available. The creation of models is done by defining the models using the DTLD. There are several ways of managing Azure Digital Twins:

- **Azure Digital Twin Explorer** – A tool to manage, view, and upload models, create instances of models, and set relationships.

- **Client App** – A custom application that makes use of available REST APIs and client SDKs to access the service.

- **Power App** – The Power Platform, a platform that provides several applications and services to build low-code/no-code solutions and allows you to create a Canvas app that can connect to an Azure Digital Twins service.

- **Azure services** – Different services such as Logic Apps, Azure Functions, and Azure Service Bus provide a way of connecting to Azure Digital Twins to manage the models. We will go into more depth about these services and how to use them in the next section.

Azure Functions

Azure Functions is a serverless compute service that runs event-triggered code. Azure Functions simplifies the development and provisioning of small applications without the concern of setting up an infrastructure. It is also called "compute on-demand." This means that it only runs when requested and automatically upscales when the number of requests increases.

Azure Functions makes it easy to handle different roles and tasks around an Azure Digital Twins instance. The following is a list of different tasks performed by Azure Functions.

Act as	Task
Input	Instantiate models to create Digital Twins based on input from other systems and services.
Input	Update properties and telemetry of Digital Twins.
Output	Act as an API to retrieve data from a Digital Twin Graph.
Managing	Set up a complete Azure Digital Twin by creating models using the DTDL notation.
Compute	Use an Azure function to compute business logic and data processing. The Azure function reads from and writes to Digital Twins within the Azure Digital Twins.

Azure IoT Hub

This is a managed service that acts as a central messaging hub between devices and IoT applications such as the Azure Digital Twins service. Any device can be connected to Azure IoT Hub. The hub provides a secure way of communicating with the attached devices. It can route device telemetry to multiple endpoints. It integrates with other services, such as Azure Event Grid, Azure Logic Apps, Azure Machine Learning, and Azure Stream Analytics.

In production solutions, you will mostly rely on Azure IoT Hub as the hub for connecting your application against real-world IoT sensors.

Azure IoT Hub Device Provisioning Service

This service, also called DPS, is a provisioning service for Azure IoT Hub. It enables the provisioning and connection of millions of devices in a secure and scalable way to Azure IoT Hub – scenarios such as zero-touch provisioning, load-balancing, and connecting devices to different IoT solutions within your IoT platform.

Azure Logic Apps

Azure Logic Apps is an Azure service that allows you to build automated, scalable workflows. It incorporates the basic elements of the Azure platform, such as user roles, security, and more. It supports hundreds of connectors to connect to any Microsoft service or external well-known service in the cloud or on-premises. And if there is no connector available, you can directly call the service or create your own custom connector. It provides a visual web interface allowing you to build these workflows quickly. These workflows can then be integrated into applications such as your Azure Digital Twins instance.

Azure Logic Apps can be used to define processes around your instance. These processes will be able to create and update Digital Twins. Processes can act upon certain values that are read from the Digital Twins.

Azure Storage

In principle, a Digital Twin can store information in its properties. But as soon as new values are set, the old values are overwritten. Data can be preserved by connecting the Azure Digital Twins service to Azure Storage. Depending on your requirements, you could choose simple storage such as storage accounts. But in most cases, your Azure Digital Twins instance depends on IoT data. Azure Data Lake provides storage for these large amounts of data.

Azure Analytics

Azure Digital Twins is part of the IoT platform. In almost any Azure Digital Twins instance, IoT data is involved such as sensor data. The IoT platform provides several ways to perform analytics.

One of them is called Azure Time Series Insights Gen2. This is an open and scalable end-to-end IoT Analytics service. The service can be used for collecting, processing, storing, and querying IoT data to analyze trends and anomalies. It does not require any lines of code to use this service.

It provides APIs to integrate it into your solution. But it can also be integrated into the workflows defined around your solution.

Together with the Azure Digital Twins service, it can be used to monitor the health, usage, and performance of devices. This allows you to optimize operational efficiency.

Azure Service Bus

Azure Service Bus is cloud **messaging as a service** (**MaaS**). It simplifies cloud messaging on an enterprise level by delivering a scalable cloud messaging solution. The service acts as a message broker by using message queues and topics. When you built large IoT solutions including Azure Digital Twins, this service can support the routing of messages from IoT devices to a Digital Twin. Backend systems can be loosely coupled by using Azure Logic Apps. Azure Functions is used to get messages from the message queue into an Azure Digital Twins instance.

Exploring the Azure Digital Twins REST API

As you will have noticed, it is possible to use a large set of different Azure services to connect to Azure Digital Twins as an input, output, management, or compute service. While in some cases, no code is required to make the connection, some services require a more tailormade approach. Azure Functions is such an example. It requires the use of the REST API of the Azure Digital Twins service to access the Digital Twin Graph.

The REST API is divided into two different API models.

REST API	Description	Operations
Control plane	These APIs are used to manage an Azure Digital Twins instance. They provide control over creating, updating, and deleting an entire instance.	Create, update, delete, get, or list Digital Twins instances.
		Create, update, delete, get, or list endpoints.
		Check for the availability of a name for a Digital Twins instance.
		List all available Digital Twins service operations.
Data plane	These APIs allow us to manage the different parts of an Azure Digital Twins instance.	Add, update, delete, and list Digital Twins models.
		Add, update, delete, and list Digital Twins.
		Add, update, delete, and list relationships between Digital Twins.
		Send telemetry.
		Query a Digital Twins instance.
		Add, delete, and list event routes.

Each of these REST APIs is available through different SDKs. This gives you the ability to manage and access the Azure Digital Twins instance from various sources.

The following SDKs are available:

- **.NET SDK** – A C# SDK provided through NuGet. This allows you to easily create Azure Functions and other Azure services via Visual Studio by adding the required NuGet packages.
- **Java SDK** – A Java SDK to support Maven projects. Maven is an automation tool that is primarily used for building Java projects.
- **JavaScript SDK** – A JavaScript SDK available to create web-based solutions that require access to Azure Digital Twins.
- **Python SDK** – A **Python Package Index** (**PyPi**) SDK that allows you to access the Azure Digital Twins instance by using Python. Python is often used when building intelligent applications with, for example, Azure Machine Learning.

In this book, we will primarily focus on code examples on the .NET SDK.

Summary

In this chapter, you have learned and understood the concept of a Digital Twin and viewed a high-level overview of a Digital Twin environment. Based on several use cases and scenarios, we have shown how a Digital Twin can be used. We followed up with Azure Digital Twins, an Azure service that allows you to build a Digital Twin. At a global level, we explained the terminology, tools, architecture, and related Azure services. The final part of this chapter went into the availability of SDKs and APIs and in which languages they are available.

In the next chapters, we will go more into detail about each of the functionalities we have discussed in this chapter. We will guide you, step by step, in designing, creating, and building your own Digital Twin.

2
Requirements and Installation

In this chapter, we will go over all the requirements to get up and running with the Azure Digital Twins service. We will explain the prerequisites of Azure Digital Twins, as well as how to install and configure the service. We will also explain how to install and use some of the tools that will help to build Digital Twins solutions with this service. After this chapter, you will have all the tools available to start building.

In the upcoming chapters, you will learn and understand how to build Digital Twins solutions using Azure Digital Twins. But before you can start, we need to have the service up and running and some additional tools to support you.

We will go through the following topics:

- Azure Digital Twins service
- Microsoft Visual Studio
- The Windows Azure CLI with Windows PowerShell
- Node.js
- Azure Digital Twins Explorer
- Creating your first Digital Twin

Technical requirements

We will require a computer with Windows 10 and its latest updates installed. It is recommended to have a 1.8 GHz or faster processor. A quad-core or better is recommended. It is also recommended to have a minimum of 8 GB of RAM. We will require around 200 GB of available disk space for all installations and room for projects.

Azure Digital Twins service

The **Azure Digital Twins service** is a service from Microsoft that provides a means of storing the definitions, models, and logic around a Digital Twin. Microsoft provides a rich set of APIs and SDKs that enable all kinds of different ways to interact with, store, retrieve, and update definitions, models, and logic.

Azure account

An Azure account is required to create and use the Azure Digital Twins service. If you don't have one available, it is possible to create an Azure account for free today. The Azure account allows you to try out several services for free over a period of 12 months. After the 12-month period, you will require a credit card to continue using paid services. Sign up using the following URL:

```
https://azure.microsoft.com/en-us/free/
```

After signing up, you will have an **Azure Active Directory** (**AAD**) account that allows you to perform several administrative tasks in the Azure portal. The Azure portal can be found using the following URL:

```
https://portal.azure.com
```

Log in with your AAD credentials to get an overview of all your services.

Resource group

In this book, we will be using several different Azure services to create Digital Twin solutions, services such as Azure Functions, Event Hub, Event Grid, Service Bus, and Logic App. More structure can be achieved by using Azure resource groups. A **resource group** is a container that contains related Azure resources for a specific purpose, in our case, the Digital Twin solutions. It will allow us to easily manage and control all the resources we will create during the book.

We begin by creating a resource group through the Azure portal. Open a web browser, enter the URL, `https://portal.azure.com`, and log in with your AAD credentials.

Perform the following steps:

1. Open the menu on the left by pressing the top-left button with the three lines. This menu is also called the **portal menu**.

2. Select **Resource groups**.

3. Press the **+ Create** button to open the dialog for creating a new resource group:

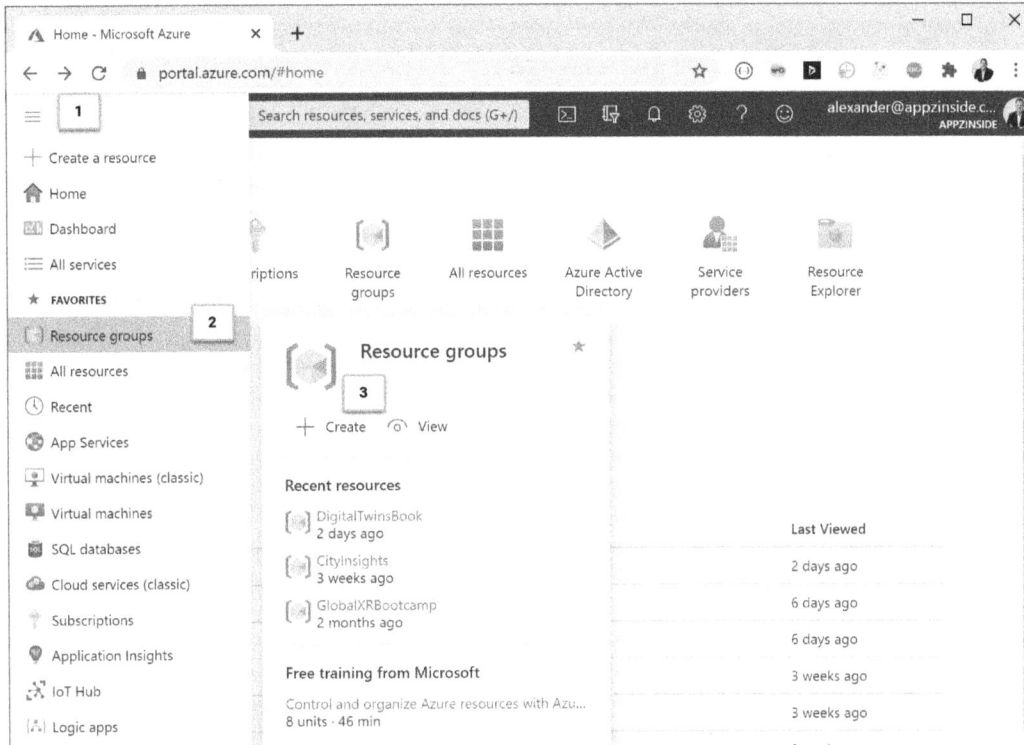

Figure 2.1 – Start creating a resource group through the Azure portal

In the next dialog, we will create a new user group. Perform the following steps:

1. Select your subscription. The name of the subscription will differ depending on whether you are using your own subscription, a trial subscription, or another subscription available to you. Having multiple subscriptions is not uncommon.

2. Enter the name `DigitalTwinsBook` as the name for your resource group.

3. Select the region of the resource group. You can select any region you want. This region can differ from the region specified when creating Azure services. At the time of writing this book, the Azure Digital Twins service is not available in every region. However, that will not be a problem since the region of the resource group does not have to be the same as the region of the service.

4. Press the **Review + create** button to review the information provided before we start creating the resource group:

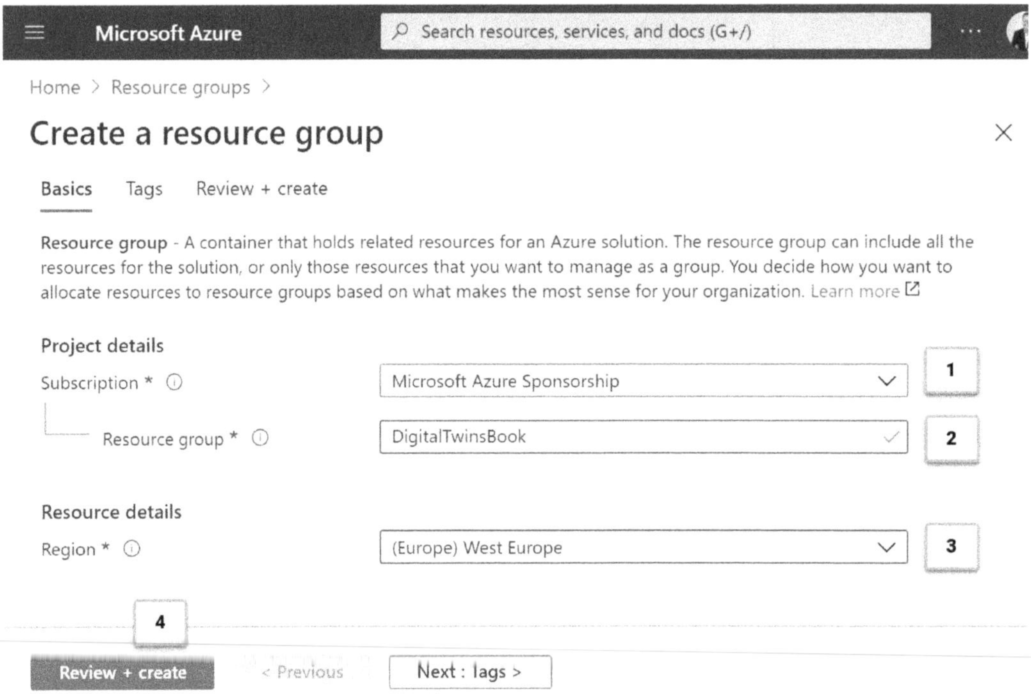

Figure 2.2 – Specifying the project details to create a resource group

Pressing the **Review + create** button will bring us the next dialog, as shown in *Figure 2.3*. Now, perform the final step.

5. Press the **Create** button to create the resource group:

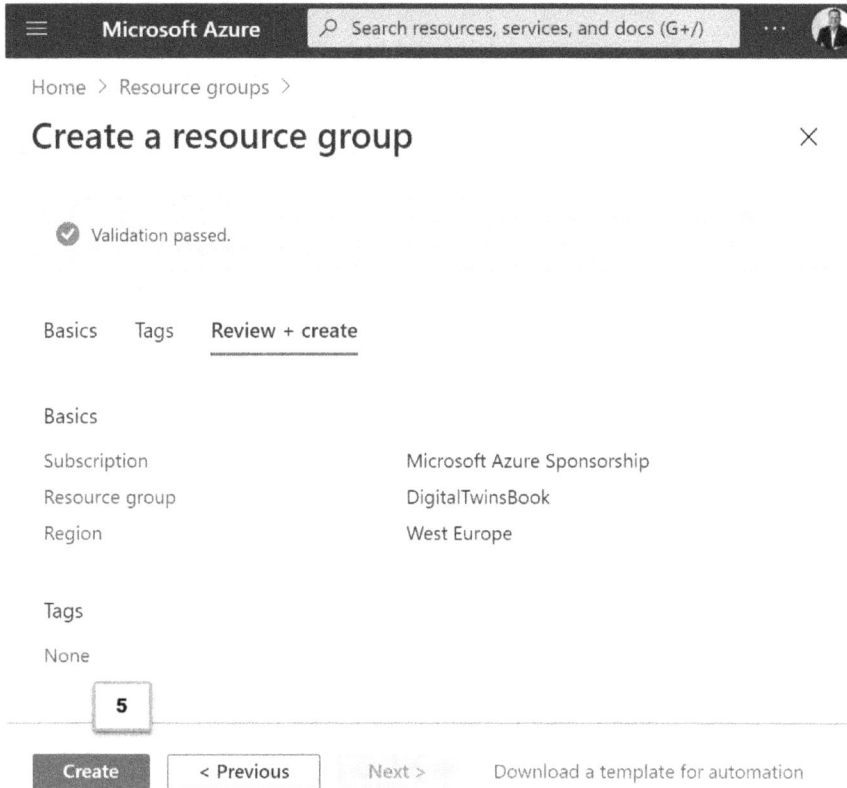

Microsoft Azure   Search resources, services, and docs (G+/)

Home > Resource groups >

Create a resource group   ✕

✔ Validation passed.

Basics   Tags   **Review + create**

Basics

Subscription   Microsoft Azure Sponsorship
Resource group   DigitalTwinsBook
Region   West Europe

Tags

None

5

Create   < Previous   Next >   Download a template for automation

Figure 2.3 – Creating the resource group

In the next step, we will create the Azure Digital Twins service and add it to the resource group we have just created.

Azure Digital Twins service

Microsoft provides a special Azure service called the Azure Digital Twins service as part of the mixed reality Azure category. This service is required to build Digital Twin solutions using Microsoft technology.

We will start installing the Azure service. Open the Azure portal if you haven't already done so. Click on the top-left button to open the portal menu again and select **Resource Groups**. This will open a list of all the resource groups created in this Azure portal. Select the **DigitalTwinsBook** resource group.

You will see an overview of the **DigitalTwinsBook** resource group. Perform the following step:

1. Press the **+ Add** button to add a new resource to the resource group:

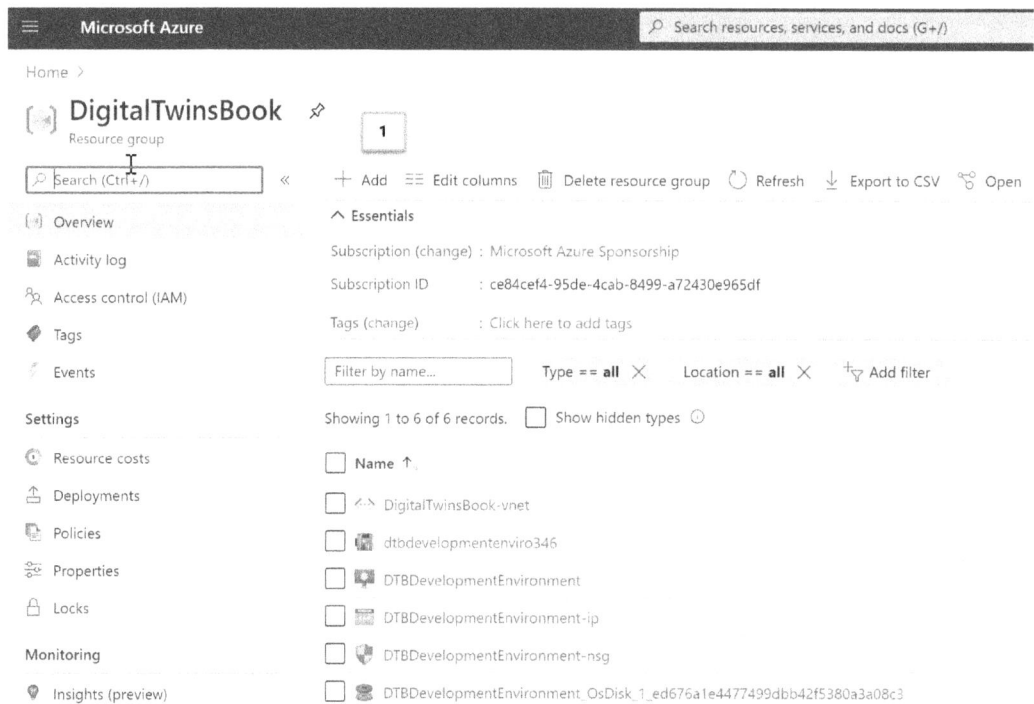

Figure 2.4 – An overview of the DigitalTwinsBook resource group

In the following screenshot, you will see the Azure Marketplace, which contains a large number of Azure services of Microsoft and third-party vendors. The Azure Digital Twins service is part of the *Mixed Reality* and the *Internet of Things* categories.

There are two options when it comes to finding the service. You can search for it by typing in Azure Digital Twins in the search field or, as in the example, via the Azure Marketplace categories. Perform the following steps:

1. Select the category called **Mixed Reality**.

2. Select the **Azure Digital Twins** service:

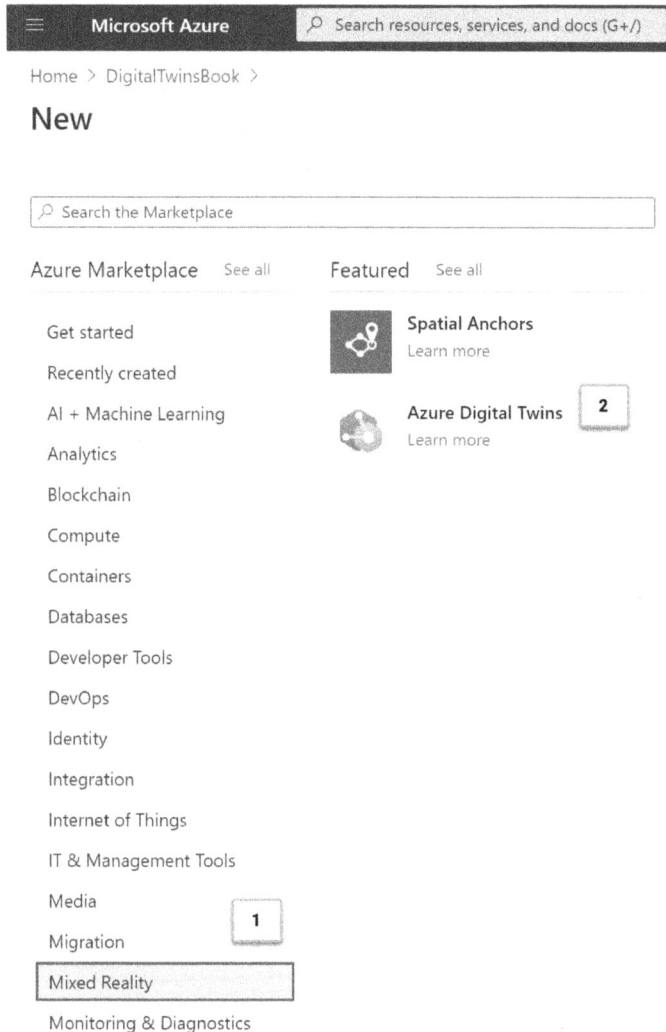

Figure 2.5 – Selecting the Azure Digital Twins service via the Mixed Reality category

We will create the Azure Digital Twins service in the next screen. Perform the following steps:

1. Since we are adding from a resource group, the **Subscription** field is already prefilled with the subscription used for the resource group. Leave this as it is.

2. Since we are adding from a resource group, the **Resource group** field is already prefilled with the resource group. Leave this as it is.

3. Select the **Location** field for the Azure Digital Twins service. To prevent any performance issues, it is always best to select the region closest to you.

    However, it is not a problem to select a region further away. In my case, the region is West Europe, since I reside in the Netherlands. More regions will become available over time. Check out the available regions in the **Region** dropdown and explore the pricing options at the following URL:

    ```
    https://azure.microsoft.com/en-us/pricing/details/digital-
    twins/
    ```

4. Enter a name for your resource. Use the name `DTBDigitalTwins`.

5. Press the **Review + create** button to review the information provided before we start creating the resource:

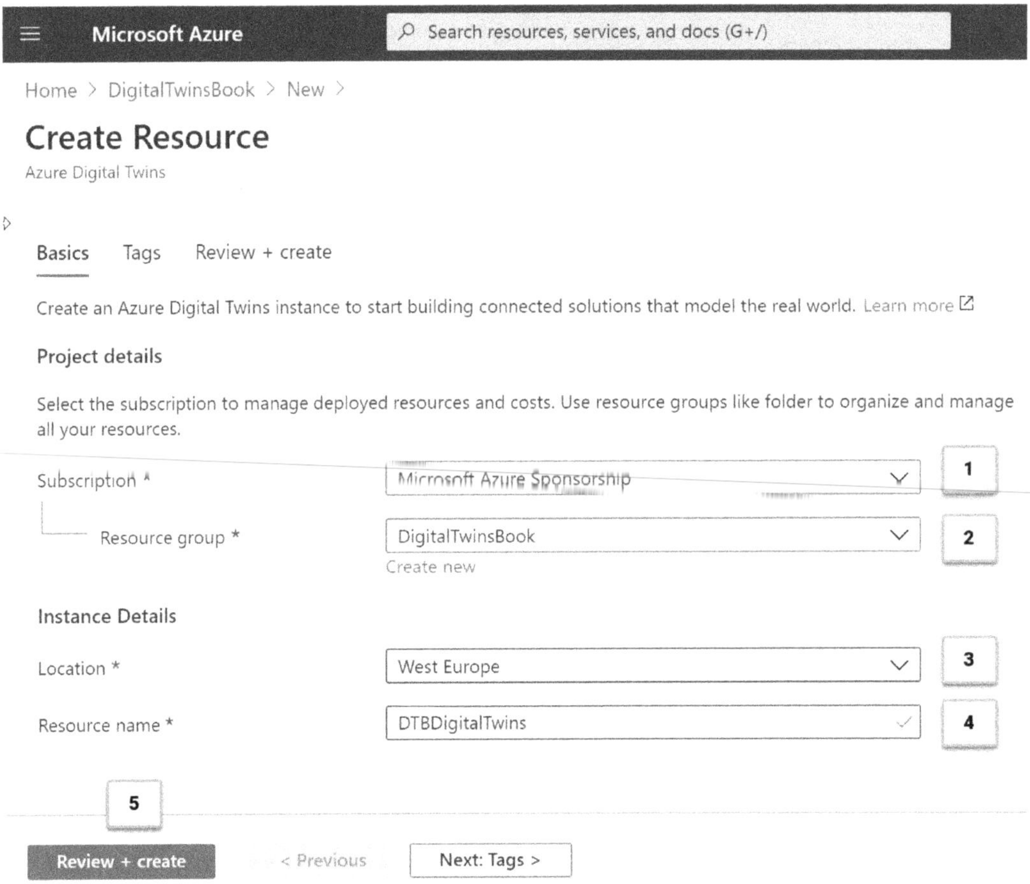

Figure 2.6 – Specifying the project details to create the Azure Digital Twins resource

6. Press the **Create** button to create the resource:

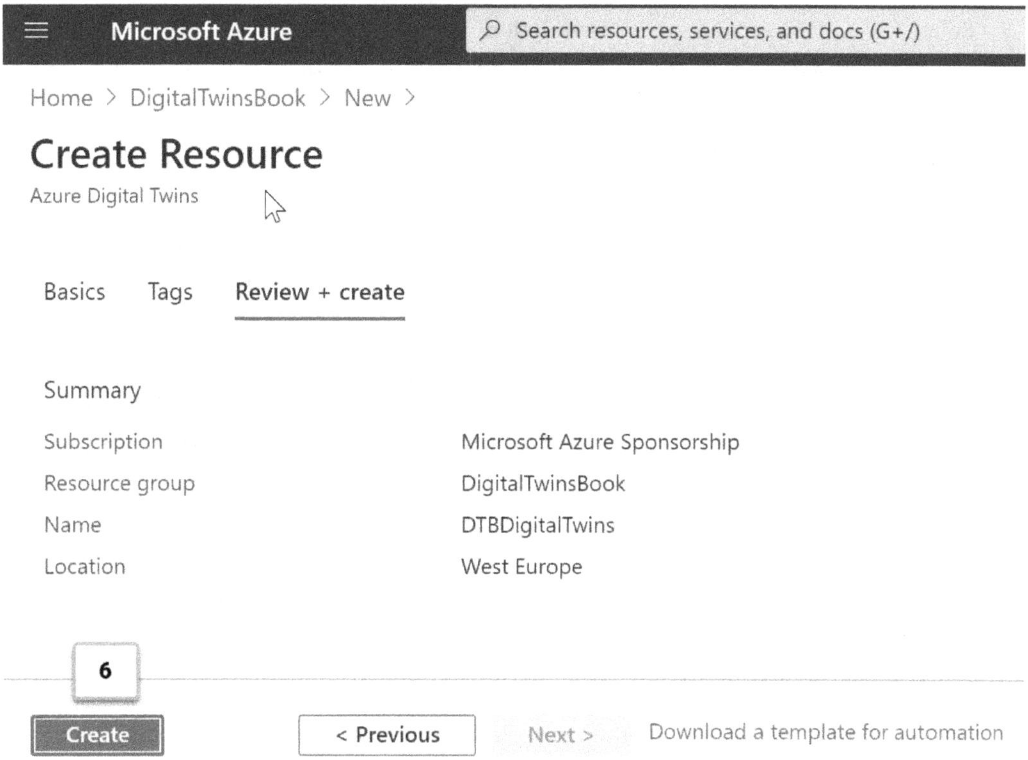

Figure 2.7 – Creating the Azure Digital Twins resource

It will take some time to create the resource. You will receive a notification when the resource has been created. From that moment, you can access the resource.

Configuring access control

We will need to configure access control for this resource. Access control is based on **RBAC. RBAC** stands for **role-based access control**. It allows us to assign a specific role to a user, application, service, or resource. This authorizes users or applications on behalf of a user to execute certain operations against the resource.

We have the following roles that can be assigned to existing users:

Role	Description
Azure Digital Twins Data Reader	This role gives read-only access to the resources of the Azure Digital Twin.
Azure Digital Twins Data Owner	This role gives full access to the resources of the Azure Digital Twin.

Determine for yourself which account you are going to use while building the examples in this book. In this example, I will be using my own work account to give it the required rights.

Since we will be creating, deleting, and updating elements from the service, we need to have the owner role, **Azure Digital Twins Data Owner**. Select the **DTBDigitalTwins** resource that has just been created from the **DigitalTwinsBook** resource group. Perform the following steps:

1. Select **Access control (IAM)** from the left-hand menu.

2. Press the + **Add** button to add a new role assignment.

3. Select the **Azure Digital Twins Data Owner** role.

4. Search for the name of the account and select it from the search results.

5. Press the **Save** button to create the new role assignment. The new role assignment for the selected user is now created:

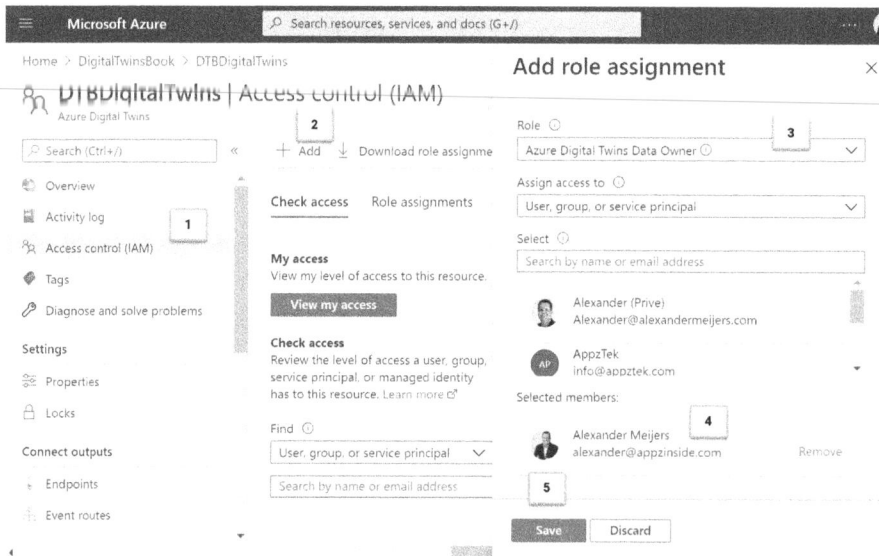

Figure 2.8 – Assigning an Azure Digital Twin role to an account

> **Important note**
>
> You need to have the Azure Digital Twins Data Reader role in order to be able to access the Azure Digital Twins service when using Azure Digital Twins Explorer.

In this section, we have set up and configured all the necessary Azure Services to run the Azure Digital Twins service. In the next section, we will begin by installing Microsoft Visual Studio as the tool used during the book to build the examples.

Microsoft Visual Studio

Several examples require us to program some code. We will be using **Microsoft Visual Studio** as the development platform. It is possible to use other applications, such as Visual Studio Code, but they will not be discussed in this book. There are several different versions of Microsoft Visual Studio available and you probably already have something installed. But if not, Microsoft has a Community Edition available. This Community Edition is free for individuals, academic users, and for use in open source projects. Microsoft Visual Studio Community can be found via the following URL: `https://visualstudio.microsoft.com/free-developer-offers/`.

When you open the URL, you will get the option to download the Microsoft Visual Studio Community Edition:

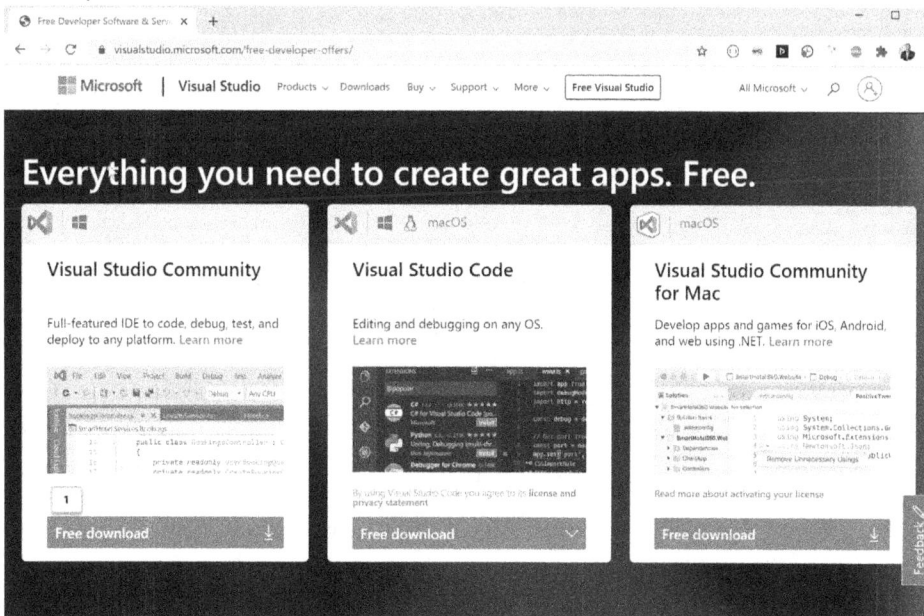

Figure 2.9 – The download website for the Microsoft Visual Studio Community Edition

The `Vs_community_[version].exe` file is downloaded into the `downloads` folder:

Figure 2.10 – The start of the Visual Studio Installer

Execute the file. This will start the Visual Studio Installer. Press the **Continue** button to continue the installation:

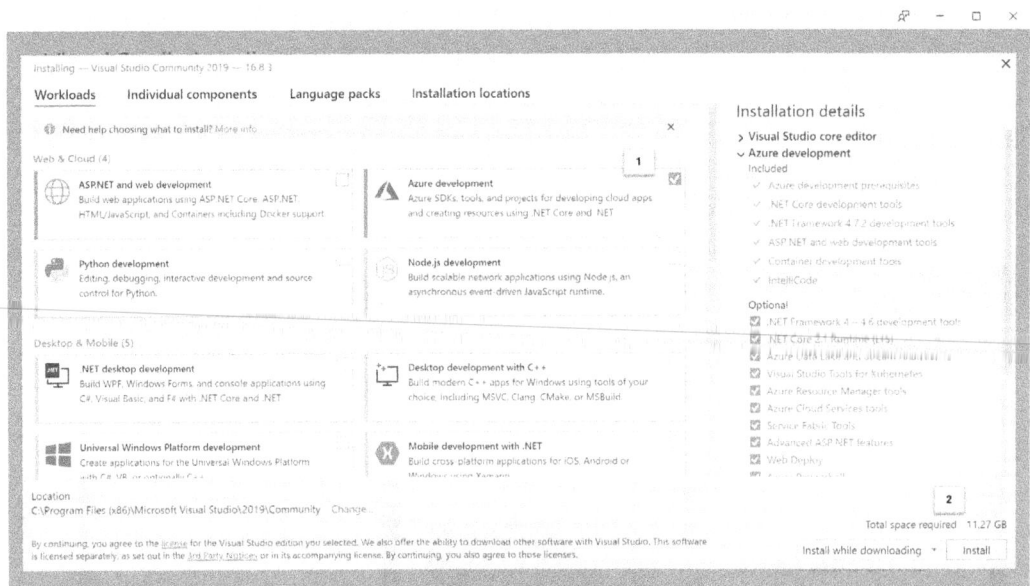

Figure 2.11 – Selecting the workloads that need to be installed

The installer will install by default most of the required packages to build a variety of applications using different languages. Separately, it is possible to select one or more workload packages that provide additional tools to build more specific applications.

In the following step, we need to select the workloads that are required to be installed. Perform the following steps (*Figure 2.11*):

1. Since we are primarily using this tool for Azure services, the Azure development workload needs to be installed. Select this workload.

2. Press the **Install** button to install Microsoft Visual Studio, including the selected workload packages.

    The installation of Microsoft Visual Studio Community will start. This can take some time, so be patient:

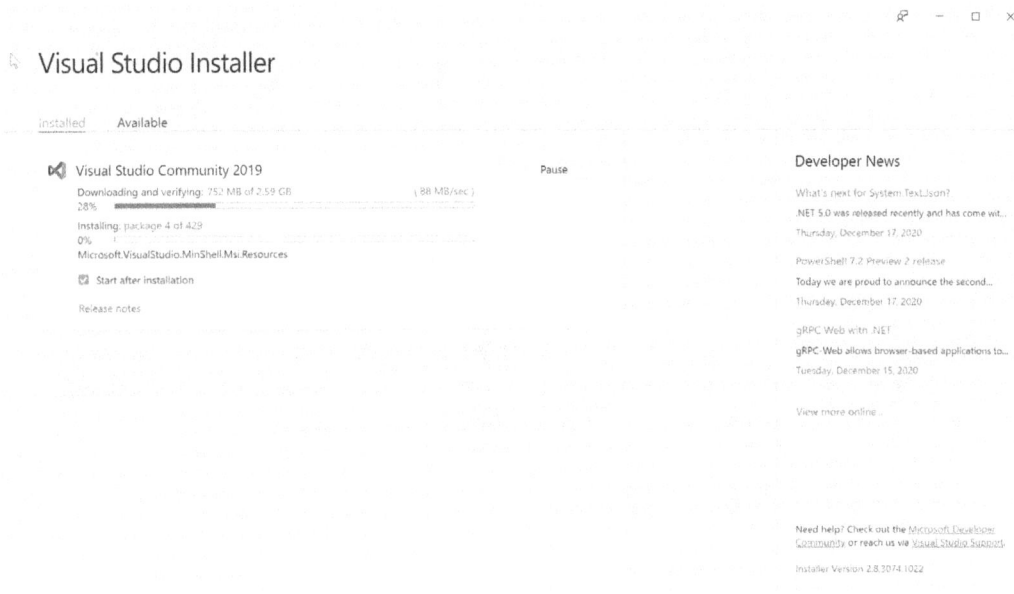

Figure 2.12 – Installation of Microsoft Visual Studio

You will see the following screen when it's finished:

Visual Studio Installer

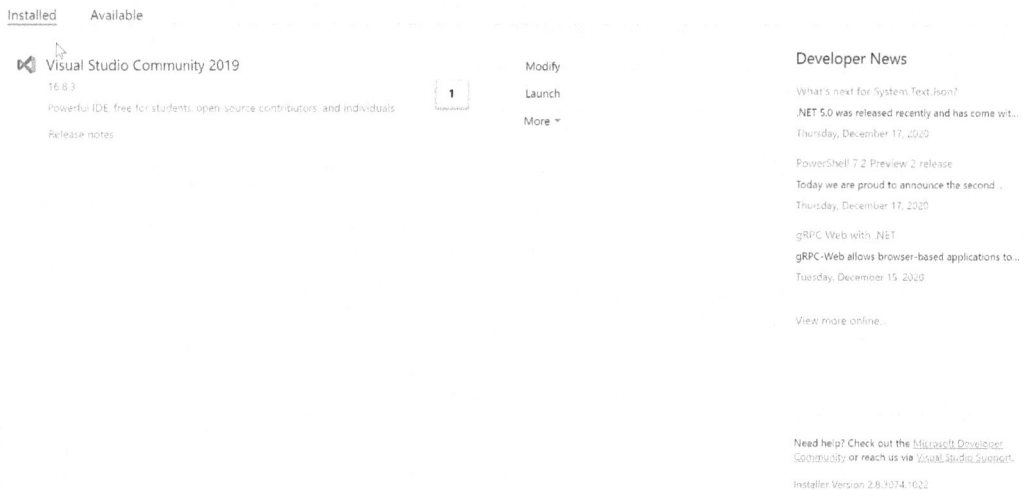

Installed   Available

Visual Studio Community 2019   Modify   Developer News

16.8.3   Launch   What's next for System.Text.Json?
Powerful IDE, free for students, open-source contributors, and individuals   More ▾   .NET 5.0 was released recently and has come wit...
Release notes   Thursday, December 17, 2020

                                                                                       PowerShell 7.2 Preview 2 release
                                                                                       Today we are proud to announce the second ...
                                                                                       Thursday, December 17, 2020

                                                                                       gRPC-Web with .NET
                                                                                       gRPC-Web allows browser-based applications to...
                                                                                       Tuesday, December 15, 2020

                                                                                       View more online...

                                                                                       Need help? Check out the Microsoft Developer
                                                                                       Community or reach us via Visual Studio Support.
                                                                                       Installer Version 2.8.3074.1022

Figure 2.13 – The installation of Microsoft Visual Studio has finished

As soon as the installer has finished, we can start Visual Studio by pressing the **Launch** button. This will open a dialog requesting you to sign in. Enter the credentials you used when creating your Azure account or use your existing Azure account credentials:

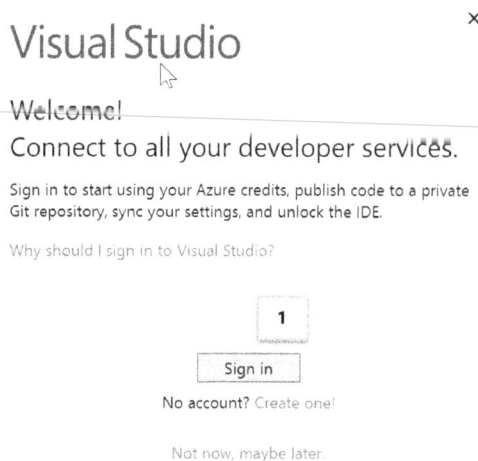

Visual Studio   ✕

Welcome!
Connect to all your developer services.

Sign in to start using your Azure credits, publish code to a private
Git repository, sync your settings, and unlock the IDE.

Why should I sign in to Visual Studio?

Sign in

No account? Create one!

Not now, maybe later.

Figure 2.14 – Logging in with your Microsoft or work account credentials

It is possible to log in later when you start Microsoft Visual Studio. The installation and configuration of Microsoft Visual Studio are now ready to create projects.

In this section, we have installed Visual Studio 2019 along with all its prerequisites. In the next section, we are going to install the Windows Azure CLI, which is part of Windows PowerShell.

The Windows Azure CLI with Windows PowerShell

We will be using Windows PowerShell to start Azure Digital Twins Explorer.

Windows PowerShell is a set of cmdlets that allows us to manage Azure resources via the PowerShell command line. Cmdlets are lightweight PowerShell commands written in a compiled .NET language.

The Azure command-line interface, also called the Azure CLI, is a set of commands for managing Azure resources. It allows us for examples to manage Azure resources using Windows Azure PowerShell.

The Azure CLI will be used to start Azure Digital Twins Explorer.

To install the Azure CLI, we will be using the MSI file, which can be found at the following URL: `https://docs.microsoft.com/en-us/cli/azure/install-azure-cli-windows?view=azure-cli-latest&tabs=azure-cli`.

To check whether the Azure CLI has been installed correctly, we need to start a Windows PowerShell instance. Perform the following steps:

1. Type in the search field next to the Windows **Start** button in your Windows 10 environment the text `Windows PowerShell`. This will show results containing the Windows PowerShell application.

2. Start the Windows PowerShell application with administrative rights.

3. Type in the following command: `az login`.

    If the Azure CLI has been installed correctly, this will open a browser window that allows you to log on with credentials. Log in with your Azure account credentials:

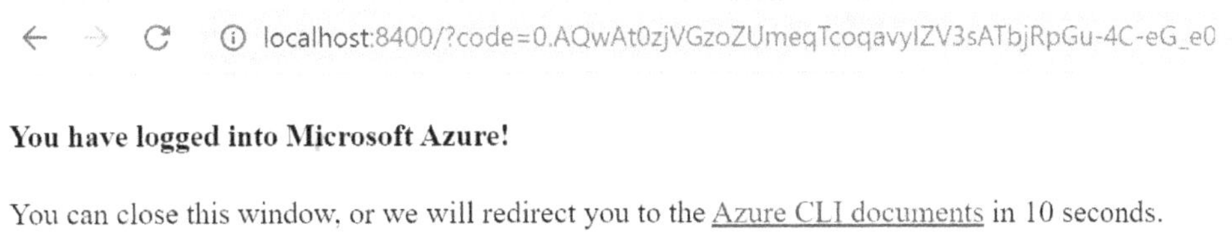

    ← → C ⓘ localhost:8400/?code=0.AQwAt0zjVGzoZUmeqTcoqavyIZV3sATbjRpGu-4C-eG_e0

    You have logged into Microsoft Azure!

    You can close this window, or we will redirect you to the Azure CLI documents in 10 seconds.

Figure 2.15 – Logged in to Microsoft Azure successfully

When the credentials are correct, you will be redirected to a page explaining that you have been logged in to Microsoft Azure. A list of available Azure subscriptions is listed through a JSON payload. Now you can use Azure CLI commands to access Azure resources.

> **Important note**
> Keep in mind that you will only be able to access and manage Azure resources for which the account you used to log on is authorized.

We have installed and configured the Windows Azure CLI, which allows us to use Azure CLI commands in Windows PowerShell. In the next section, we will install and configure Node.js.

Node.js

Node.js is an open source JavaScript runtime environment that allows us to execute JavaScript code outside a web browser. It is mostly used for creating event-driven servers to create websites and API services and operates on a single-thread architecture. It allows you to create a local web server with ease. We will be using Node.js to create the local web server in which Azure Digital Twins Explorer will run:

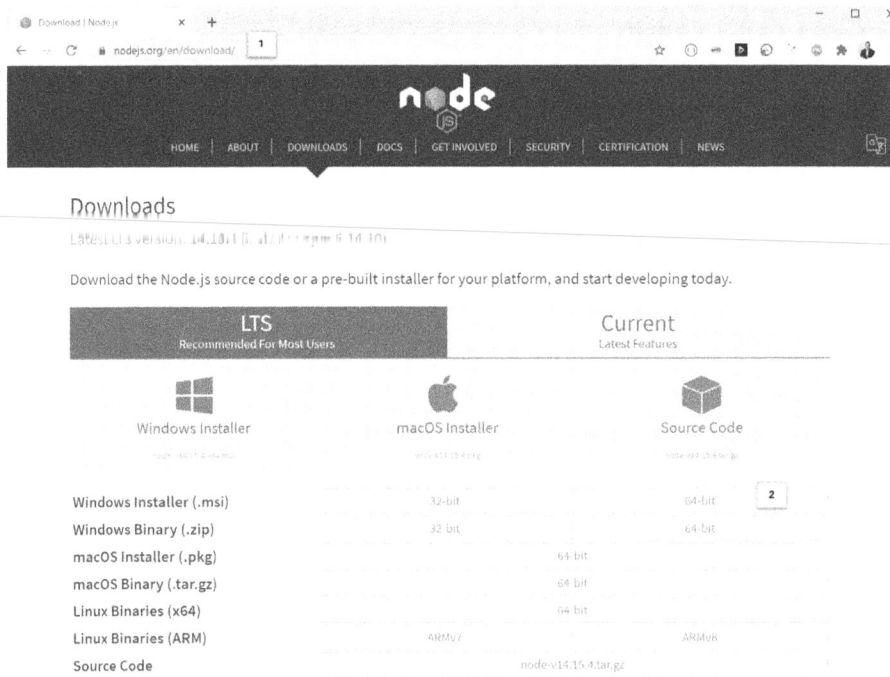

Figure 2.16 – The download location of Node.js

We begin by installing Node.js. Perform the following steps:

1. Open a web browser and go to the following URL: `https://nodejs.org/en/download`. This will open a download page for Node.js.

2. Select the 32-bit or 64-bit version of the Windows Installer file depending on the Windows 10 operating system that is installed.

This will download the MSI file into the downloads folder. Run the installer. Select the following installation path: `C:\Program Files\nodejs`.

We need to make sure that the installation location of Node.js is part of the standard PATH variable. Node.js is required for running the `npm` command in PowerShell. `npm` is a package manager for JavaScript and allows us to compile, build, and run a JavaScript project. And this is exactly what we use to build and start Azure Digital Twins Explorer.

We have installed and configured Node.js, which is required to get Azure Digital Twins Explorer installed and running in the next section.

Azure Digital Twins Explorer

While the Azure Digital Twins service stores Digital Twin-related elements, it has no way of visualizing it. We will need to use the APIs and SDKs to see what is defined therein. For that, Microsoft has released Azure Digital Twins Explorer. This explorer is one of the Azure samples found at GitHub. It is a JavaScript-based website that accesses the Azure Digital Twins service via the JavaScript SDK to visualize and manage an Azure Digital Twins instance.

Download

We will be getting the Azure Digital Twins Explorer source code from GitHub in the following steps. This requires having an account on GitHub. If you do not have an account, create one accordingly. Open a web browser and go to the corresponding URL, `https://github.com`, to create an account.

If you have an account, perform the following steps:

1. Open a web browser and use the following URL: `https://github.com/Azure-Samples/digital-twins-explorer`. Make sure you are logged in to GitHub.

2. Select the **Code** button. This will reveal a drop-down dialog.

3. From the drop-down dialog, select **Download ZIP**. This will download the complete package locally to the Downloads folder:

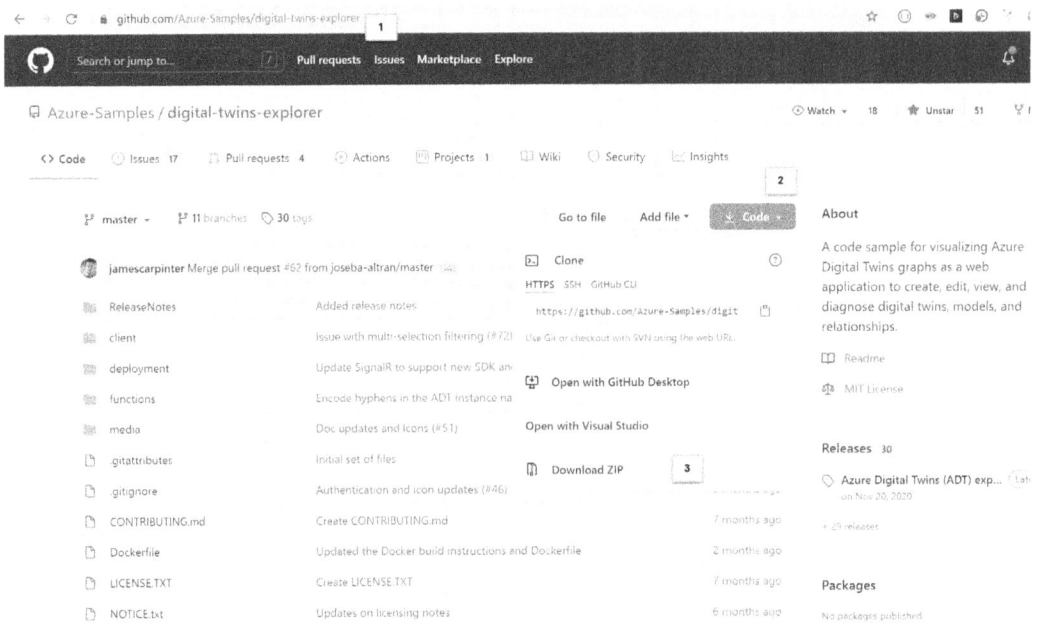

Figure 2.17 – GitHub location of the Azure Digital Twins Explorer source code

4. We need to unpack the downloaded package into a folder. Open Windows File Explorer.

5. Create a `Github` directory under `C:\`.

6. Unpack the downloaded file into this folder. This will create a `digital-twins-explorer-master` folder containing all the files for the Azure Digital Twins Explorer application:

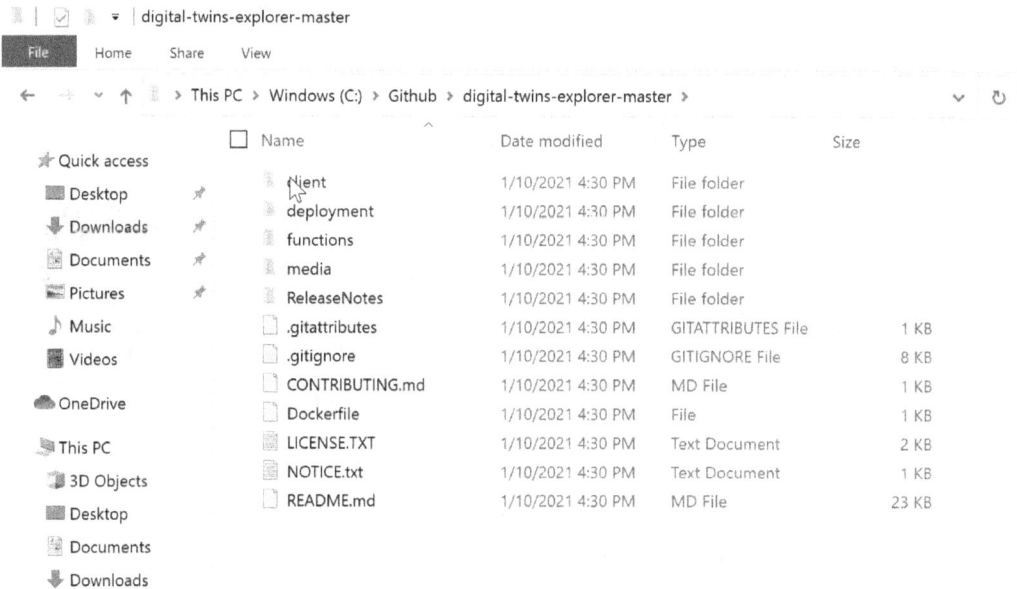

Figure 2.18 – Azure Digital Twins Explorer source code unpacked in the C:\Github folder

Now we're ready for installation.

Installation

In this next stage, we will install Azure Digital Twins Explorer. We will be using npm. npm is a package manager for JavaScript and allows us to compile, build, and run a JavaScript project:

1. Open a Windows PowerShell using administrative rights. Change the current directory to the source directory of the Azure Digital Twins Explorer project using the following command:

    ```
    cd c:\github\digital-twins-explorer-main\client\src
    ```

2. Start the installation by using the following command:

```
npm install
```

Here's what the output will look like:

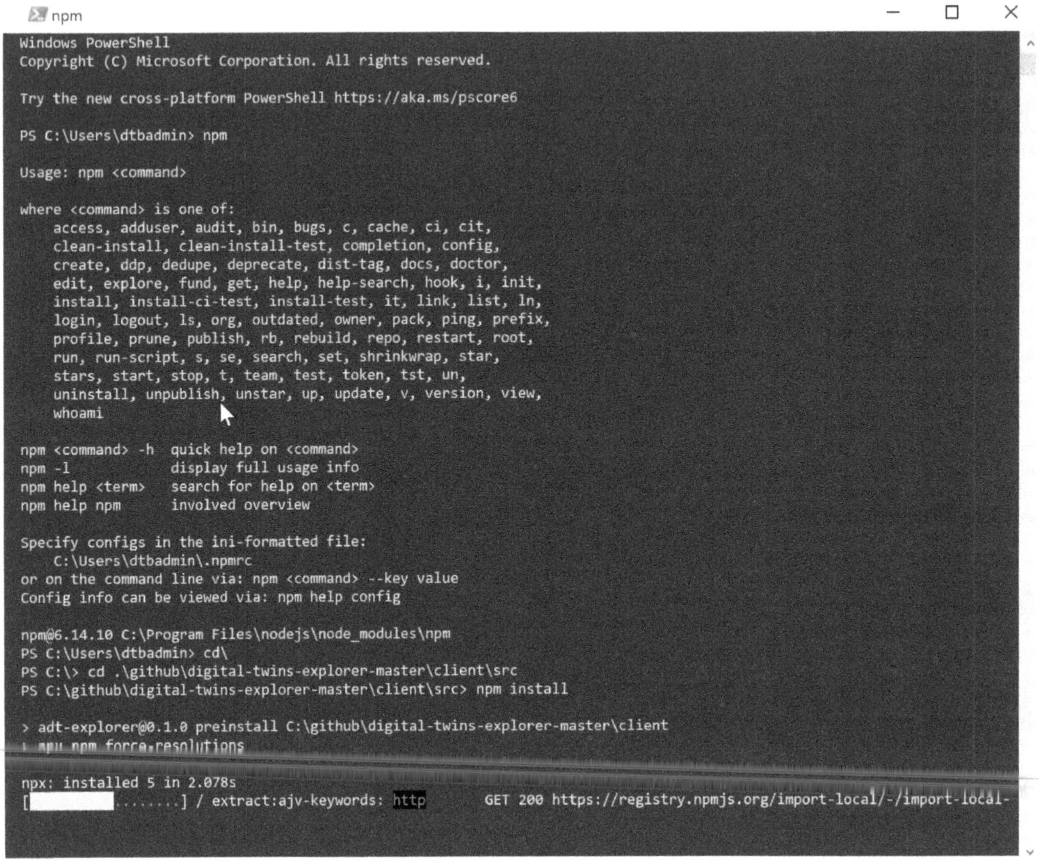

Figure 2.19 – npm is installing all the requisite packages and the packages that these packages depend on

This command will install all the requisite packages and the packages that these packages depend on.

Compile and run

At this stage, we will compile and run Azure Digital Twins Explorer. It compiles the JavaScript, loads the packages, and creates a web server. You will need to do this whenever you want to use Azure Digital Twins Explorer.

Azure Digital Twins requires you to be logged in. Use the following command:

```
az login
```

It will open your browser to let you log in:

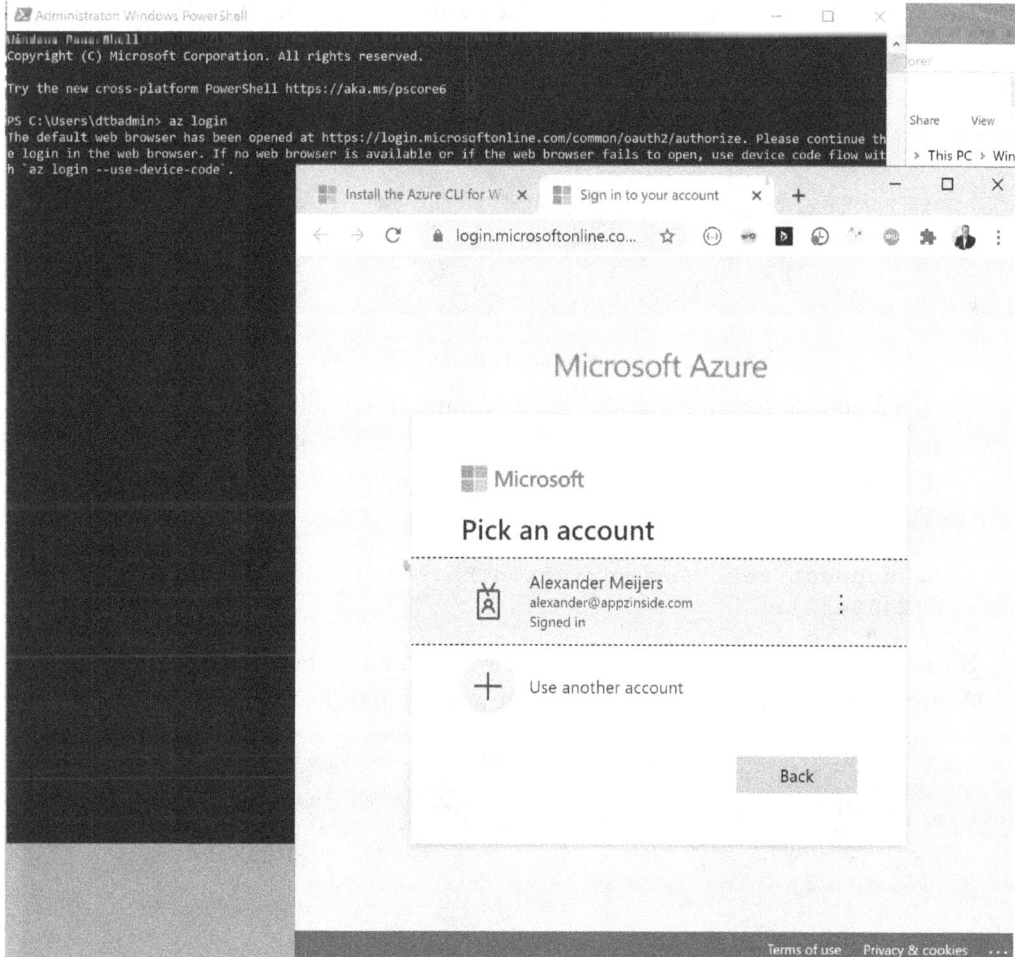

Figure 2.20 – Logging in to Azure using your Azure account credentials

Use an Azure account that has the **Azure Digital Twins Data Owner** role. When you log in successfully, it will list the available subscriptions. In most cases, this will be a single subscription, which is the default subscription. In some cases, however, like mine, I have several more subscriptions.

The subscription that contains the Azure Digital Twins service must be the default subscription. Azure Digital Twins Explorer will use the default subscription by default.

Perform the following steps to change the default subscription:

1. Begin with the following command:

    ```
    az account list --output table
    ```

    This will list a table containing all available Azure subscriptions, the ID of the Azure subscription, and whether the Azure subscription is the default subscription:

Figure 2.21 – List of available Azure subscriptions

2. Use the following command to change the default Azure subscription by using the Azure subscription ID. The following ID is a fake one. It is always best to use the ID since in some cases, the name of Azure subscriptions can be the same as you see above:

    ```
    az account set --subscription bb84cef4-9de4-4cab-1232-
    a72430e321ae
    ```

    If you are logged in to Azure and have the wrong Azure subscription set to default, Azure Digital Twins Explorer will generate an error like the following:

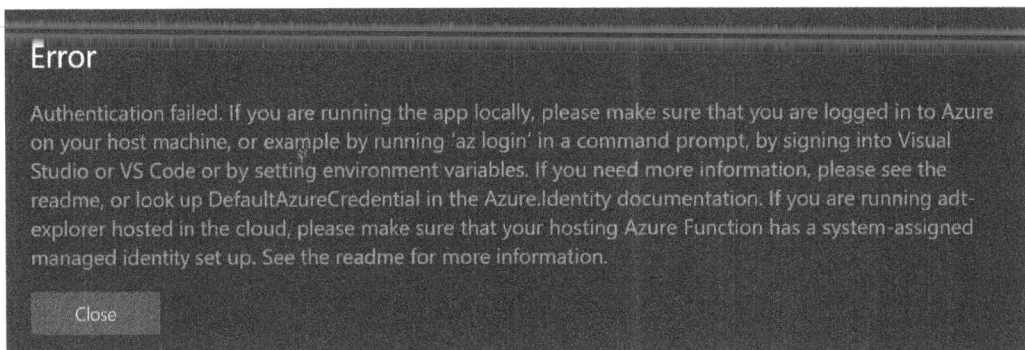

Figure 2.22 – Error produced by Azure Digital Twins Explorer

3.   Start Azure Digital Twins Explorer by using the following command:

```
npm run start
```

4.   Azure Digital Twins Explorer will ask you to provide the Azure Digital Twins URL. This URL can be found by going to the Azure Digital Twins service resource in the Azure portal:

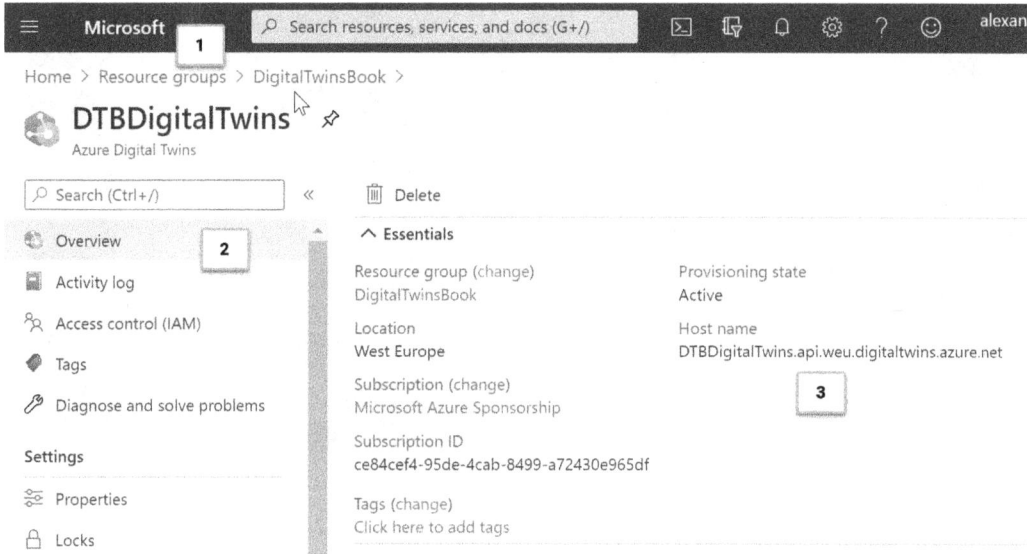

Figure 2.23 – Looking up the hostname of the Azure Digital Twins service

Now, let's perform the following steps to retrieve the hostname of your Azure Digital Twins service:

1.   Open the **DTBDigitalTwins** resource.

2.   Select **Overview** from the left-hand menu.

3.   Copy the hostname.

Perform the following steps in Azure Digital Twins Explorer to enter the Azure Digital Twins URL:

1. Enter the URL by combining `https://` with the copied hostname.

2. Press the **Save** button to store the URL:

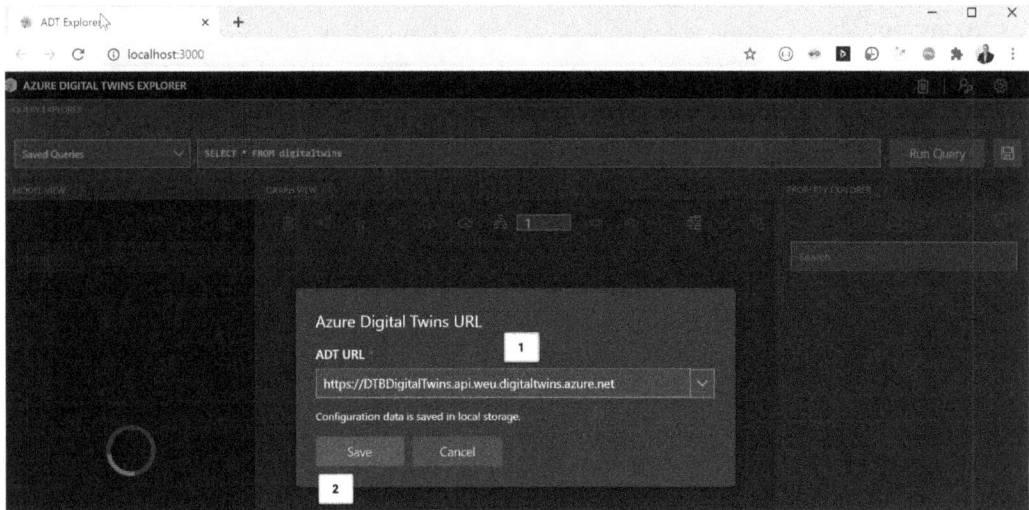

Figure 2.24 – Configuring the Azure Digital Twins URL in Azure Digital Twins Explorer

You can always change the URL later via the settings icon in the top-right corner.

We have now installed all the necessary tools for building examples and solutions using the Azure Digital Twins service. In the final section of this chapter, we will be creating our first Digital Twin using these tools.

Creating your first Digital Twin

This final section will help you to create your first Digital Twin using Azure Digital Twins Explorer. The source code of Azure Digital Twins Explorer comes with a rich set of example files that we can use to create the first Digital Twin graph. The example files can be found here: `C:\Github\digital-twins-explorer-master\client\examples`.

The structure of Azure Digital Twins will be explained in depth in the upcoming chapters. For now, we need to understand that an Azure Digital Twins instance contains one or more Digital Twins. Each Digital Twin, which represents an entity, is based on a model. A model defines the properties of the Digital Twin and the relationships to other Digital Twins. A **Digital Twin Graph** is a visual representation of Digital Twins and their relationships.

Uploading models

A Digital Twin is always created from a model. Models are defined by JSON files. This means that we need to upload the models first. Perform the following steps:

1. Select the **import model** icon at the top of the model view section.

2. Go to the examples folder: `C:\Github\digital-twins-explorer-master\client\examples`.

3. Select all JSON files.

4. Press the **Open** button:

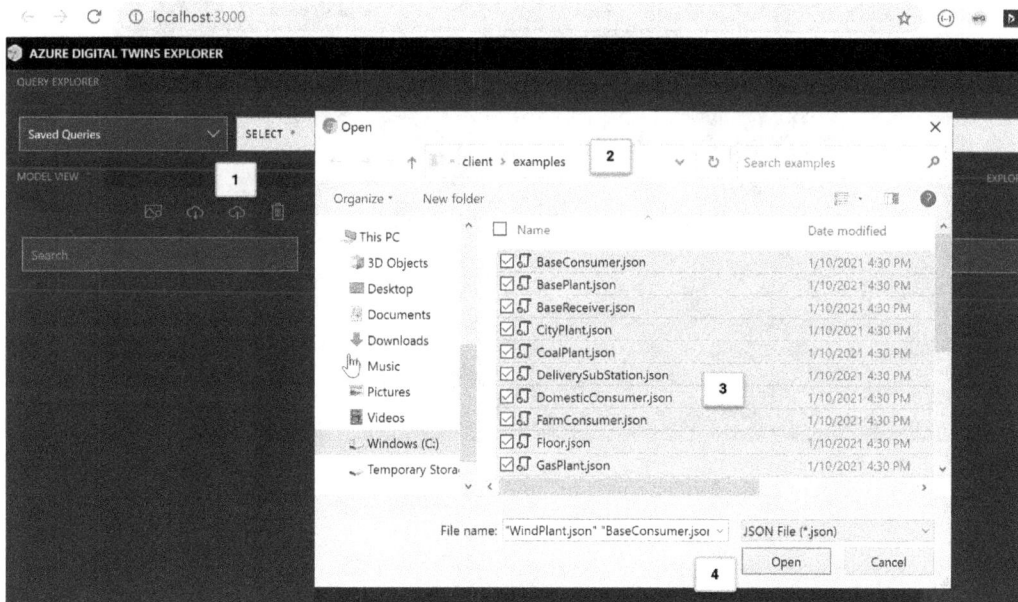

Figure 2.25 – Uploading Digital Twin models into Azure Digital Twin Explorer

This will create the models based on the information from the JSON files. We will go into more detail regarding the definition of models in the next chapters. The result of the import can be seen here:

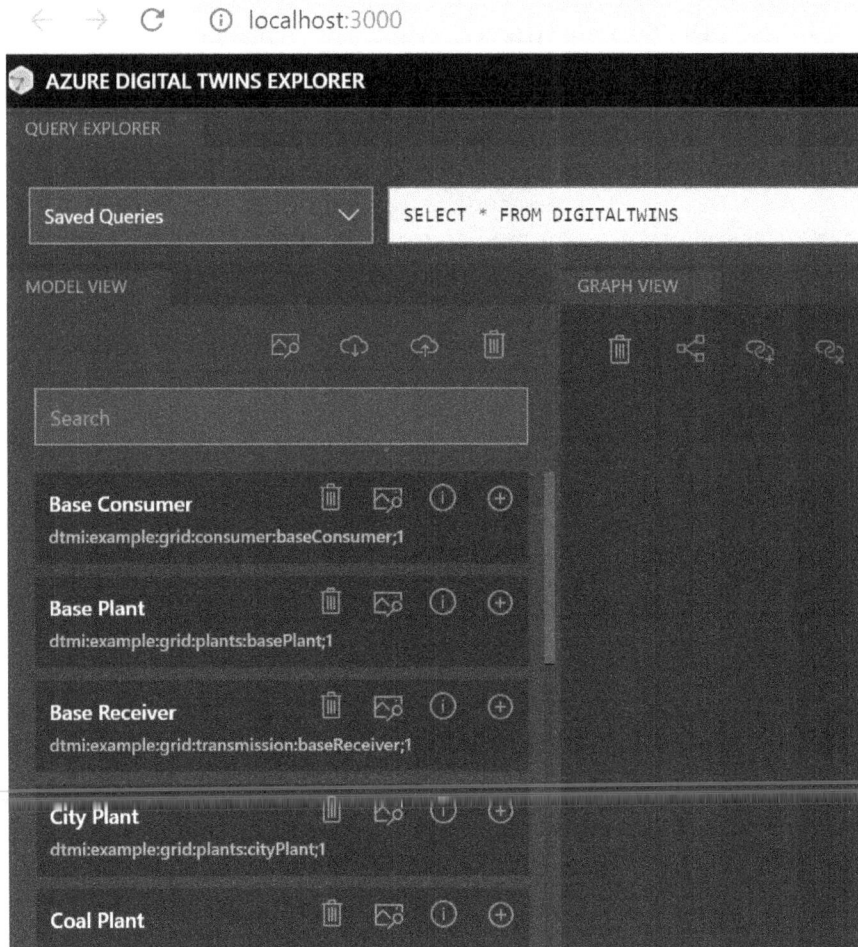

Figure 2.26 – Models uploaded to Azure Digital Twins Explorer

In the next step, we need to import the Azure Digital Twins graph. A graph describes the Digital Twins and their relationships. Each Digital Twin is based on a Digital Twin model:

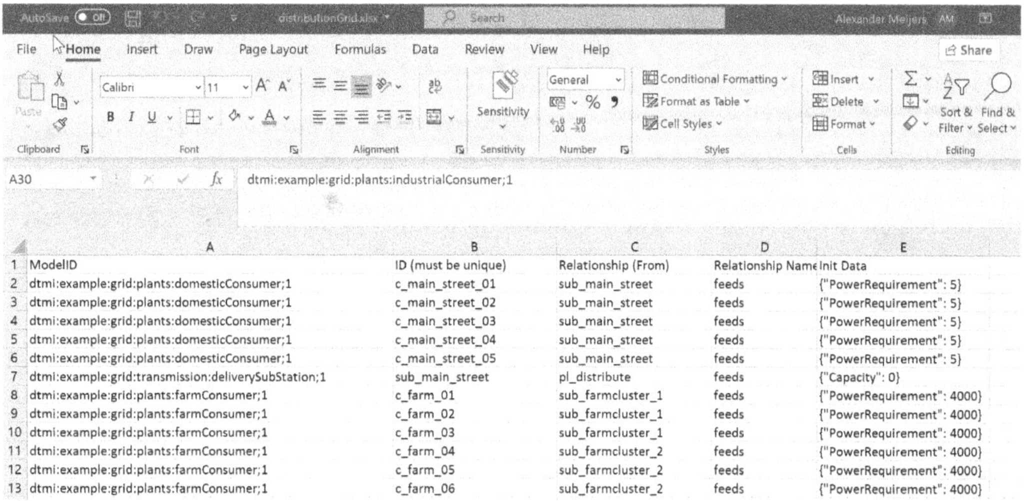

	A	B	C	D	E
1	ModelID	ID (must be unique)	Relationship (From)	Relationship Name	Init Data
2	dtmi:example:grid:plants:domesticConsumer;1	c_main_street_01	sub_main_street	feeds	{"PowerRequirement": 5}
3	dtmi:example:grid:plants:domesticConsumer;1	c_main_street_02	sub_main_street	feeds	{"PowerRequirement": 5}
4	dtmi:example:grid:plants:domesticConsumer;1	c_main_street_03	sub_main_street	feeds	{"PowerRequirement": 5}
5	dtmi:example:grid:plants:domesticConsumer;1	c_main_street_04	sub_main_street	feeds	{"PowerRequirement": 5}
6	dtmi:example:grid:plants:domesticConsumer;1	c_main_street_05	sub_main_street	feeds	{"PowerRequirement": 5}
7	dtmi:example:grid:transmission:deliverySubStation;1	sub_main_street	pl_distribute	feeds	{"Capacity": 0}
8	dtmi:example:grid:plants:farmConsumer;1	c_farm_01	sub_farmcluster_1	feeds	{"PowerRequirement": 4000}
9	dtmi:example:grid:plants:farmConsumer;1	c_farm_02	sub_farmcluster_1	feeds	{"PowerRequirement": 4000}
10	dtmi:example:grid:plants:farmConsumer;1	c_farm_03	sub_farmcluster_1	feeds	{"PowerRequirement": 4000}
11	dtmi:example:grid:plants:farmConsumer;1	c_farm_04	sub_farmcluster_2	feeds	{"PowerRequirement": 4000}
12	dtmi:example:grid:plants:farmConsumer;1	c_farm_05	sub_farmcluster_2	feeds	{"PowerRequirement": 4000}
13	dtmi:example:grid:plants:farmConsumer;1	c_farm_06	sub_farmcluster_2	feeds	{"PowerRequirement": 4000}

Figure 2.27 – An Excel file describing the Azure Digital Twin graph

To import this file describing the Azure Digital Twin graph, perform the following steps:

1. Click on the import graph symbol at the top of the graph view.
2. Go to the examples folder: `C:\Github\digital-twins-explorer-master\client\examples`.
3. Select the `distributionGrid.xlsx` file.

4.   Press the **Open** button:

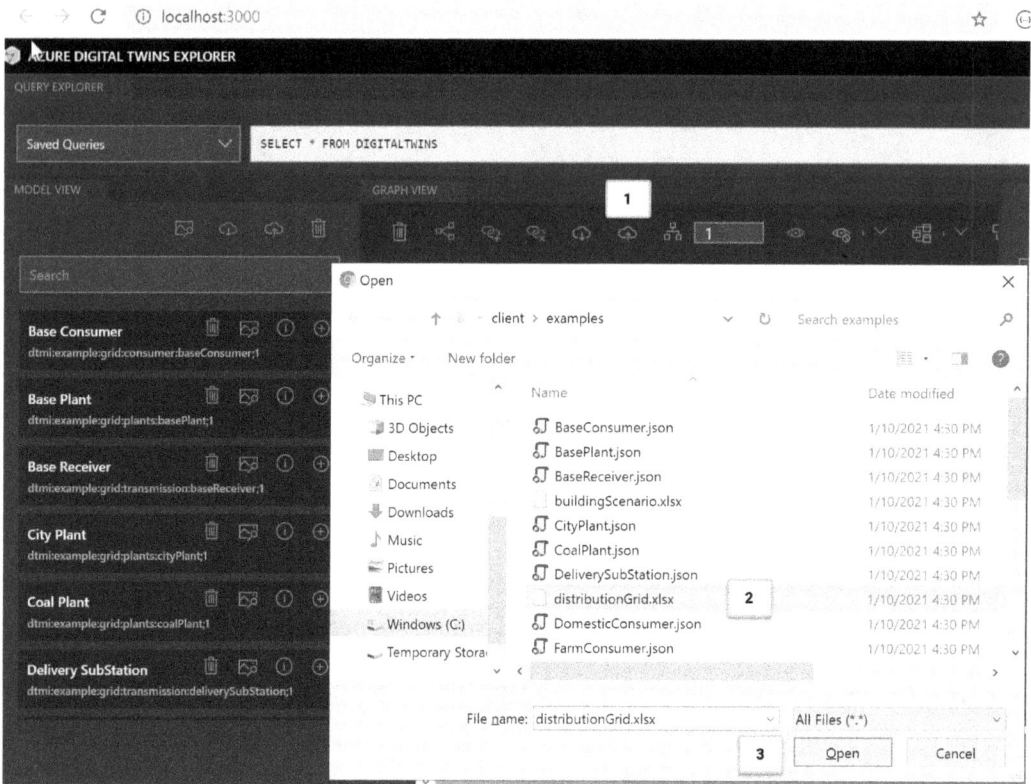

Figure 2.28 – Importing the Digital Twin graph defined by an Excel file

This will start the import of the Azure Digital Twin graph. Initially, you get to see a preview of that graph. Perform the following steps to import the graph into the Azure Digital Twins instance:

1.   Press the **Save** symbol in the top-right corner in the **Import** tab.

2.   After a short time, a dialog will appear with the number of twins and relationships imported. Press the **Close** button to close the dialog:

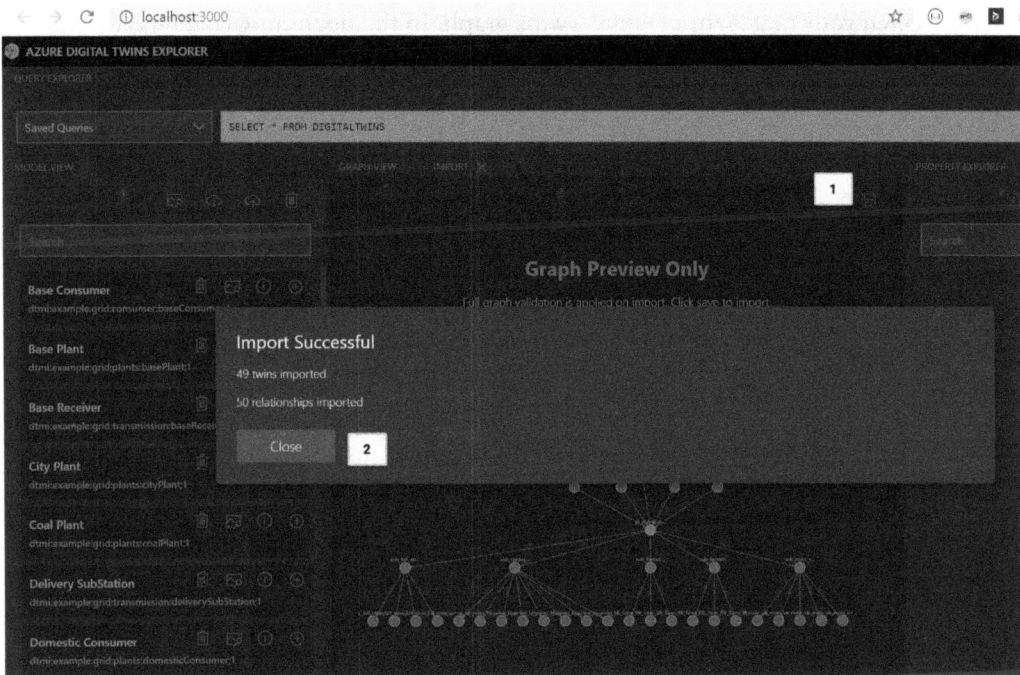

Figure 2.29 – Importing Azure Digital Twins and relationships

The result shows the Azure Digital Twins graph based on Digital Twins and relationships:

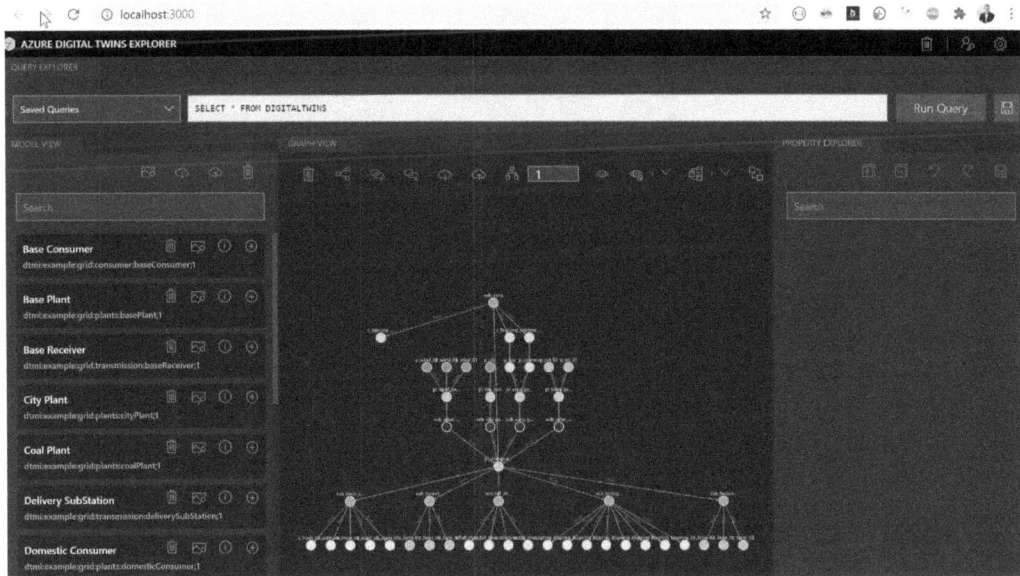

Figure 2.30 – Result of an imported Azure Digital Twins graph

You have created your first Azure Digital Twins graph. In the upcoming chapters, we will go into more detail regarding how to create your own models and relationships. As part of that, we will also explain more about how you can edit, the different types of visualization, and running queries using Azure Digital Twins Explorer.

Summary

During this chapter, you have learned and understood all the prerequisites and requirements for using the Azure Digital Twins service. You have installed and configured this service. You have also learned about the necessary tools that allow you to build solutions around Azure Digital Twins. This incorporated development includes tools such as Microsoft Visual Studio and viewer tools such as Azure Digital Twins Explorer. You have created your first Digital Twins graph, containing instances of predefined models and underlying relationships.

In the next chapter, we will go into more detail regarding the definition model of Azure Digital Twins.

Section 2: Getting Started with Azure Digital Twins

This next section will help you to understand the foundation of a digital twin specified by the Digital Twin definition model. We will go through each of the elements and how these apply to the elements of the Azure Digital Twins service. Several examples are used to build models, properties, and relationships, and query the data.

This part of the book comprises the following chapters:

- *Chapter 3, Digital Twin Definition Model*
- *Chapter 4, Understanding Models*
- *Chapter 5, Model Elements*
- *Chapter 6, Creating Relationships between Azure Digital Twin Models*
- *Chapter 7, Querying Digital Twins*
- *Chapter 8, Building Models Using Ontologies*

3
Digital Twin Definition Model

This chapter will look at the **Digital Twin Definition Language** in depth. The Digital Twin Definition Language is used to describe the meta models we use to create custom models to instantiate Digital Twins. First, we will explain the different meta models such as interfaces, properties, and components. Then, we will learn about the Digital Twin Model identifier, its available schemas, and its semantic types. Some examples will be used to explain it more thoroughly.

While this chapter is more theoretical than the other chapters, it will pave the way for building Digital Twins using the Azure Digital Twins service. You will use each of these definitions in the upcoming chapters to understand their examples, which allow you to start building Digital Twins using the service.

In this chapter, we'll cover the following topics:

- Digital Twins Definition Language
- Interface content
- Schemas
- Primitive schemas
- Complex schemas

- Geospatial schemas
- Semantic type
- Validating a model

Technical requirements

This chapter will talk about validating models using the DTDL Validator tool, which can be found at `https://docs.microsoft.com/en-us/samples/azure-samples/dtdl-validator/dtdl-validator/`. We will also be using some of the tools that we installed and configured in the previous chapter.

Digital Twins Definition Language

The models in the Azure Digital Twins service are based on the **Digital Twins Definition Language** (**DTDL**). DTDL is a language that's used to describe models. These models are used with IoT plug and play devices and Digital Twins. DTDL allows us to use semantics and definitions uniformly to define these models, which are used across IoT platforms and IoT solutions. DTDL enables us to define several elements, behaviors, and abilities of a Digital Twin. Digital Twins' behaviors are modeled using meta model classes with DTDL. Several examples of these meta models are listed here and will be explained in more detail in the next part of this chapter:

- **Interfaces**: A descriptive language for describing models.
- **Properties**: The simple types of metadata of a model.
- **Telemetry**: The telemetry data that comes from a sensor. It can be used as metadata in a model.
- **Relationships**: Relationship definitions between models.
- **Components**: A way of grouping the properties in models as part of a Digital Twin.
- **Data Types such as Array, Enum, Map, Object, and Field**: Complex types of metadata in a model.

Some SDKs and API libraries, such as the .NET SDK, implement these behaviors to allow you to create models based on these meta models with ease. This allows you to create your own custom models. These custom models are then used to create Digital Twins, which represent instances of these custom models. The **JavaScript Object Notation** (**JSON**) payload can be used to load these custom models into the Azure Digital Twins service using Azure Digital Twins Explorer.

JSON-LD

DTDL is based on **JavaScript Object Notation for Linked Data (JSON-LD)**. JSON-LD is a way of formatting data into a uniform structure. JSON-LD is based on JSON, which makes it easy to read for everyone.

JSON-LD allows you to structure data analogously by using a list of pairs of attributes and values on top of what you can normally do with JSON. It can be compared to multi-dimensional arrays.

JSON-LD was selected by Microsoft because it can be used as JSON and can be used in **Resource Description Framework (RDF)** systems. RDF is a widely known standard that is used to describe data in a distributed and extensible way. This framework also provides semantic annotations. Semantic annotations allow us to give meaning to the behavior of a model. JSON-LD allows us to use semantic annotations without having to understand RDF.

Versioning

The current Azure Digital Twins service is based on version 2 of DTDL. DTDL has been extended thoroughly in version 2 and now has several new model elements and semantic changes made to it. Several elements have been removed. In some cases, elements have been deprecated to prevent regression errors.

In this section, we learned what the DTDL is and how it can be used to create Digital Twins. In the next section, we will learn about the Digital Twin interface and describe several meta models.

Digital Twin interface

Each digital twin is described by an interface. This interface contains a description of the content of a digital twin. Let's explain this better with an example. Here, we will create an interface describing a room. This room contains several sensors, such as temperature, light, and movement. In this case, the interface describes the room and defines the content of that room by its sensors. This interface is shown here in JSON:

```
{
    "@id": "dtmi:com:contoso:room;1",
    "@type": "Interface",
    "displayName": "Room",
    "contents": [
        {
```

```
                "@type": "Telemetry",
                "name": "temperature",
                "schema": "double"
        },
        {
                "@type": "Telemetry",
                "name": "lights",
                "schema": "boolean"
        },        {
                "@type": "Telemetry",
                "name": "hasmovement",
                "schema": "boolean"
        },
        {
                "@type": "Property",
                "name": "setlight",
                "writable": true,
                "schema": "boolean"
        }
    ],
    "@context": "dtmi:dtdl:context;2"
}
```

There are several parts in this example. The first part is the @id property. This property is the **digital twin model Identifier** (DTMI). Each element, such as interfaces, properties, telemetry, commands, components, relationships, and even complex schemas within a digital twin model, must have a DTMI.

A DTMI consists of the following parts:

```
[scheme]:[path];[version]
```

Note the second semicolon between [path] and [version] instead of a colon. The following table explains the different parts more thoroughly:

Part	Description
[scheme]	This is a string literal. The string literal in this example is dtmi.
[path]	The path is a sequence of several segments. Each segment is a string containing only letters, digits, and underscores. Segments can't start with a digit or end with an underscore. Segments can be identified as a system DTMI or as a user DTMI. System DMITs always begin with an underscore, while user DTMIs always begin with a letter. In this example, the path is com:contoso:Room.
[version]	This part describes the version of your interface. Versions are limited to nine digits and need to be specified and treated as integers while they are part of a piece of text. In this example, the version is 1.

DTMIs are case-sensitive. Be aware of making mistakes when using lowercase and uppercase letters. My recommendation would be to always use lowercase letter so that you don't make any mistakes.

> **Tip**
> The best approach for creating a DTMI is looking at your organization. As an example, we will use the organization Contoso, which has adomain of www. contoso.com, and department analytics as an example. As a best practice, the domain of the organization is used in the path, which creates a path such as com:contoso:analytics:Room. This makes the complete DTMI dtmi:com:contoso:analytics:Room;1.

The next required property is called @type and needs to be set to *Interface*. The final required property is called @context. @context determines how the interface is being processed by the system. This also contains the version that must be used by the DTDL. In this example, it is defined as dtmi:dtdl:context;2, which specifies version 2.

An interface does not necessarily need to contain content. An Interface could just be a description of a digital twin within the model, as shown here:

```
{
    "@id": "dtmi:com:contoso:door;1",
    "@type": "Interface",
    "displayName": "Door",
    "@context": "dtmi:dtdl:context;2"
}
```

For that reason, the `contents` property is optional. The `contents` property allows you to use a collection containing up to 300 different object types based on `Property`, `Telemetry`, `Relationship`, and `Component`.

In this section, we learned that each digital twin is described by an interface. Then, we looked at the structure of an interface. In the next section, we will learn about different object types, which are used as the content of the interface.

Interface content

Interface content is all about defining the digital twin. We can identify the following object types, which can be used as content:

- `Property`
- `Telemetry`
- `Relationship`
- `Component`

Let's look at each of these object types.

Properties

Properties define values within a digital twin. These values can be read-only or have read and write states. Properties have a backing storage. This allows us to read the value of the property at any time. However, the property can also be writable if we set the `writable` property to `true`. This allows us to store a value in the property.

As you already know, each digital twin has a representation in the real world. This means that the property describes the following:

- **State**: This property describes a part of the state of a digital twin.
- **Synchronization**: This property describes the synchronization between the digital twin and the real-world object.

Every property contains synchronization information that facilitates the connection between the digital twin and the real-world object. Since this is the same for all properties, it is not included in the definition of the digital twin.

The following properties are required:

Required Property	Description
@type	This needs to specify `Property`. You can also use a semantic type. Semantic types will be explained later in this chapter.
name	A unique name within the content of the interface.
schema	This can be any primitive or complex schema. Complex schemas are not allowed to have `Array` in one of their recursive levels. Schemas will be explained in more detail later in this chapter.

The following example shows a property called settemperature. This property has its read/write state enabled. This means that it can synchronize with a real-world object:

```
{
    "@type": "Property",
    "name": "settemperature",
    "schema": "double",
    "writable": true
}
```

You will be using properties in most cases when you're designing a model in Azure Digital Twins. This allows you to use its backing storage and gives you the ability to read and query values.

Telemetry

Telemetry is a stream of events and can be compared to events in C#. Since telemetry is mostly used in combination with IoT devices, it can be compared to a stream of measurement values floating from the IoT device. The reason for this is that most IoT devices are not capable of storing their measurement values. Due to this, these measurement values will be lost. Telemetry allows us to view the latest value of that measurement.

Data ingress, which is done through APIs, is mostly handled by combining properties and telemetry. This allows us to real-time connect to the IoT device and provides its measurements, which we can then store in the Azure Digital Twin.

Telemetry has the following required properties:

Required Property	Description
@type	This needs to specify `Telemetry`. It can also use a semantic type. Semantic types will be explained later in this chapter.
name	A unique name within the content of the interface.
schema	This is the data type of the telemetry based on a schema.

The following example shows the definition of telemetry for a IoT temperature device:

```
{
    "@type": "Telemetry",
    "name": "temperature",
    "unit": "degreeCelcius",
    "schema": "double"
}
```

Relationships

A relationship describes the link between two digital twins. This relationship can be used to build a graph of digital twins. This is something you can see happening in the Azure Digital Twins Explorer tool.

Relationships have the following required properties:

The best way of explaining relationships is by looking at an example. In the following example, we have two relationships defined to other interfaces:

Required Property	Description
@type	This needs to specify `Relationship`.
name	A unique name within the content of the interface.

```
{
    "@id": "dtmi:com:contoso:room;1",
    "@type": "Interface",
    "displayName": "Room",
    "contents": [
        {
            "@type": "Relationship",
```

```
            "name": "floor",
            "target": "dtmi:com:contoso:floor;1"
        },
        {
            "@type": "Relationship",
            "name": "inspectedby",
            "target": "dtmi:com:contoso:employee;1"
        }
    ],
    "@context": "dtmi:dtdl:context;2"
}
```

The interface represents a room. A room belongs to a certain floor. Therefore, we have defined a relationship called `floor` that targets the `floor` interface. The target is the `digital win model identifier` property of the floor interface.

It is important to use a name that describes the relationship. The second relationship is a good example of this. This relationship is called `inspectedby` and describes that the room is inspected by a facility employee. These describing names will also help us quickly identify relationships when we're using the Azure Digital Twins Explorer tool.

A relationship can also have properties. In the following example, we have specified a property that describes the relationship between the facilitator employee and room with regards to the inspection that was carried out:

```
{
    "@type": "Relationship",
    "name": "cleanedby",
    "target": "dtmi:com:contoso:employee;1",
    "properties" : [
        {
            "@type": "Property",
            "name": "lastchecked",
            "schema": "dateTime"
        }
    ]
}
```

Component

Components allow us to group a set of related properties. While components reference an interface that contains the relevant grouping, the interface does not require a separate identity. In other words, the interface that's referenced by the component does not need to be instantiated or deleted separately, since it only has meaning in the interface where the component is used.

In the following example, we have defined two interfaces. The first interface, Ceiling, is used to describe the ceiling of a room. The second interface, Room, is used to describe a room and contains a component that references the ceiling interface:

```
[{
    "@id": "dtmi:com:contoso:ceiling;1",
    "@type": "Interface",
    "displayName": "Ceiling",
    "contents": [
        {
            "@type": "Property",
            "name": "height",
            "schema": "float"
        }
    ],
    "@context": "dtmi:dtdl:context;2"
},
{
    "@id": "dtmi:com:contoso:room;1",
    "@type": "Interface",
    "displayName": "Room",
    "contents": [
        {
            "@type": "Component",
            "name": "ceiling",
            "schema": "dtmi:com:contoso:ceiling;1"
        }
    ],
    "@context": "dtmi:dtdl:context;2"
}]
```

In principle, you will never instantiate a digital twin based on the `ceiling` interface. As soon as we instantiate a digital twin based on the room, it will instantiate the ceiling as part of the `Room` interface. Therefore, the ceiling is also deleted when the digital twin for the room is deleted.

Here, we learned about the different object types we can use to define content in an interface. Content is defined by schemas. In the next section, you will learn what a schema is.

Schemas

Schemas are used to describe the format of the data in a digital twin interface. This format can be serialized using JSON. We can identify the following schema types:

Schema Type	Description
Primitive schema	A primitive schema uses primitive data types such as `boolean`, `datetime`, `double`, `float`, and more.
Complex schema	A complex schema allows us to create more complex types by using `Arrays`, `Enums`, `Maps`, and `Objects`.
Geospatial schema	Geospatial schemas allow us to model several different forms of geographical data structures, such as points, lines, and polygons in a single way or multiple ways.

In this section, you learned what a schema is, and which different types of schemas are used. In the next section, we will discuss primitive schemas.

Primitive schemas

There is a clear list of available primitive schemas. These schemas can be directly applied as values. This allows you to define values in all formats within the digital twin interface. The following table shows a list of all the available primitives for version 2 of the DTDL:

Primitive Schema	Format	Description
`boolean`	N/A	Boolean value
`date, dateTime, time`	RFC 3339	Definition of date, date/time, or time
`duration`	ISO 8601	Duration
`double`	IEEE	8-byte floating point
`float`	IEEE	4-byte floating point
`integer`	N/A	Signed 4-byte integer
`long`	N/A	Signed 8-byte integer
`string`	UTF-8	String text

RFC 3339 is a widely used standard for date-time formats. All primitive schemas related to date and time use this standard. The exception is `duration`. `duration` uses the **ISO 8601** standard, which represents the format of dates and time in the Gregorian calendar.

The **IEEE** Standard for Floating-Point Arithmetic is used with the `double` and `float` primitive schemas. **UTF-8** is used for `string`. This is a widely used standard for variable width encoded character formats.

With that, you have learned about primitive schemas. In the next section, you will learn about complex schemas and their underlaying data types.

Complex schemas

Complex schemas allow us to model more complex data types that are based on primitive data types, as mentioned by the primitive schemas. These complex data types support recursive up to five levels deep. The following complex schemas are available:

- `Object`
- `Enum`
- `Array`
- `Map`

Complex schemas have required and optional properties. Required properties are properties that define, in most cases, the purpose of the complex schema, while optional properties provide additional information that describes the complex schema in more detail:

	Description	Examples of Properties
Required	These are the required properties. Properties differ per complex schema.	name, schema, @type, fields, mapKey, mapValue, enumValue, enumValues, valueSchema, elementSchema
Optional	These properties are optional but available in every complex schema.	@id, comment, description, displayName

Each of these complex schemas will be described here, and their required properties will be explained.

Field

Fields are the basic data types for building a meta model using the complex Object schema type. A field data type describes a named field in an Object. A field has required properties called name and schema. The name property describes the name of the field, while the schema property describes the field type. There are several other properties available.

A field type is always described by a primitive schema or a complex schema. Both were discussed earlier in this chapter. Fields are always part of an object and are never used separately.

Object

An Object data type is used to describe a data type based on named fields. It contains the @type and fields required properties. @type is set to Object, while fields contains the named fields of the object. It allows up to 30 fields with a maximum depth of 5 levels. An example of an Object is as follows:

```
{
    "@type":"Object",
    "name":"temperature",
    "field":[
        {
            "name":"temperature",
            "schema":"double"
        }
    ]
}
```

Map

A Map is a complex schema that can be best compared to a dictionary in C#. It represents a list of key-value pairs. Map contains two required properties that refer to the following schemas:

Required Property	Description
mapKey	This property describes the key part of the key-value pair. This schema requires that we use the string primitive schema. No other type is allowed.
mapValue	This property describes the value part of the key-value pair. The schema for a value needs to be the same for each key-value pair. However, the schema can be of any type; that is, either primitive or complex.

The following example defines a property in an interface called rooms. The property is using a complex schema based on a Map that defines the room name as the key and the room temperature as the value:

```
{
    "@type": "Property",
    "name": "rooms",
    "writable": true,
    "schema": {
        "@type": "Map",
        "mapKey": {
            "name": "roomname",
            "schema": "string"
        },
        "mapValue": {
            "name": "roomtemperature",
            "schema": "float"
        }
    }
}
```

This example could become even more complex if the value is a complex Object schema type with multiple layers of values represented by primitive and complex schema types.

Array

This complex schema type can be compared to an array in C#. It represents an indexable data type. Each value in the array is of the same primitive or complex schema. The required @type property needs to be set to Array. The required elementSchema property contains the schema type that represents the value definition.

The following example describes a property in a digital twin interface called EquipmentState. It is based on an array of equipment that describes their state. The value of the array is based on a complex schema, which describes the name, state, and the active radius of the equipment:

```
{
    "@type": "Property",
    "name": "equipmentstate",
    "schema": {
        "@type": "Array",
        "elementSchema": {
            "@type":"Object",
            "name":"equipmentstate",
            "field":[
                {
                    "name":"name",
                    "schema":"string"
                },
                {
                    "name":"active",
                    "schema":"boolean"
                },
                {
                    "name":"radius",
                    "schema":"long"
                }
            ]
        }
    }
}
```

This example can also be modeled in a different way. We could treat all the equipment as separate digital twin instances in our digital twin. However, this completely depends on your data and model requirements.

Enum

This complex schema type represents an enum, similar to what you would use in C#. It is a data type with named labels mapped to values. An enum has a required @type property that needs to be set to Enum. There are two more required properties that describe these labels and values:

Required property	Description
valueSchema	This describes the format of the value of an Enum. It can only be a primitive integer or string schema type.
enumValues	This contains the available values for our Enum. They are based on the EnumValue schema. The EnumValue schema has two required properties called name and enumValue. name is the label of the Enum value, while enumValue is the value of the label represented by an integer or a string value.

In the following example, we have a property defined in the interface that describes a state. This property is based on the complex Enum schema type and contains three labels, each of which are represented by an integer value. These labels are inactive, active, and onhold:

```
{
    "@type": "Property",
    "name": "state",
    "schema": {
        "@type": "Enum",
        "valueSchema": "integer",
        "enumValues": [
            {
                "name": "inactive",
                "enumValue": 1
            },
            {
                "name": "active",
                "enumValue": 2
```

```
            },
            {
                "name": "onhold",
                "enumValue": 3
            }
        ]
    }
}
```

In this section, we learned about complex schemas and their underlaying data types. In the next section, we will learn about a special schema called the geospatial schema.

Geospatial schemas

Geospatial schemas allow us to model several different geographic types. These geospatial schemas are based on **GeoJSON**. GeoJSON is a standard format for encoding several geographical data structures using JSON.

The following geometry types are available:

DTDL term	Geometry Type for a Single Definition	Geometry Type for a Multi-Definition
point, multiPoint	Point	MultiPoint
lineString, multiLineString	LineString	MultiLineString
polygon, multiPolygon	Polygon	MultiPolygon

The DTDL geospatial schema's DTMI can be derived from the DTDL term:

```
dtmi:standard:schema:geospatial:[DTDL term];2
```

The following example shows a `Telemetry` that defines the location of a vehicle. The location is defined by a point:

```
{
    "@type": "Telemetry",
    "name": "vehiclelocation",
    "schema": "point"
}
```

This allows us to send a message containing the longitude and latitude coordinates of the vehicle to a digital twin instance using a model containing this object type.

> **Important note**
>
> Geospatial types cannot be used in `Property` schemas. The reason for this is that geospatial types are based on arrays, which are can't be used in a `Property` schema. However, you can use geospatial types in the `Telemetry` schema.

In this section, we learned about geospatial schemas. In the next section, we will learn about semantic types.

Semantic types

Semantic types are specific types of annotations that give meaning to a behavior. This allows machine learning and other computational systems to *understand and reason* about this behavior. This reasoning allows these systems to interpret semantic types into something meaningful. Think of using the `Distance` semantic type in a meta model called `Silo`. Instead of handling the behavior as a schema, it can be reasoned as the distance to that Silo.

The list of semantic types is rather large and can be found via one of the links in the *Further reading* section of this chapter.

With that, you have learned what semantic types are and how they are used. In the last part of this chapter, we will learn about how to validate a model.

Validating a model

It is important to prevent failures by validating a model upfront before the model is uploaded to the Azure Digital Twins instance via an API, or via the Azure Digital Twins Explorer tool. Microsoft provides a .NET client-side DTDL parser library to support this process.

There are two ways to use the validator. Since the parser library is available as a NuGet package called `Microsoft.Azure.DigitalTwins.Parser`, you can create your own custom parser or make it part of a service that dynamically uploads models to your Azure Digital Twins instance.

Microsoft also provides a command-line parser example as a .NET project. This project can be downloaded from `https://docs.microsoft.com/en-us/samples/azure-samples/dtdl-validator/dtdl-validator/`.

Download the ZIP file and extract the contents of it to `c:\github\DTDL_Validator`. Start Visual Studio, as shown in the following screenshot:

1. Open the `DTDLValidator.sln` solution, which can be found in the `C:\Github\DTDL_Validator\DTDLValidator-Sample` folder.

2. Set the build to **Release**.

3. Right-click the solution and select **Build Solution**:

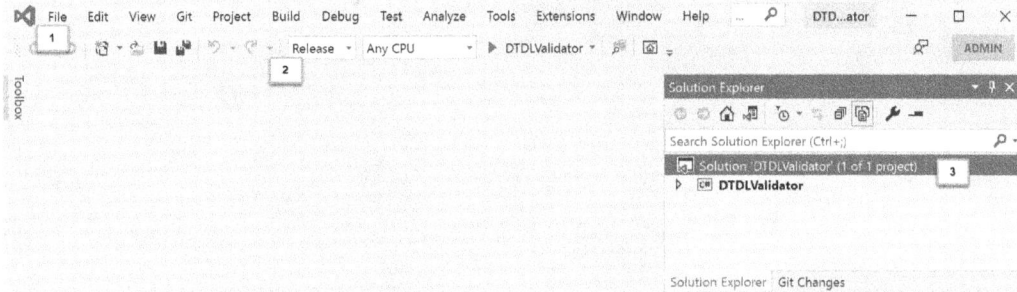

Figure 3.1 – Creating the command-line validator executable

We want to use this DTDL parser in our command line in any folder. Therefore, we need to add the path to the DTDL parser in the `Path` environment variable.

Open the window settings of your computer. Search for *system environment variables*. This will open **SystemProperties** with the **Advanced** tab selected. Execute the following steps, as shown in the following screenshot:

1. Select **Environment Variables**. This will open a new dialog for **Environment Variables**.

2. Double-click on the **Path** user variable. This will open a new dialog for **Edit environment variable**.

3. Select **New**.

4. Add the path to the validator in the new row. The path to the validator is the folder that the `DTDLValidator.exe` file can found be. In my case, this is `C:\Github\DTDL_Validator\DTDLValidator-Sample\DTDLValidator\bin\Release\netcoreapp3.1`.

5. Click **OK** to close **Environment Variables**.

6. Click **OK** to close **Edit environment variable**.

7. Click **OK** to close **System Properties**:

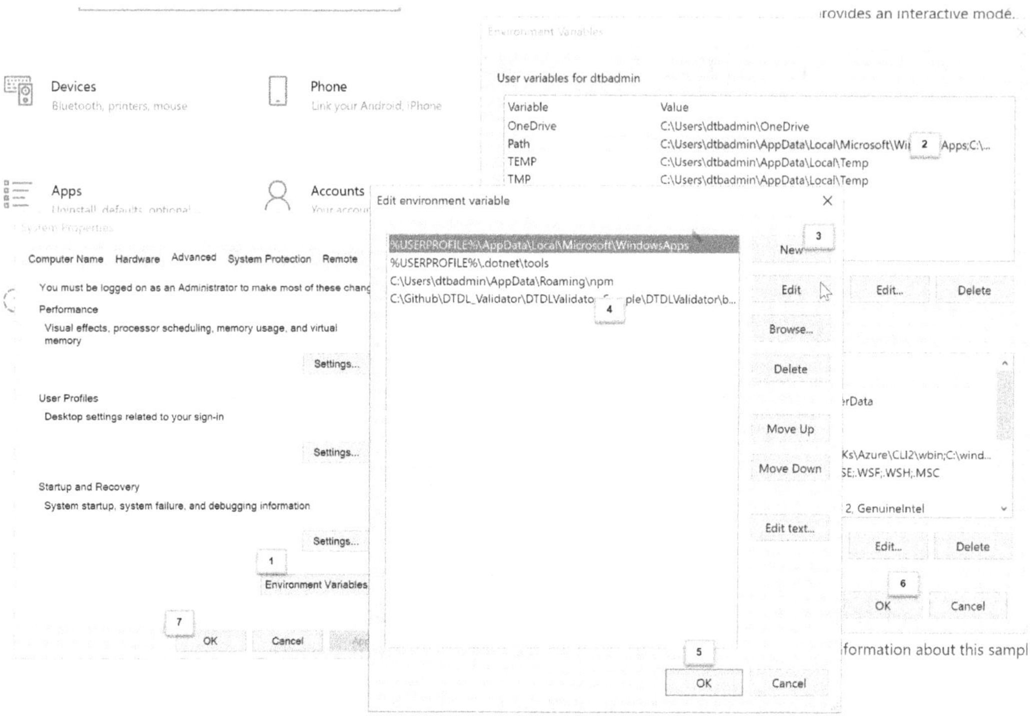

Figure 3.2 – Adding the path of the validator to the environment variables

Next, we will test a DTDL file. Perform the following steps:

- Open File Explorer and create a folder. In my case, I created a folder called `DTDL` under `C:\Github`.

- Create a file with a `json` extension that contains the JSON payload of the Digital Twin Interface example, called `dtmi:com:contoso:room;1`, and place it in that folder.

- Open a command line. Make sure to run it as Administrator.

- Change the folder in the command line to `C:\Github\DTDL` using the `cd C:\github\dtdl` command.

- Run the `DTDLValidator` command. This will validate all the DTDL files that are found in that folder:

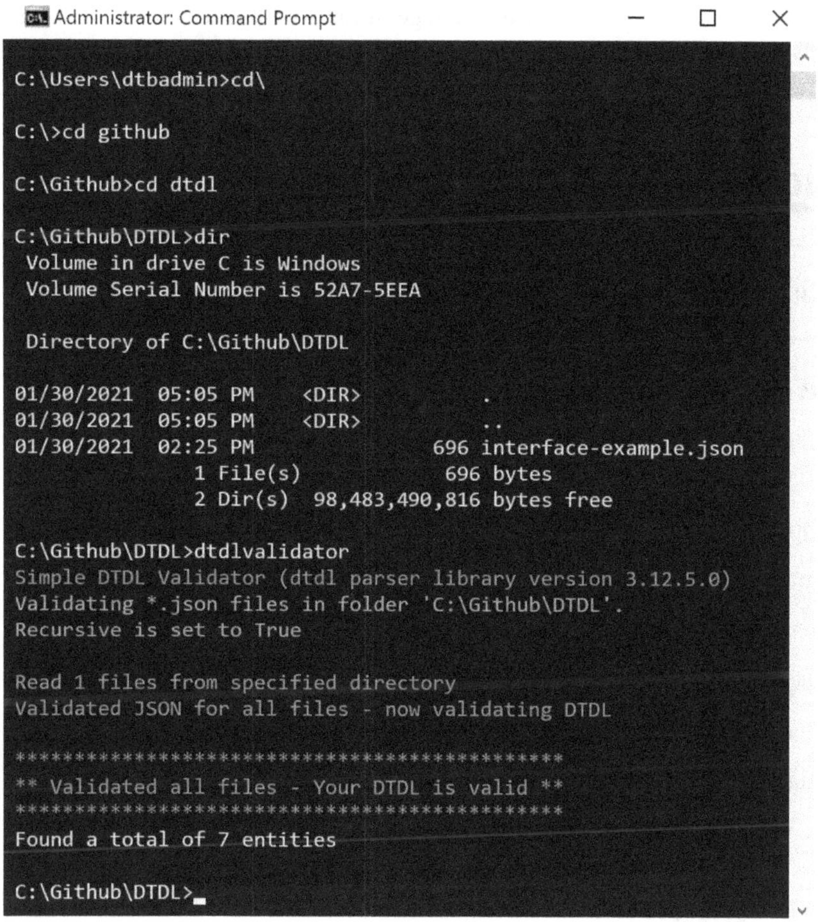

Figure 3.3 – Testing the validator to validate a DTDL file

The result of `DTDLValidator` can be seen in the preceding screenshot. We can see that it has validated a single file in the directory. The file has been validated.

Summary

In this chapter, we explored the *Digital Twins Definition Language* and how it is structured. We looked at its interface and its content, such as its properties, telemetry, relationships, and components. This content relies heavily on schemas. Due to this, you learned about primitive and complex schemas and viewed examples that show how to use them. In the upcoming chapters, we will start using this basic understanding to create our first models, digital twins, and many other examples around the Azure Digital Twins service.

In the next chapter, we will learn how to manage models with Azure Digital Twins. These models are based on the *Digital Twins Definition Language*, which we learned about in this chapter.

Questions

1.   What is the maximum depth allowed for a field in an `Object`?

     A. One level

     B. Three levels

     C. Five levels

2.   What specific data types can be used to allow machine learning to reason about a type?

     A. Property

     B. Semantic

     C. Telemetry

3.   Why is it important to validate a model?

     A. You can prevent failures when uploading the model

     B. It makes it easier to upload the model

     C. The validation process creates an overview of your model

Further reading

The Digital Twins Definition Language is still something that will change in the future. New versions of DTDL will arise. The following links provide more information about DTDL, along with Azure Digital Twins:

- **Understanding twin models in Azure Digital Twins**: `https://docs.microsoft.com/en-us/azure/digital-twins/concepts-models`

- **Digital Twins Definition Language (DTDL)**: `https://github.com/Azure/opendigitaltwins-dtdl/blob/master/DTDL/v2/dtdlv2.md`

- **Parsing and validating models using the DTDL parser library**: `https://docs.microsoft.com/en-us/azure/digital-twins/how-to-parse-models`

4
Understanding Models

This chapter will explain models in Azure Digital Twins. Models are the templates that define an Azure Digital Twin. We will learn how to manage models by creating, retrieving, updating, and removing models. We will also learn how to instantiate Azure Digital Twins based on these models via the Azure Digital Twins Explorer and by using a .NET console application. A smart building concept will be used to give you a better understanding of the model and what is required to model it.

By the end of this chapter, you will be able to design and manage your own models and create Digital Twins based on them.

In this chapter, we'll cover the following topics:

- Designing models
- Creating an application
- Managing models
- Managing Digital Twins

Technical requirements

We will be using a .NET console application and the .NET SDK to build an Azure Digital Twins instance using models. The Azure Digital Twins Explorer will be used to view the result of the .NET calls made by the console application. While the SDK supports both synchronous and asynchronous calls, the examples in this chapter will be based on synchronous coding so that the focus is mainly on the SDK instead of .NET.

Designing models

As we explained in *Chapter 1*, *About Digital Twins*, an Azure Digital Twins instance contains Azure Digital Twins. Azure Digital Twins are instances of a model that are used to represent a certain entity. This means that a model can be seen as a template for creating Azure Digital Twins. Models are based on the **Digital Twins Definition Language** and are specified using **JSON-LD**.

Modeling a smart building

The best way of explaining models and creating instances is by using an example that is based on the smart building concept. This concept allows us to model a building and its sensors into an Azure Digital Twins instance. We will keep this simple by using the simplest structure for this concept. We have the following entities:

- **Campus**: A company's location containing one or more buildings.
- **Building**: A building located on the *Campus* of a company.
- **Floor**: Buildings contain one or more floors. Each floor is defined by *Floor*.
- **Room**: Each floor contains multiple rooms. Each room is defined by *Room*.
- **Sensor**: A sensor that can be found in a *Room*.

As you already know, this can be applied to any company that uses a physical location and buildings for their organization. This concept will be used in this chapter and the next to explain each of the elements that are required to build an Azure Digital Twins instance.

Designing models

An important aspect of working with models is the design process of modeling the required models. To create the right models, we will need to have a clear understanding of our scenario. As stated in *Figure 1.1* of *Chapter 1*, *About Digital Twins*, we have *Input* and *Output*. These two are related to each other. What we want to achieve in the *Output* will define what is required for the *Input*.

Let's explain this. In our example, the smart building concept is described by several entities that represent the structure of a building containing floors, rooms, and sensors. In this example, the *Input* describes which sensors we want to monitor. The *Output* gives us the current state of sensors per entity in a structured way, allowing us to build a visualization that shows an actual building where sensor values are shown inside each of the rooms.

In the next section, we will learn about several modeling recommendations you need to consider when modeling models.

Modeling recommendations

There are some recommendations you should take into consideration when you start to design models for your Digital Twins:

- **Required models**: Think about what needs to be modeled for your Azure Digital Twins instance. It depends on what your requirements are, and which content of the real-life world needs to be modeled.

- **Splitting into multiple models**: This involves splitting a real-life asset into different models that have an underlying relationship. This is not always needed but can simplify your understanding of the model.

- **Model query implications**: Defined models and their underlaying relationships can have an impact on querying the data. This could, for example, cause you to require multiple queries to get the right data. The recommendation here would be to define our query results and, based on that, define the required models and their metadata. In some cases, the number of queries and their complexity can be simplified by adding metadata to the relationships between the models.

In the upcoming chapters, it will become clearer how to incorporate these recommendations when you design a Azure Digital Twins instance.

In this section, we learned about what is required to design and model an Azure Digital Twin Model. In the next section, we will create a basic .NET console application that allows us to test the different .NET examples for managing Models and Digital Twins.

Creating an application

This chapter contains several scripts for managing Models and Digital Twins. We will be using a .NET Console application to test these scripts against the Azure Digital Twins instance.

Start Visual Studio and select the **Create a new project** option, as shown in the following screenshot:

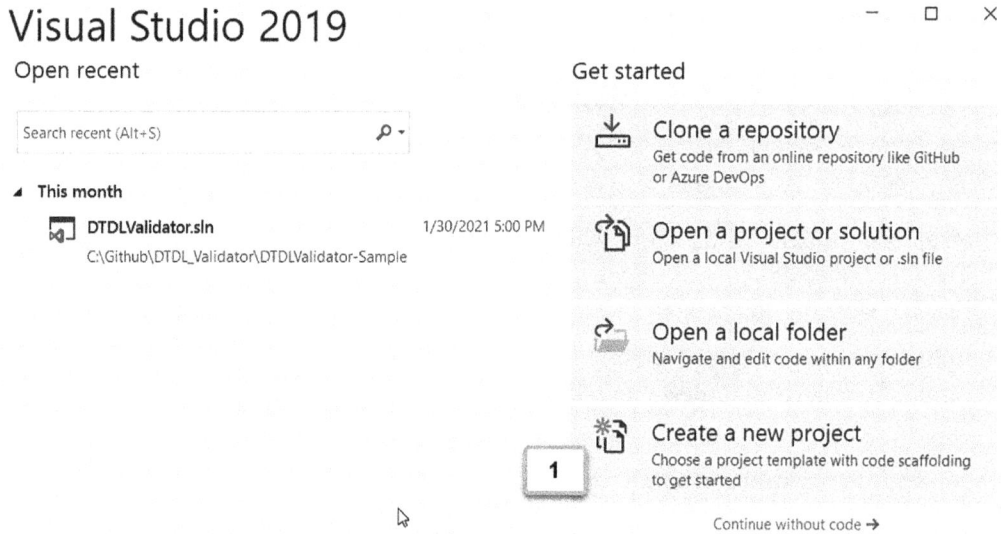

Figure 4.1 – Opening Visual Studio

We need to use the **Console App (.NET Core)** template here, which allows us to create a .NET Console application. If the template is not visible as one of the default selections, use the search box to find it. Execute the following steps, as shown in the following screenshot:

1. Select the **Console App (.NET Core)** template.

2. Press the **Next** button:

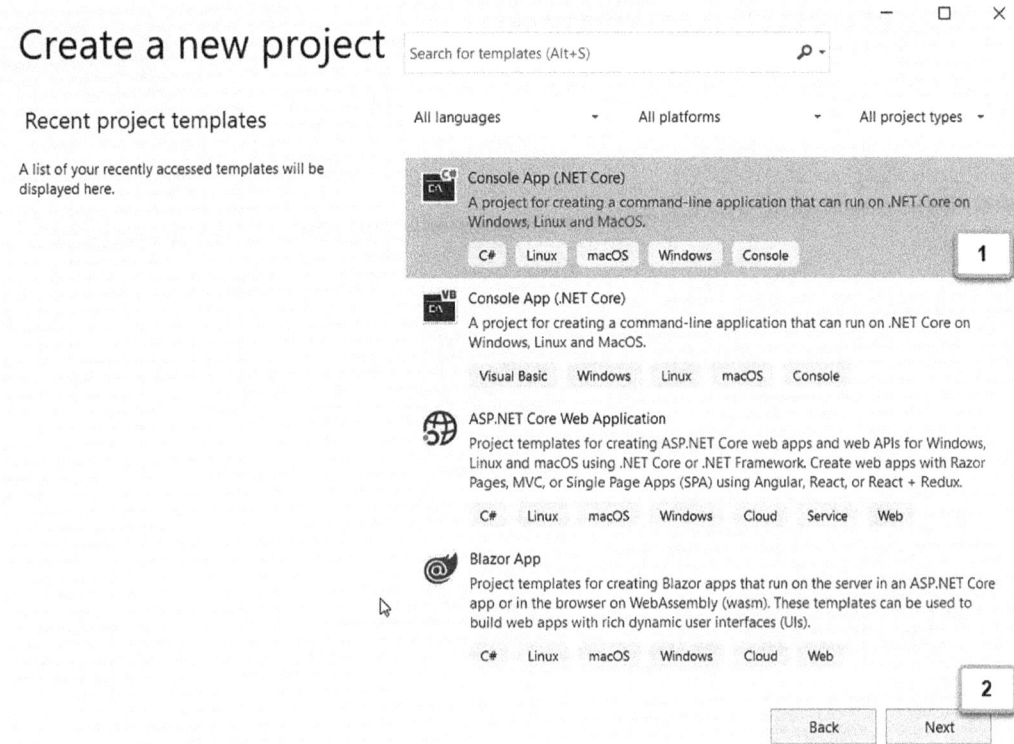

Figure 4.2 – Creating a new project in Visual Studio

Next, as shown in the following screenshot, we need to configure our new project. Execute the following steps:

1. Enter SmartBuildingConsoleApp for the name.

2. Enter the location where the folder for the application will be created. In my case, I used c:\Github.

3. Enter SmartBuildingConsoleApp as the solution name.

4.   Press the **Create** button to create the project:

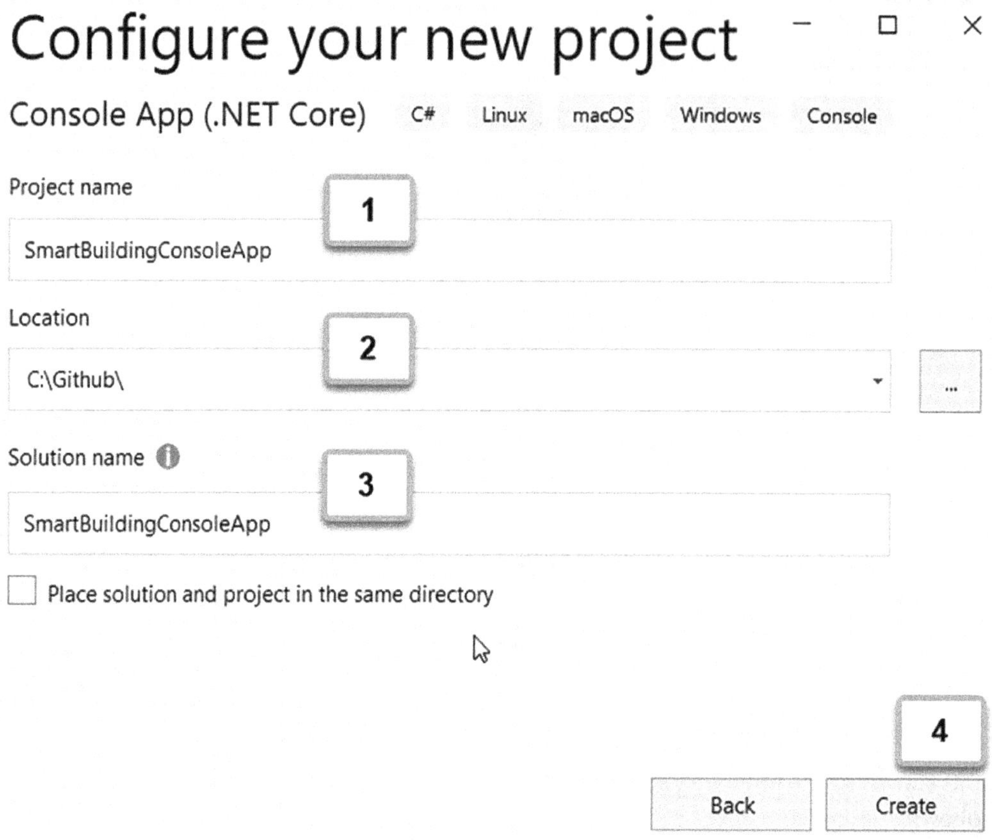

Configure your new project   —   ☐   ✕

Console App (.NET Core)   C#   Linux   macOS   Windows   Console

Project name

1

SmartBuildingConsoleApp

Location

2

C:\Github\   ▾   ...

Solution name ⓘ

3

SmartBuildingConsoleApp

☐ Place solution and project in the same directory

                                                4

                                        Back   Create

Figure 4.3 – Setting the required fields for creating a new Console App

With that, Visual Studio has created a .NET Console application for you. Right-click on the SmartBuildingConsoleApp node in the Solution Explorer to open the menu. Select **Manage NuGet packages**. We will need to add the following two NuGet packages:

- Azure.Identity: This is an implementation of the Azure SDK Client Library for Azure Identity and is used by the application to connect to the Azure Digital Twins instance.

- Azure.DigitalTwins.Core: This is an SDK for accessing a variety of functions to manage and build the Azure Digital Twins service.

Each of these packages can be installed through the window shown in the following screenshot. Follow these steps for each of the packages:

1. Select the **Browse** tab.

2. Search for the package by entering its name.

3. Select the package from the list of packages. This will update the window on the right.

4. Select the latest stable version of the package. Press the **Install** button to start installing the package. During the installation, two dialogs will appear. One will be for previewing the changes, while the second will be for accepting the license. Press the **OK** button for the preview changes and press the **Accept** button for the license acceptance:

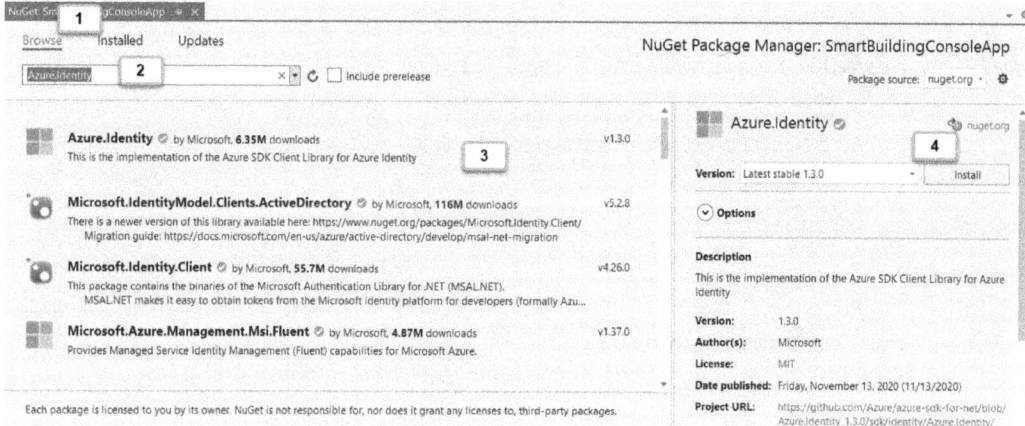

Figure 4.4 – Adding NuGet packages to the project

We need to have a folder under the `SmartBuildingConsoleApp` node in the Solution Explorer that contains the code for managing the Digital Twins instance. Follow these steps:

1. Right-click **SmartBuildingConsoleApp** and select **Add | New Folder**.

2. Name the folder `DigitalTwins`.

3. Right-click the `DigitalTwins` folder and select **Add | New class**.

4. Select the C# class template.

5. Name the file of the class `DigitalTwinsManager.cs`.

Copy the following code into the `DigitalTwinsManager.cs` file:

```csharp
using Azure.DigitalTwins.Core;
using Azure.Identity;
using System;
using System.IO;

namespace SmartBuildingConsoleApp.DigitalTwins
{
    public class DigitalTwinsManager
    {
        private static readonly string adtInstanceUrl = "<Azure
Digital Twins Instance URL>";

        private DigitalTwinsClient client;

        public DigitalTwinsManager()
        {
            Connect();
        }

        public void Connect()
        {
        }
    }
}
```

This block of code is our base for managing the Azure Digital Twins instance. It is a standard class that calls the `Connect()` method when it is initiated. It contains a reference to the `DigitalTwinsClient` class, which gives us access to the Digital Twin in Azure.

We need to replace <Azure Digital Twins Instance URL> with the actual URL of our Azure Digital Twins instance. This URL can be found by performing the following steps, as shown in the following screenshot:

1. Open the Azure Digital Twins service called **DTBDigitalTwins** in Azure via `https://portal.azure.com`.

2. Copy the **Host name** details by clicking on the copy symbol beside the URL and paste it into <Azure Digital Twins Instance URL>:

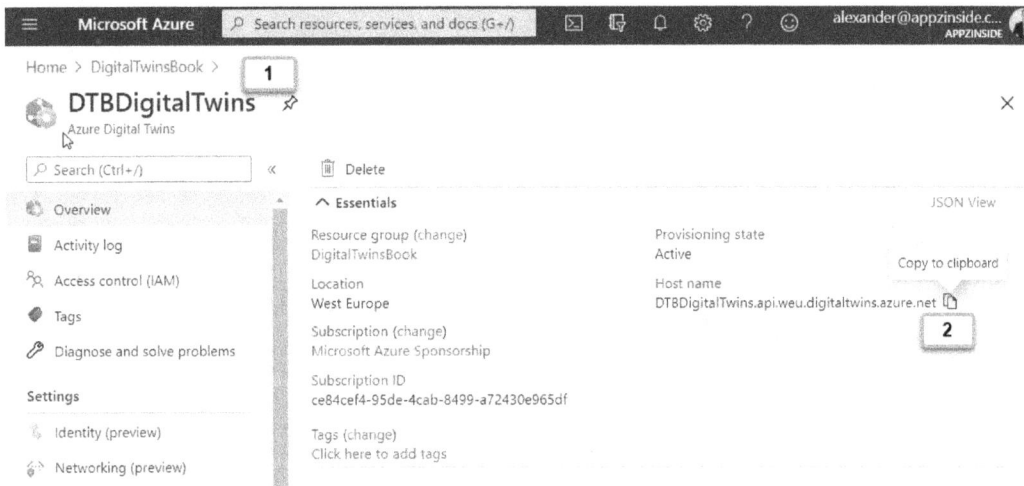

Figure 4.5 – Retrieving the Azure Digital Twins instance URL from the Azure portal

Now, we need to connect to the Azure Digital Twins instance. We will use the `Azure.Identity.DefaultAzureCredential()` class for this. This class will sequentially call several different credential classes to retrieve the credentials. One of them is the credentials set in Visual Studio.

Replace the `Connect()` method with the following code:

```
public void Connect()
{
    var cred = new DefaultAzureCredential();
    client = new DigitalTwinsClient(new Uri(adtInstanceUrl),
cred);
}
```

Make sure that you have used credentials in Visual Studio that have access to the Azure Digital Twins instance. This will automatically give us access and a valid `DigitalTwinsClient` client.

Now, we need to instantiate the `DigitalTwinsManager` class. Open the `Program` class by double-clicking on `Program.cs` in the Solution Explorer. Replace the contents of the file with the following code:

```
using SmartBuildingConsoleApp.DigitalTwins;
using System;

namespace SmartBuildingConsoleApp
{
    class Program
    {
        static void Main(string[] args)
        {
            DigitalTwinsManager dtHelper = new
DigitalTwinsManager();

            // TODO
        }
    }
}
```

You will notice a `// TODO` remark. Later, this will be replaced with the code in each step that follows to test different functions for managing Models and Digital Twins.

With that, we have set up our base application support for testing Models and Digital Twins. In the next section, we will start managing models using the .NET SDK.

Managing models

The first part of this chapter is all about managing models. Models are the templates for Azure Digital Twins. We will be exploring how to create, inherit, delete, and get a Model by using the .NET SDK.

Creating a model

Let's start by creating a model for a Room. We need to create a folder called `Models` under the `SmartBuildingConsoleApp` node in the Solution Explorer. Follow these steps:

1. Right-click on the SmartBuildingConsoleApp node and select Add | New Folder

2. Name the folder Models

3. Right-click on the Models folder and select Add|New folder

4. Name the folder chapter4

5. Right-click on the chapter4 folder and select Add|New Item

6. Look for a JSON file type

7. Name the file room.json.

Replace the contents of the `room.json` file with the following code:

```
{
    "@id": "dtmi:com:smartbuilding:Room;1",
    "@type": "Interface",
    "@context": "dtmi:dtdl:context;2",
    "displayName": "Room"
}
```

This definition file only defines the Room by using an `@id` and `displayName` to keep it simple. In the next chapter, we will go further by adding additional properties and metadata.

We want to open the file while executing the console application. We need to make sure that the file is copied, along with the output of the build. We can accomplish this by changing a property of the file in the Solution Explorer, as shown in the following screenshot. Follow these steps:

1. Click on the `room.json` node in the Solution Explorer. This will open the properties of that file.

2. Change the **Copy to Output Directory** property to `Copy always`:

> **Important note**
>
> We need to set the **Copy to Output Directory** property to `Copy always` for every future JSON file representing a model. Otherwise, the model files won't be found when the application is run.

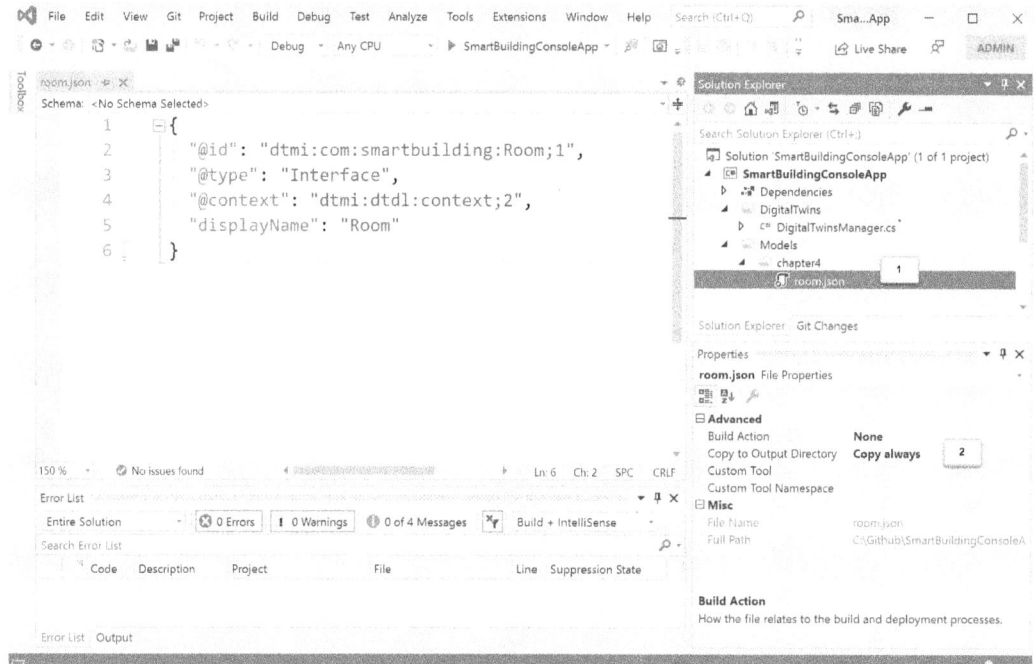

Figure 4.6 – Defining that the model file is copied to the output directory

Let's create our first piece of code for creating a model based on a model definition stored in a JSON file. Add the `CreateModel(string path)` method to the `DigitalTwinsManager` class, like so:

```
public void CreateModel(string path)
{

    using var modelStreamReader = new StreamReader(path);
    string dtdl = modelStreamReader.ReadToEnd();
    string[] dtdls = new string[] { dtdl };


    client.CreateModels(dtdls);
}
```

> **Important note**
> The preceding code allows us to add more than one model. This will be
> explained later in this chapter, with an example containing multiple models.

We need to call this method when our application is running. Add the following code
below the `// TODO` remark field in the `Program.Main` method:

```
string path = "Models/Chapter4/room.json";
dtHelper.CreateModel(path);
```

Before we run our application, we need to have Azure Digital Twins Explorer in place
and ready. Azure Digital Twins Explorer will be used to view the results of our code.

Open the Azure Digital Twins Explorer, as shown in the following screenshot. You will
notice that there are already a lot of models present due to what we did in *Chapter 2,
Requirements and Installation*. Let's remove them all. Click on the recycle bin icon to
remove them:

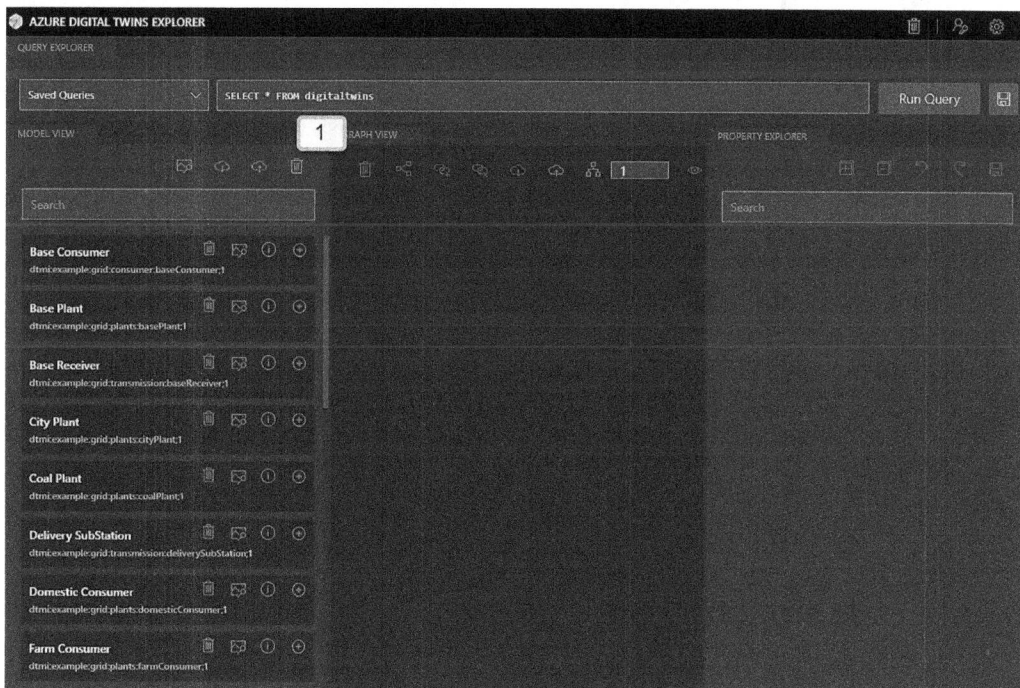

Figure 4.7 – Azure Digital Twins Explorer

Run the application via Visual Studio. Check the Azure Digital Twins Explorer, as shown in the following screenshot, to see whether the new model has been added:

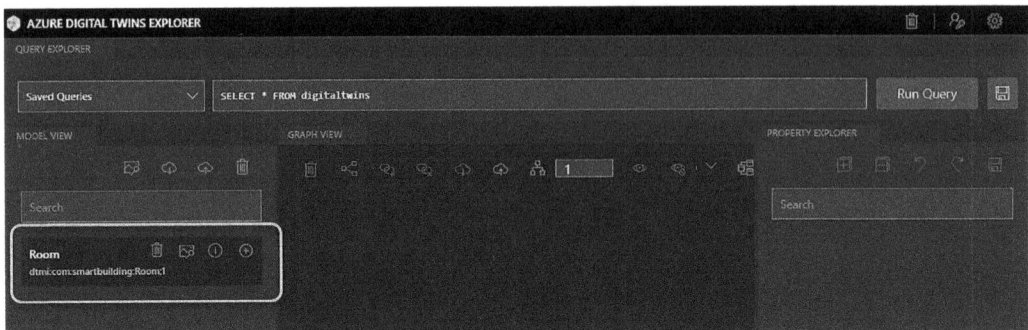

Figure 4.8 – Azure Digital Twins Explorer containing our first created model

In the next section, we will learn how to create multiple models simultaneously.

Creating multiple models

It is possible to create several models together with the same `DigitalTwinsClient.CreateModels()` method.

Model creation will fail when they inherit from a model that hasn't been created yet. Using this method will make sure that the models are created in the right order to prevent this.

Replace the `CreateModel()` method in the `DigitalTwinsManager` class with the following code:

```
public bool CreateModel(string path)
{
    return CreateModels(new string[] { path });
}

public bool CreateModels(string[] path)
{
    List<string> dtdls = new List<string>();

    foreach (string p in path)
    {
        using var modelStreamReader = new StreamReader(p);
        string dtdl = modelStreamReader.ReadToEnd();
```

```
            dtdls.Add(dtdl);
    }

    try
    {
        DigitalTwinsModelData[] models = client.
CreateModels(dtdls.ToArray());
    }
    catch (RequestFailedException)
    {
        return false;
    }

    return true;
}
```

This block of code has split the process of creating the models into two methods called `CreateModel()` and `CreateModels()`. The `CreateModels()` method contains a parameter that allows you to specify more than one model. Each model is loaded and then all the loaded models are created by the `DigitalTwinsClient.CreateModels()` method.

> **Note**
>
> Before we execute the code we need to have each of the mentioned model files in our project. Copy the remaining content of SmartBuildingConsoleApp/Models/chapter4 on GitHub into our project.

Replace the code below the `// TODO` remark field in the `Program.Main` method with the following code:

```
string[] paths = new string[] {
    "Models/chapter4/campus.json",
    "Models/chapter4/building.json",
    "Models/chapter4/floor.json",
    "Models/chapter4/workarea.json",
    "Models/chapter4/sensor.json"
};
dtHelper.CreateModels(paths);
```

You will notice that the list does not contain the `room.json` file. The method will fail to upload any model when one of the models has already been uploaded.

Run the application and check the result in the Azure Digital Twins Explorer. You will see that all the models have been created:

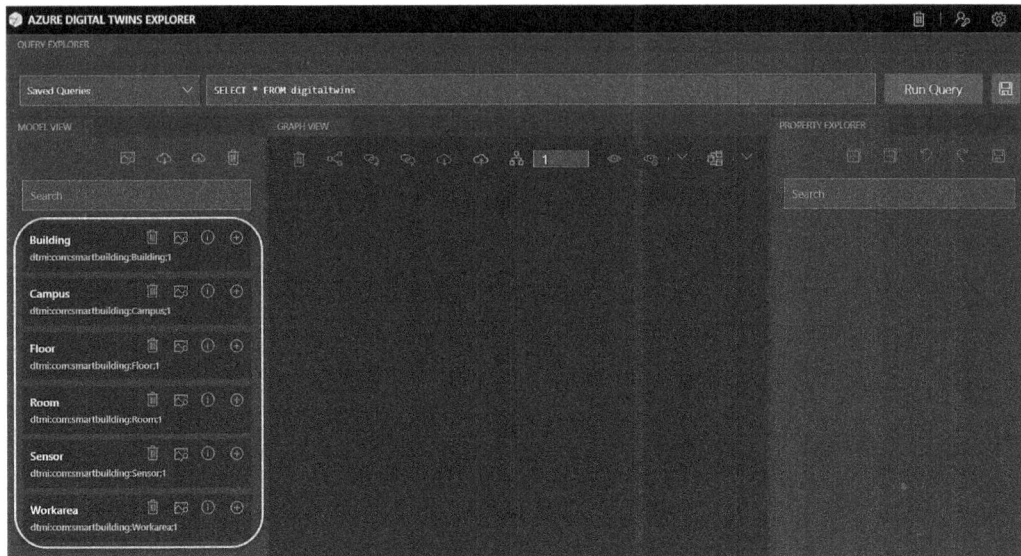

Figure 4.9 – An overview of the models created

Inheriting from a model

We can inherit from an existing model. This can be useful when you want to extend an existing model with additional properties, metadata, and relationships. In the following example, we have a `Meetingroom` that inherits from `Room`. `Meetingroom` could have, for example, an additional property called `occupied` that is specific for a meeting room. Still, `Meetingroom` will inherit all the properties of `Room`.

Create a new JSON file called `meetingroom.json` under the `Models/Chapter4` folder. Make sure that the file is copied to the output folder, as explained earlier.

Copy the following code in the `meetingroom.json` file:

```
{
    "@id": "dtmi:com:smartbuilding:Meetingroom;1",
    "@type": "Interface",
    "@context": "dtmi:dtdl:context;2",
    "extends": [ "dtmi:com:smartbuilding:Room;1" ],
```

```
  "displayName": "Meetingroom",
  "occupied": false
}
```

To indicate that this model inherits from another one, add the `extends` property. This property contains one or more model IDs that this model inherits from.

Now, we need to create this model. Replace the code below the `// TODO` remark field in the `Program.Main` method with the following code:

```
string path = "Models/Chapter4/meetingroom.json";
dtHelper.CreateModel(path);
```

Run the application to create the model. Check the result with the Azure Digital Twins Explorer:

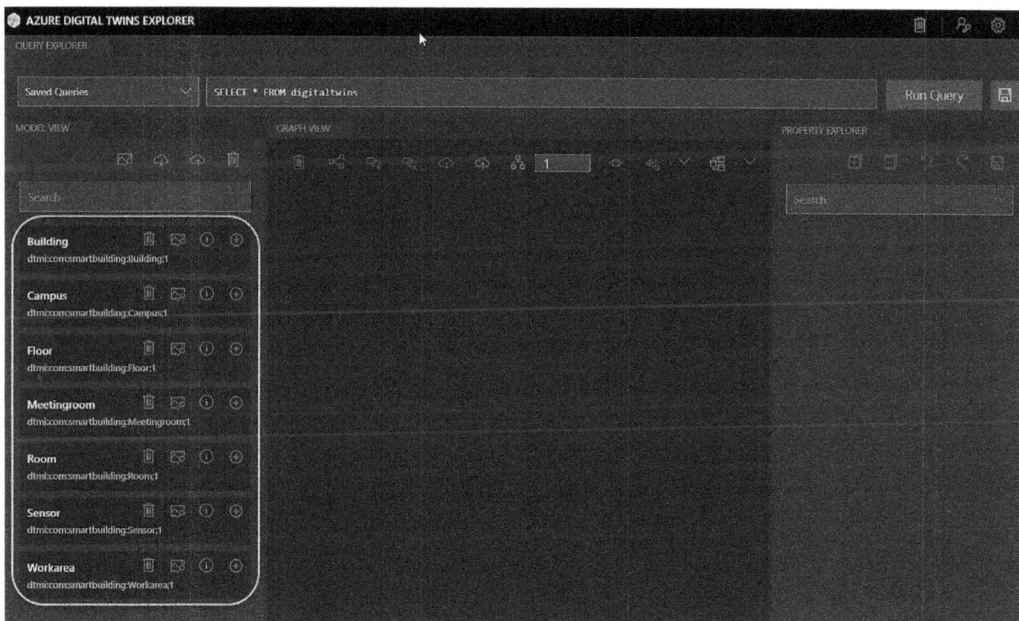

Figure 4.10 – An overview of the models created

Deleting a model

It is also possible to delete a model through code. Add the following method to the `DigitalTwinsManager` class:

```
public void DeleteModel(string modelId)
{
    client.DeleteModel(modelId);
}
```

Replace the code below the `// TODO` remark field in the `Program.Main` method with the following code:

```
dtHelper.DeleteModel("dtmi:com:smartbuilding:Meetingroom;1");
```

Run the application. Check the Azure Digital Twins Explorer to check that the `Meetingroom` model has been removed from the list of models:

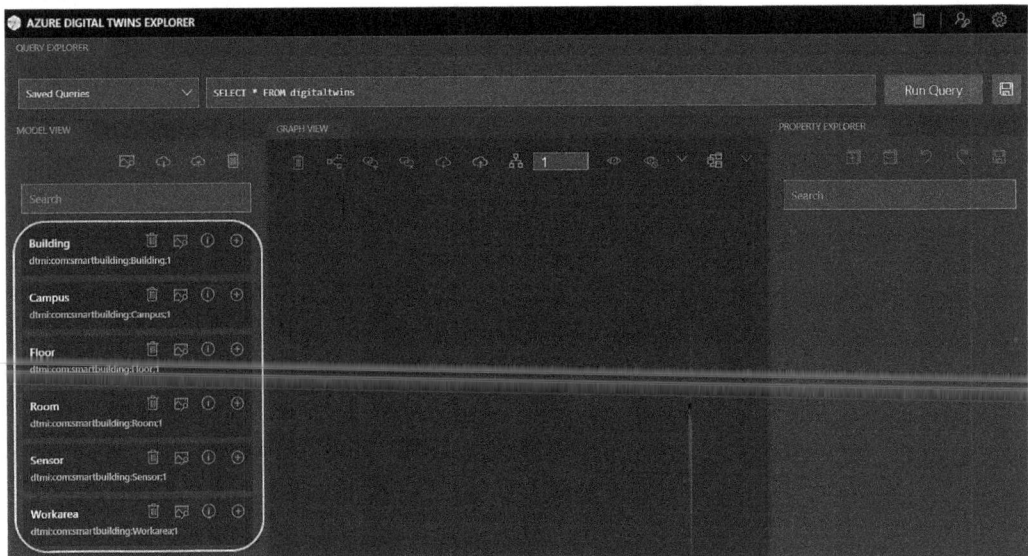

Figure 4.11 – An overview of the created models

Next, we'll learn how to get a model in Azure.

Getting a model

We can get a model by using the `DigitalTwinsClient.GetModel()` method. Copy the following code in the `DigitalTwinsManager` class:

```
public DigitalTwinsModelData GetModel(string modelId)
{
    try
    {
        return client.GetModel(modelId);
    }
    catch (RequestFailedException)
    {
        return null;
    }
}
```

Replace the code below the `// TODO` remark field in the `Program.Main` method with the following code:

```
DigitalTwinsModelData model = dtHelper.
GetModel("dtmi:com:smartbuilding:Room;1");

Console.WriteLine(model.Id);

Console.WriteLine(model.LanguageDisplayNames["en"]);
```

Run the application and check the output in the command console. It will display the `@id` and `displayName` properties. `displayName` is stored in a collection with the language as the key:

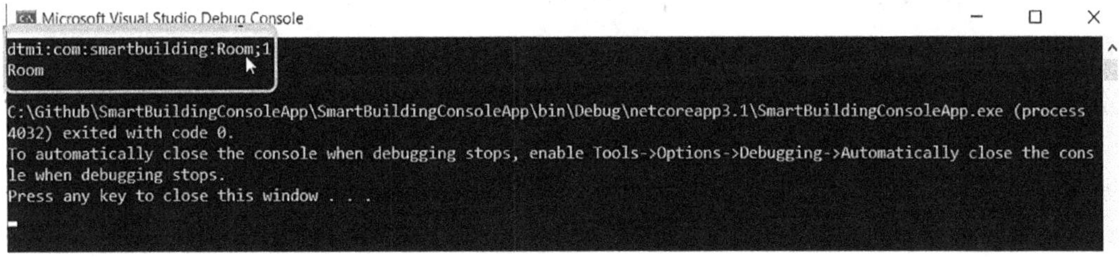

Figure 4.12 – Viewing the ID and name of a single model

Now, let's learn how to get multiple models in an Azure Digital Twin.

Getting models

It is also possible to retrieve all the models in the Azure Digital Twins instance by using the `DigitalTwinsClient.GetModels()` method. Add the following code to the `DigitalTwinsManager` class:

```
public Pageable<DigitalTwinsModelData> GetModels()
{
    GetModelsOptions options = new GetModelsOptions();

    return client.GetModels(options);
}
```

This method has an optional `GetModelsOptions` parameter that allows you to have more control over the result. The following properties can be used:

- `DependenciesFor`: This allows us to specify a list of model IDs so that we can retrieve their dependencies.

- `IncludeModelDefinition`: This property, when set to `true`, will include the model definition in the result.

Replace the code below the `// TODO` remark field in the `Program.Main` method with the following code:

```
var models = dtHelper.GetModels();
foreach (DigitalTwinsModelData model in models)
{
    Console.WriteLine(model.Id);
    Console.WriteLine(model.LanguageDisplayNames["en"]);
}
```

Run the application and check the output in the command console. It will display the `@id` and `displayName` properties of each model that has been defined within the Azure Digital Twins instance:

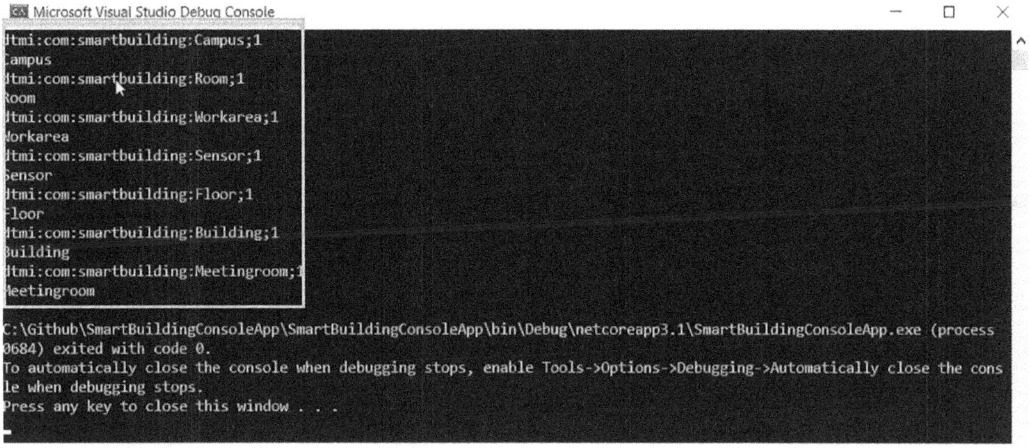

Figure 4.13 – Viewing the IDs and names of all the models

With that, we have learned how to create, delete, get, and inherit Models. In the next section, we will learn how to instantiate these models to create Digital Twins.

Managing Digital Twins

The second part of this chapter is all about managing Digital Twins. Digital Twins are instances of Models. We will be exploring how to create, update, delete, and get a Digital Twin by using the .NET SDK.

Creating a Digital Twin

The `BasicDigitalTwin` class is a helper class that can serialize or deserialize a Digital Twin. This class is used to create a Digital Twin based on metadata and content. The metadata specifies the ID of the model by setting `BasicDigitalTwin. Metadata.ModelId`.

Copy the following code in the `DigitalTwinsManager` class:

```
public bool CreateDigitalTwin(string twinId, string modelId)
{
    BasicDigitalTwin digitalTwin = new BasicDigitalTwin();
    digitalTwin.Metadata = new DigitalTwinMetadata();
    digitalTwin.Metadata.ModelId = modelId;
    digitalTwin.Id = twinId;
```

```
    try
    {
        client.
CreateOrReplaceDigitalTwin<BasicDigitalTwin>(twinId,
digitalTwin);
    }
    catch (RequestFailedException)
    {
        return false;
    }

    return true;
}
```

Now, create the `DigitalTwinsClient.CreateOrReplaceDigitalTwin` call, with `BasicDigitalTwin` set as a strongly typed object type class, to create the Digital Twin.

Replace the code below the `// TODO` remark field in the `Program.Main` method with the following code:

```
dtHelper.CreateDigitalTwin("Campus",
"dtmi:com:smartbuilding:Campus;1");
dtHelper.CreateDigitalTwin("MainBuilding",
"dtmi:com:smartbuilding:Building;1");
dtHelper.CreateDigitalTwin("GroundFloor",
"dtmi:com:smartbuilding:Floor;1");
dtHelper.CreateDigitalTwin("MeetingRoom1_01",
"dtmi:com:smartbuilding:Meetingroom;1");
```

It is not allowed to use spaces in the ID of the Digital Twin.

Run the application and use the Azure Digital Twins Explorer to see the Digital Twin you created. Digital Twins appear in the Graph View. Follow these steps, as shown in the following screenshot, to view the Digital Twins:

1. Enter the `SELECT * FROM digitaltwins` query in the query field.

2. Press the **Run Query** button.

3. The Digital Twins will appear in the graph view:

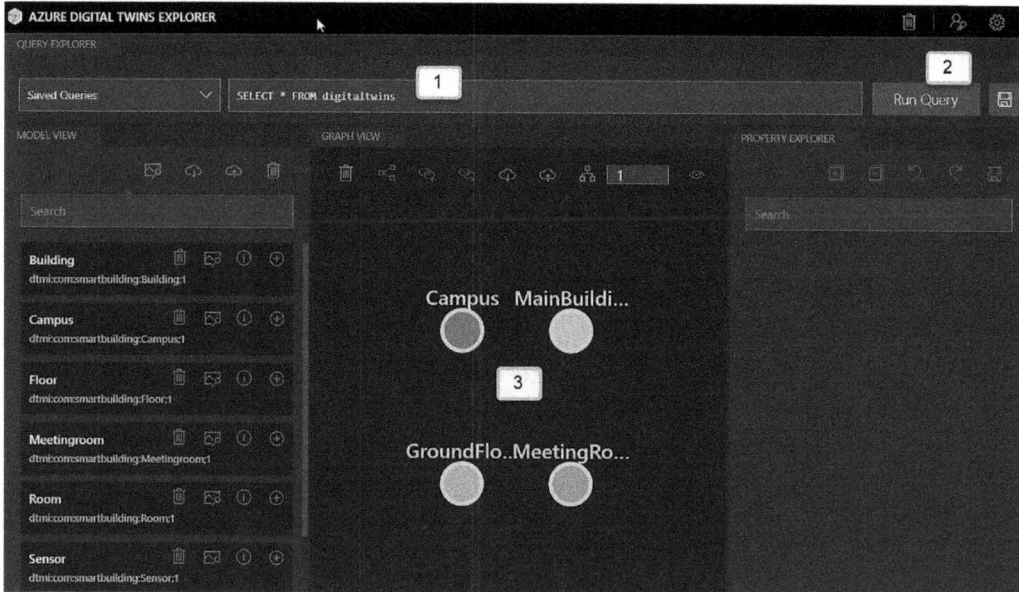

Figure 4.14 – Several Digital Twins created and visible in Azure Digital Twins Explorer

In the next section, we will learn how to update a Digital Twin.

Updating a Digital Twin

A Digital Twin contains values that are stored in the content part of the Digital Twin. The content part consists of properties, telemetry, and more. Updating a Digital Twin is important if you wish to set the value of these elements.

Copy the following code in the `DigitalTwinsManager` class:

```
public bool UpdateDigitalTwin(string twinId, string property,
object value)
{
    try
    {
        BasicDigitalTwin digitalTwin = client.
GetDigitalTwin<BasicDigitalTwin>(twinId);
        digitalTwin.Contents[property] = value;
        client.
CreateOrReplaceDigitalTwin<BasicDigitalTwin>(twinId,
digitalTwin);
    }
```

```
    catch (RequestFailedException)
    {
        return false;
    }
    return true;
}
```

The preceding example retrieves the Digital Twin based on its ID as a strongly typed object of the `BasicDigitalTwin` class.

We use the `Contents` property to set the value of a certain property. We use `object` as the type of the parameter. This is very important since the value needs to be set based on the schema of the property. We will explain this more in depth in the next chapter.

The `DigitalTwinsClient.CreateOrReplaceDigitalTwin()` method updates the Digital Twin with the new set of values.

Replace the code below the `// TODO` remark field in the `Program.Main` method with the following code:

```
dtHelper.UpdateDigitalTwin("MeetingRoom1.01", "occupied",
true);
```

Run the application and check, via the Azure Digital Twins Explorer, whether the property of the Digital Twin has been updated. You will need to run the query again to view the updated value. Follow these steps, as shown in the following screenshot:

1. Press the **Run Query** button to run the query again.

2. Select the Digital Twin called `Meetingroom1.01`.

3. View the properties of the Digital Twin. You will see that the `occupied` property has been updated to `true`:

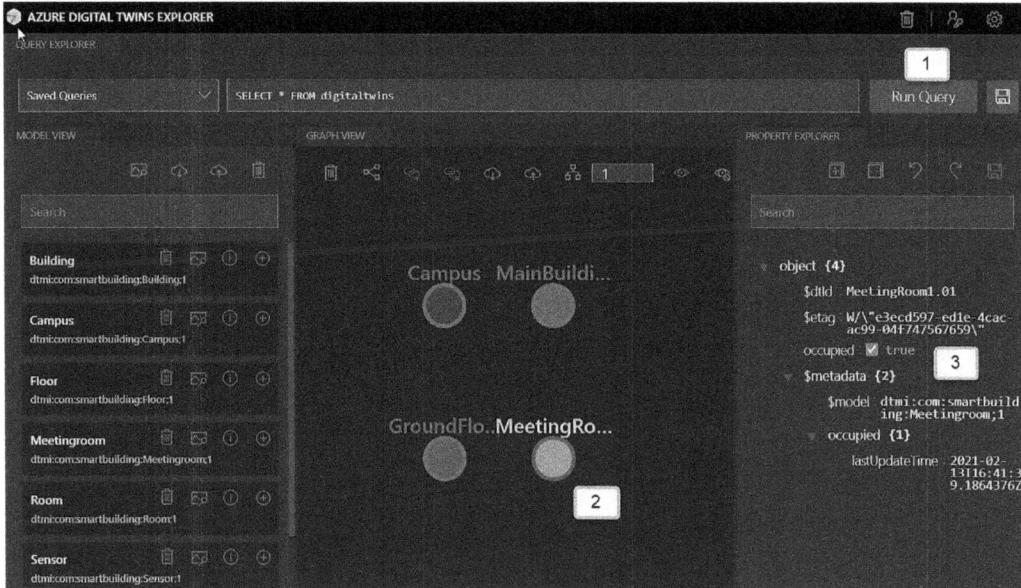

Figure 4.15 – Viewing the updated properties of the MeetingRoom1.01

In the next section, we will learn how to delete a Digital Twin.

Deleting a Digital Twin

A Digital Twin can be deleted based on its ID. We can use `DigitalTwinsClient.DeleteDigitalTwin()` for this.

Copy the following code in the `DigitalTwinsManager` class:

```
public bool DeleteDigitalTwin(string twinId)
{
    try
    {
        client.DeleteDigitalTwin(twinId);
    }
    catch (RequestFailedException)
    {
        return false;
    }

    return true;
}
```

Replace the code below the `// TODO` remark field in the `Program.Main` method with the following code:

```
dtHelper.DeleteDigitalTwin("Campus");
```

Run the application to remove the Digital Twin. Use the Azure Digital Twins Explorer to check the results.

Getting a Digital Twin

We can retrieve a Digital Twin based on its ID by using the `DigitalTwinsClient.GetDigitalTwin()` method as a strongly typed object type of the `BasicDigitalTwin` class.

Copy the following code in the `DigitalTwinsManager` class:

```
public BasicDigitalTwin GetDigitalTwin(string twinId)
{
    try
    {
        return client.GetDigitalTwin<BasicDigitalTwin>(twinId);
    }
    catch (RequestFailedException)
    {
        return null;
    }
}
```

Replace the code below the `// TODO` remark field in the `Program.Main` method with the following code:

```
BasicDigitalTwin twin = dtHelper.
GetDigitalTwin("MeetingRoom1.01");
bool occupied = false;
Boolean.TryParse(twin.Contents["occupied"].ToString(), out
occupied);
if (occupied)
{
    Console.WriteLine("the meeting room is occupied");
}
```

Here, the Digital Twin is returned as a `BasicDigitalTwin`. Then, we try to parse the value of the `occupied` property and set it to a Boolean. Based on the result, we print a message if the meeting room is occupied.

Run the application and check the command console for the message. Since we have set the `occupied` property to `true`, it should return a message stating that the room is occupied.

In *Chapter 7, Querying Digital Twins*, we will learn how to use queries to query Digital Twins in more detail. This is a far easier way of retrieving a set of Digital Twins.

Summary

In this chapter, you learned what it takes to design and model Models. We created a .NET console application to test our examples around managing Models and Digital Twins. Several examples for creating, deleting, updating, and getting Models and Digital Twins were explained. All the examples were based on the synchronous methods of the `DigitalTwinsClient` class. However, it is also possible to use asynchronous calls. This will be explained in a later chapter of this book.

The next chapter will talk about the structure of Digital Twins. We will explain and test properties, telemetry, and components. We will also extend the models that we used in this chapter that contain these elements.

Questions

As we conclude, here is a list of questions for you to test your knowledge regarding this chapter's material. You will find the answers in the *Assessments section* of the *Appendix*:

1. What is the base class for managing Models and Digital Twins?

    a. `DigitalTwins`

    b. `DigitalTwinsClient`

    c. `AzureDigitalTwinsClient`

2. What do we need to do to make sure that the models are part of the distribution of the application?

    a. Set **Copy to Output Directory** to **Always copy**.

    b. Set **Build Action** to **None**.

    c. Set **Build Action** to **Resource**.

3.   Which of the following should we use to update a Digital Twin?

a. `DigitalTwinsClient.UpdateDigitalTwin<T>`

b. `DigitalTwinsClient.AddOrReplaceDigitalTwin<T>`

c. `DigitalTwinsClient.CreateOrReplaceDigitalTwin<T>`

Further reading

You can learn more about creating your own custom models for Azure Digital Twins in the online Microsoft documentation: `https://docs.microsoft.com/en-us/azure/digital-twins/concepts-models`.

5
Model Elements

In this chapter, we will look at several examples that show how to add elements such as properties, telemetry, and components to a Digital Twin. We will extend our models from the previous chapter and add some additional models to them. We will also learn how to apply these elements and how to program them using the console application. We will then add the different elements.

In this chapter, we'll cover the following topics:

- Primitive properties
- Complex properties
- Telemetry
- Components

Let's get started!

Technical requirements

We will be using the previous .NET console application using the .NET SDK to build up our Azure Digital Twins instances using models. The Azure Digital Twins Explorer will be used to view the results of the .NET calls that are made by the console application.

Primitive properties

We will start by creating some primitive properties. **Primitive properties** are properties of a Digital Twin such as `boolean`, `double`, `float`, `integer`, `string`, and others. They are used to define the different metadata fields of your model. All the available primitive schemas were listed in *Chapter 3, Digital Twin Definition Model*.

We will extend the *Floor* model with two new properties.

Create a new version of the model file for a Floor by executing the following steps:

- Create a new folder called `chapter5` under the `Models` folder of `SmartbuildingConsoleApp`. Right-click on the `Models` folder and select **Add → New folder**.

- Create a new `.json` file called `floor.json` in the `chapter5` folder that contains the following code:

```
{
    "@id": "dtmi:com:smartbuilding:Floor;1",
    "@type": "Interface",
    "@context": "dtmi:dtdl:context;2",
    "displayName": "Floor",
    "contents": [
        {
            "@type": "Property",
            "name": "floornumber",
            "schema": "integer",
            "writable": false
        },
        {
            "@type": "Property",
            "name": "lightson",
            "schema": "boolean",
            "writable": true
        }
    ]
}
```

Here, we have added two properties. The `floornumber` property is a property of the `integer` type and represents the floor number in the building. This property has been set to *readonly* by us setting the `writable` attribute to `false`, since the floor number never changes.

The `lightson` property defines whether the lights are switched on or off. Since this can change, the property is writable.

To update the model, we must remove the model and create the model again based on the new file. Replace the code in the `Program.Main` method in the `Program.cs` file with the following code:

```
string[] paths = new string[] {
    "Models/chapter5/floor.json"
};

dtHelper.DeleteModel("dtmi:com:smartbuilding:Floor;1");
dtHelper.CreateModels(paths);
Console.WriteLine("Model updated");
```

While we already have a Digital Twin instantiated from the *Floor* model, it will not be removed and will still work when the model is removed and created again.

Open Azure Digital Twins Explorer and execute the following steps, as shown in the following screenshot:

1. Run the query to refresh the **Graph View** area.
2. Click on the Digital Twin node called **GroundFloor**.

3. You will see the new properties defined under the `metadata` specification:

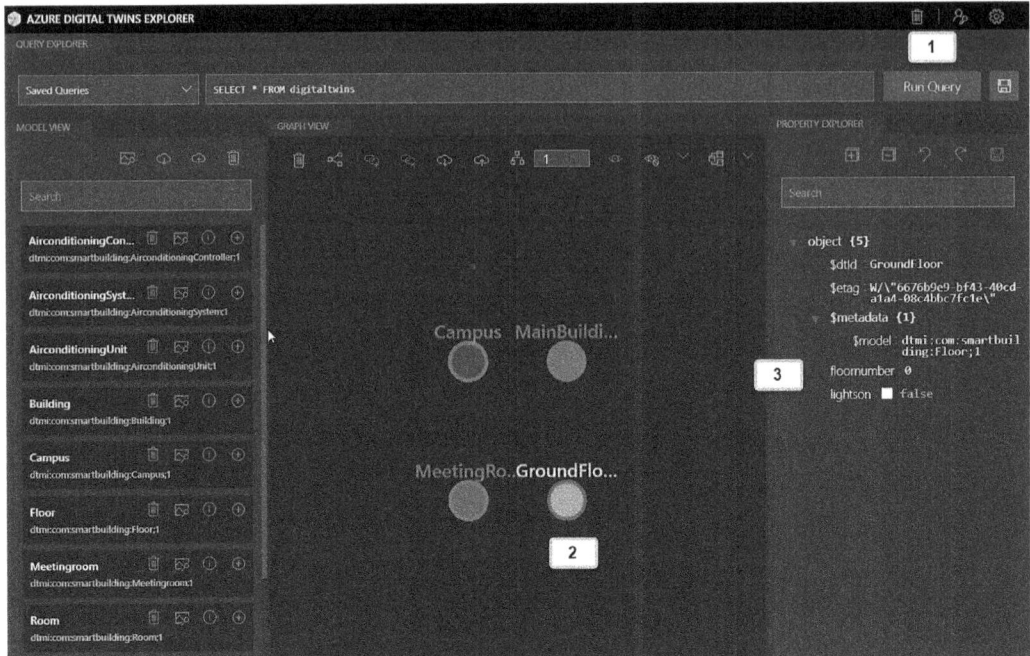

Figure 5.1 – Viewing the properties of the GroundFloor Digital Twin

You will notice that the new properties only appear under the `metadata` specification. Since the Digital Twin was initiated from the previous version, its properties haven't been defined yet.

> **Tip**
> You will receive an error when you try to read the properties of a Digital Twin when they only appear under the metadata specification. This can happen when you update the model after the model has been used for creating Digital Twins. You will need to add or update the property first.

Let's run the following code to see what happens with the view containing the properties of the Digital Twin. Replace the code in the `Program.Main` method in the `Program.cs` file with the following code:

```
DigitalTwinsManager dtHelper = new DigitalTwinsManager();

dtHelper.UpdateDigitalTwin("GroundFloor", "floornumber", 1);
dtHelper.UpdateDigitalTwin("GroundFloor", "lightson", true);
```

Run the application in Visual Studio. The code will fail because our update method tries to replace the Digital Twin with the property that was set. Since the property hasn't been added yet, we receive an error.

This means we must add properties to a Digital Twin in a different way. Add the following code to the `DigitalTwinsManager` class:

```
public void UpdateDigitalTwinProperty(string twinId, string
property, object value)
{
    JsonPatchDocument patch = null;
    try
    {
        patch = new JsonPatchDocument();
        patch.AppendAdd("/" + property, value);
        client.UpdateDigitalTwin(twinId, patch);
    }
    catch (RequestFailedException)
    {
    }

    patch = new JsonPatchDocument();
    patch.AppendReplace("/" + property, value);
    client.UpdateDigitalTwin(twinId, patch);
}
```

The preceding code uses the `JsonPatchDocument` class to update the Digital Twin. First, the code tries to add the property. Then, it replaces the value of the property.

Replace the code in the `Program.Main` method in the `Program.cs` file with the following code:

```
DigitalTwinsManager dtHelper = new DigitalTwinsManager();

dtHelper.UpdateDigitalTwinProperty("GroundFloor",
"floornumber", 1);
dtHelper. UpdateDigitalTwinProperty ("GroundFloor",
"lightson", true);
```

Run the application in Visual Studio:

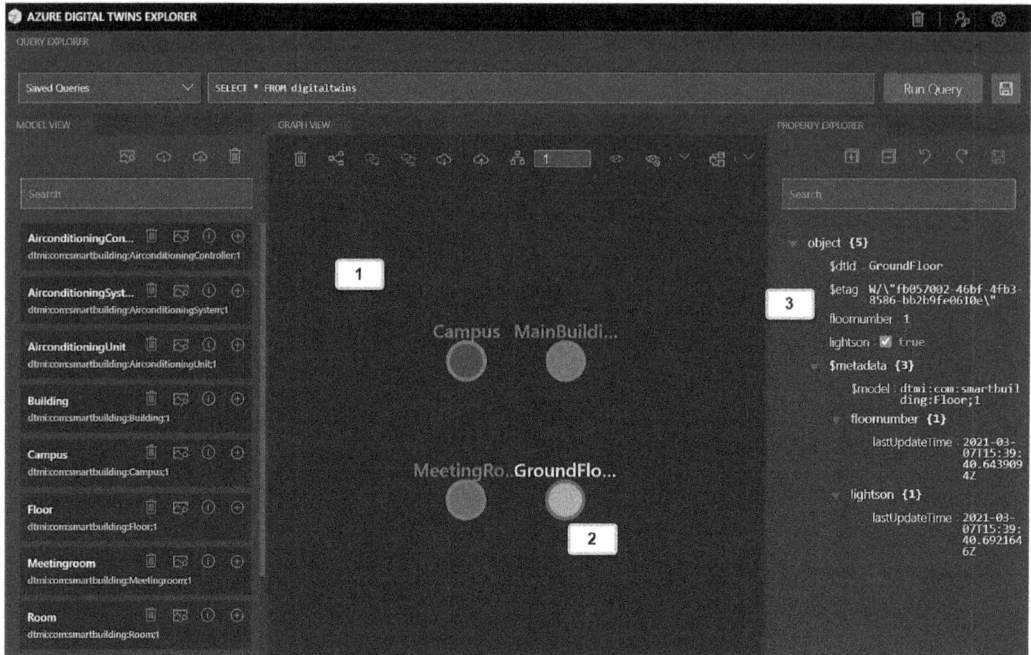

Figure 5.2 – Viewing the properties of the GroundFloor Digital Twin

To see the result, we need to execute the following steps in the Azure Digital Twins Explorer, as shown in the preceding screenshot:

1. Click somewhere outside the nodes in the **Graph View** area.
2. Select the Digital Twin node of **GroundFloor**.
3. View the properties of the Digital Twin.

The properties of the Digital Twin are extended with `floornumber` and `lightson`.

Here, we can see that both properties have been set. However, one of the properties has been set to read only, so it shouldn't be overwritten. There is a good reason for this behavior. The Digital Twins Definition Language is used to describe the definition of a Digital Twin. The `writable` property is part of that. However, the `writable` property is not used within the Azure Digital Twins SDK. It is mainly used to determine the capabilities of the model for **IoT Plug and Play** purposes. **IoT Plug and Play** allows us to integrate smart devices without manual configuration. These configurations will obey the `writable` property. This means that **IoT Hub**, an Azure cloud service that acts as a central message hub between these smart devices and the Azure Digital Twin, will not be able to write to the read-only property since it complies with the capabilities.

So far, we have extended some models with primitive properties and learned how to create and update them through programming. In the next section, you will learn about complex properties.

Complex properties

It is also possible to use complex properties. These properties use complex schemas such as *Object*, *Enum*, *Array*, and *Map*. All of these were explained in *Chapter 3*, *Digital Twin Definition Model*. In this section, we will focus on the complex schema type called *Map*. This complex schema type can be best compared to a dictionary in **C#**.

This example will demonstrate how to extend the Building model with a map containing all the temperatures of different rooms. This allows us to query the temperatures of the complete building easily in one go.

We need to create a new `.json` file called `building.json` under `Models/chapter5`. Replace the content of the file with the following code:

```
{
  "@id": "dtmi:com:smartbuilding:Building;1",
  "@type": "Interface",
  "@context": "dtmi:dtdl:context;2",
  "displayName": "Building",
  "contents": [
    {
      "@type": "Property",
      "name": "rooms",
      "writable": true,
      "schema": {
        "@type": "Map",
        "mapKey": {
          "name": "roomname",
          "schema": "string"
        },
        "mapValue": {
          "name": "roomtemperature",
          "schema": "float"
        }
      }
    }
```

```
        }
    ]
}
```

The schema of the property consists of a definition for the key and the value. In this case, the key is a string specifying the name of the room, while the value is a float containing the temperature of that room.

Just like we did previously, we will update the model using code. Replace the code in the `Program.Main` method in the `Program.cs` file with the following code:

```
DigitalTwinsManager dtHelper = new DigitalTwinsManager();

string[] paths = new string[] {
    "Models/chapter5/building.json"
};

dtHelper.DeleteModel("dtmi:com:smartbuilding:Building;1");
dtHelper.CreateModels(paths);
Console.WriteLine("Model updated");
```

Now, run the application in Visual Studio.

A map can be updated by using a dictionary **C#** class. Replace the code in the `Program.Main` method in the `Program.cs` file with the following code:

```
DigitalTwinsManager dtHelper = new DigitalTwinsManager();

var map = new Dictionary<string, float>();
map.Add("MeetingRoom101", 21.3f);
map.Add("MeetingRoom102", 22.1f);
map.Add("MeetingRoom103", 20.9f);

dtHelper.UpdateDigitalTwinProperty
("MainBuilding", "rooms", map);
```

The preceding code creates a dictionary by using the `Dictionary<>` class based on a `string` key and a `float` value. We added several rooms to the list before updating the Digital Twin using the `UpdateDigitalTwinProperty` code.

Now, run the application in Visual Studio.

Important note

While our rooms have punctuation in their names, we are not allowed to use punctuation in the map key. Therefore, we removed it from the names of the rooms.

The result can be seen by using the Azure Digital Twins Explorer, as shown in the following screenshot:

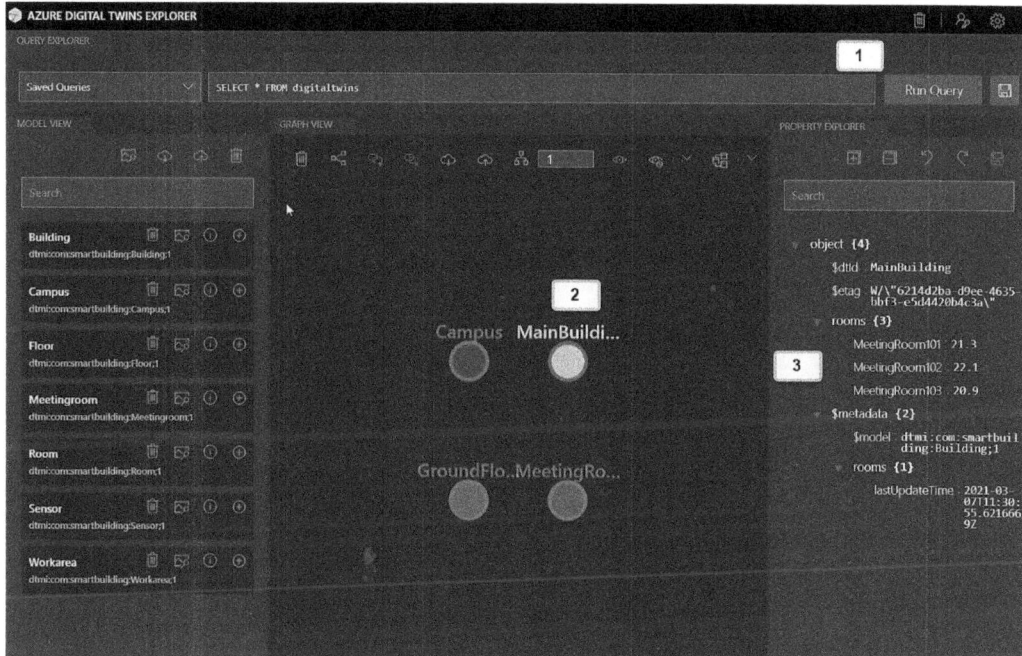

Figure 5.3 – Viewing the complex schema of a Map

Execute the following steps to view the results:

1. Run the query again to update the **Graph View** area.
2. Select the Digital Twin node called **MainBuilding**.
3. View the properties of **MainBuilding**.

With that, we have extended the **MainBuilding** model with the complex property called Map and learned how to create and update them through programming. In the next section, you will learn about telemetry.

Telemetry

Telemetry is a special type of property. It doesn't store any data and acts like a **C#** event. More information can be found in *Chapter 3, Digital Twin Definition Model.*

It is not possible to update the telemetry field of a Digital Twin. It is part of the schema and matches the actual IoT sensor. An additional property is used to store the value in the Digital Twin. This field is normally updated with the latest measured value and is part of the **ingress** process. **Ingress** is a way of binding the output of a sensor to the Azure Digital Twin. This process will be explained more thoroughly later in this book.

Since we aren't focusing on the **ingress** process just yet, we will be using a method that generates temperature data and pushes it to the Digital Twin:

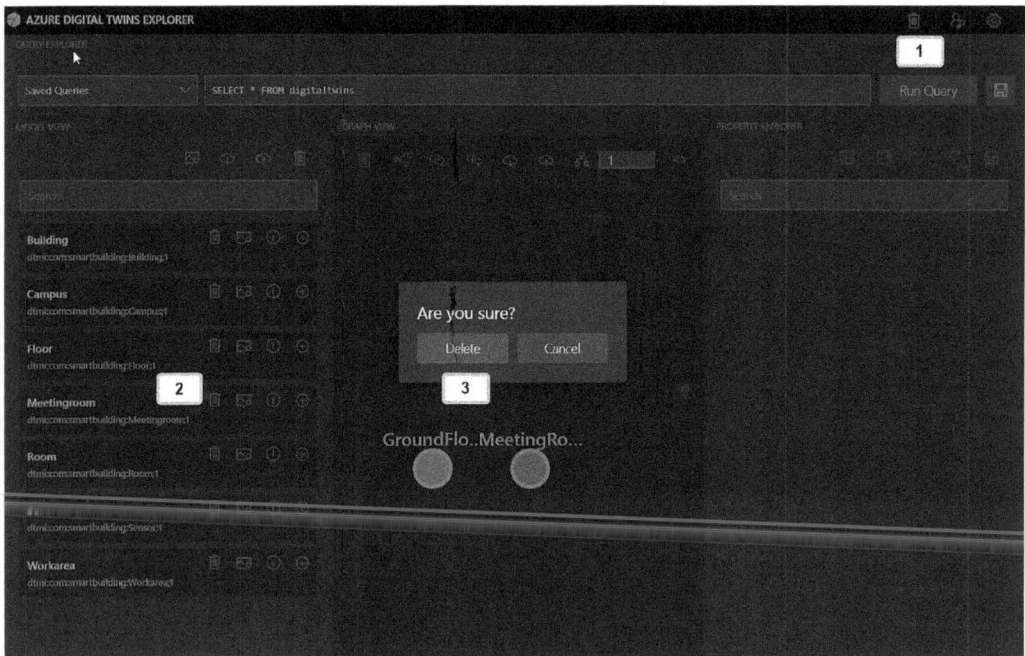

Figure 5.4 – Removing the Meetingroom model via the Azure Digital Twins Explorer

We will start by removing the **MeetingRoom** model via the Azure Digital Twins Explorer, as shown in the preceding screenshot. Execute the following steps:

1. Refresh the page of the Azure Digital Twins Explorer browser.
2. Press the recycle bin icon next to the **MeetingRoom** model.
3. Press the **Delete** button in the dialog.

Create a new `.json` file called `Meetingroom.json` under `Models/chapter5`.
Replace the content of the file with the following code:

```json
{
    "@id": "dtmi:com:smartbuilding:Meetingroom;1",
    "@type": "Interface",
    "@context": "dtmi:dtdl:context;2",
    "extends": [ "dtmi:com:smartbuilding:Room;1" ],
    "displayName": "Meetingroom",
    "contents": [
        {
            "@type": "Property",
            "name": "occupied",
            "schema": "boolean",
            "writable": true
        },
        {
            "@type": "Property",
            "name": "temperaturevalue",
            "schema": "double",
            "writable":   true
        },
        {
            "@type": "Telemetry",
            "name": "temperature",
            "schema": "double"
        }
    ]
}
```

In this model, we have defined a telemetry property called `temperature` and
a `temperaturevalue` property to store the actual value of the latest measurement.

Replace the code in the `Program.Main` method in the `Program.cs` file with the following code:

```
DigitalTwinsManager dtHelper = new DigitalTwinsManager();

string[] paths = new string[] {
    "Models/chapter5/Meetingroom.json"
};

dtHelper.DeleteModel("dtmi:com:smartbuilding:Meetingroom;1");
dtHelper.CreateModels(paths);
Console.WriteLine("Model updated");
```

Run the application in Visual Studio. This will replace the **MeetingRoom** model with the new schema:

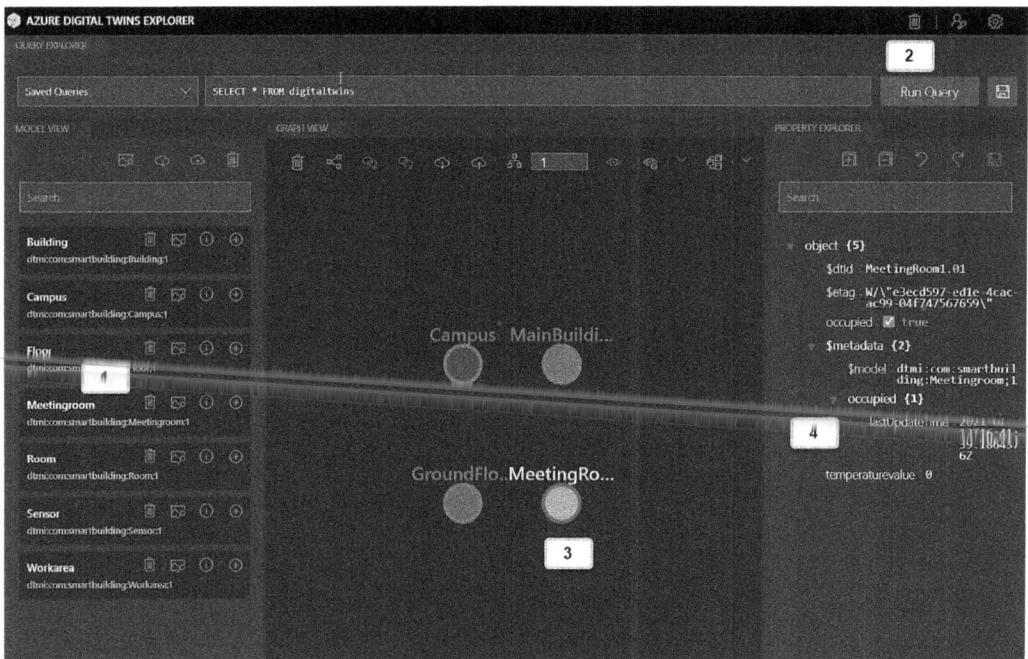

Figure 5.5 – Viewing the properties of the MeetingRoom Digital Twin instance

Use the Azure Digital Twins Explorer to see the results of the update. Execute the following steps:

1. Refresh the page to see the updated models.
2. Select **Run Query** to get an overview of the Digital Twins in the **Graph View** area.
3. Select the Digital Twin node named **MeetingRoom**.
4. View the properties of the Digital Twin.

Again, you will notice that the Digital Twin instance that was created does not reflect the `temperature` property yet. This is because it hasn't been set.

Create a new folder named Sensor in our SmartBuildingConsoleApp project. Copy the file TemperatureSensor.cs from the folder SmartBuildingConsoleApp/Sensor on GitHub to this new created folder. This class will generate random temperature values which will be used in this example.

Create a new static method called `GenerateSensorData()` by copying the following code into the `Program` class in `program.cs`:

```
public static void GenerateSensorData()
{
    DigitalTwinsManager dtHelper = new DigitalTwinsManager();

    // generate sensor data
    TemperatureSensor sensor = new TemperatureSensor();
    Console.WriteLine("Temperature sensor");
    while (true)
    {
        double temperature = sensor.GetMeasurement();
        Console.WriteLine(string.Format("{0} degrees",
temperature));

        dtHelper.UpdateDigitalTwinProperty("MeetingRoom1.01",
"temperaturevalue", temperature);

        System.Threading.Thread.Sleep(1000);
    }
}
```

We need to call this method. To imitate the idea that it is from a different application, we can use **threading** to execute the method in a separate thread. Replace the code in the `Program.Main` method in the `Program.cs` file with the following code:

```
static void Main(string[] args)
{
    // generate sensor data
    System.Threading.Thread sensorThread = new System.
Threading.Thread(GenerateSensorData);
    sensorThread.Start();
}
```

Run the application. The result is shown in the following screenshot:

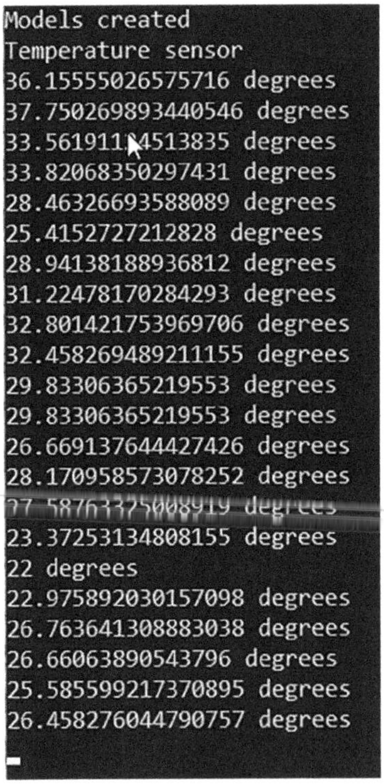

Figure 5.6 – The result of the GenerateSensorData method running in a separate thread

Now, while that application is running, check the value of the `temperature` field of the meeting room using the Azure Digital Twins Explorer, as shown in the following screenshot:

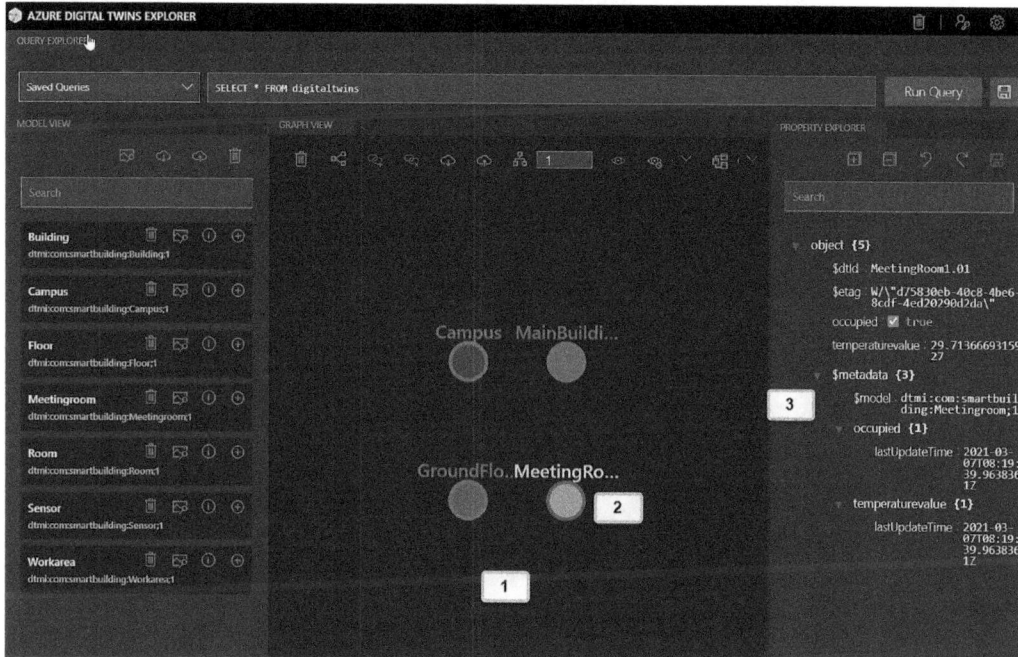

Figure 5.7 – The result of the temperature stored in the temperature property

To view the result, you will need to follow these steps:

1. Click somewhere in the **Graph View** area.

2. Select the Digital Twin node named **MeetingRoom**.

3. View the properties of the Digital Twin.

You will notice that the `temperature` property contains the latest temperature data from the method that's generating temperature values. We can also see that the telemetry field is not shown in the property explorer. This is normal since it is not a viewable field.

In this section, we have learned how to add telemetry to a model. We saw that the value is stored using a property and used a method to generate random sensor data. In the next section, you will learn how to use components.

Components

Components are mostly used in scenarios where you want to model the existing situation in parts. Components are interfaces that do not normally exist on their own. They are an integrated part of the interface where they are referred from. Therefore, they differ from relationships. Relationships are more focused on the definition of the relationship between two interfaces, while interfaces can co-exist and act separately from each other. This was explained in *Chapter 3*, *Digital Twin Definition Model*:

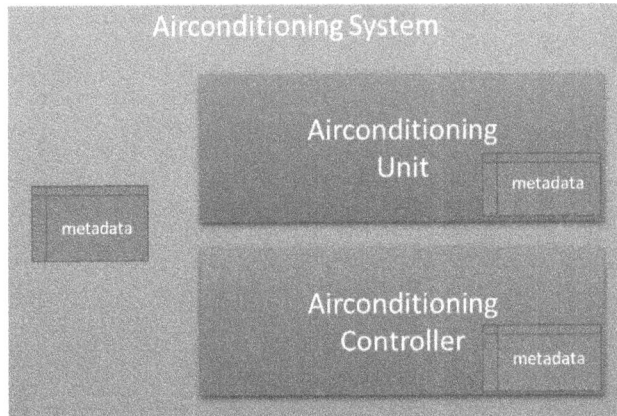

Figure 5.8 – A high-level view of a model that contains two components

To explain and show such components, we can look at the preceding diagram. Here, we have an air conditioning system. This system consists of a unit and a controller. The unit and controller are intrinsically bound to each other. Their existence defines the air condition system. Therefore, the unit and controller are components of the system, each with their own defined metadata.

We will start by defining the air conditioning unit. Create a new `.json` file under `Models/chapter5` called `airconditioningunit.json`. Replace the contents of that file with the following code:

```
{
    "@id": "dtmi:com:smartbuilding:AirconditioningUnit;1",
    "@type": "Interface",
    "@context": "dtmi:dtdl:context;2",
    "displayName": "AirconditioningUnit",
    "contents": [
        {
            "@type": "Property",
            "name": "poweron",
```

```
      "schema": "boolean",
      "writable": true
    },
    {
      "@type": "Property",
      "name": "level",
      "schema": "double",
      "writable": true
    }
  ]
}
```

Now, let's create the air conditioning controller. Now, create a new .json file under Models/chapter5 called airconditioningcontroller.json. Replace the contents of that file with the following code:

```
{
  "@id": "dtmi:com:smartbuilding:AirconditioningController;1",
  "@type": "Interface",
  "@context": "dtmi:dtdl:context;2",
  "displayName": "AirconditioningController",
  "contents": [
    {
      "@type": "Property",
      "name": "temperatureset",
      "schema": "float",
      "writable": true
    }
  ]
}
```

Finally, let's create the air conditioning system. Create a new .json file under Models/
chapter5 called airconditioningsystem.json. Replace the contents of that file
with the following code:

```
{
  "@id": "dtmi:com:smartbuilding:AirconditioningSystem;1",
  "@type": "Interface",
  "@context": "dtmi:dtdl:context;2",
  "displayName": "AirconditioningSystem",
  "contents": [
    {
      "@type": "Component",
      "name": "airconditioningunity",
      "schema": "dtmi:com:smartbuilding:AirconditioningUnit;1"
    },
    {
      "@type": "Component",
      "name": "airconditioningcontroller",
      "schema":
"dtmi:com:smartbuilding:AirconditioningController;1"
    }
  ]
}
```

Replace the code in the Program.Main method in the Program.cs file with the
following code:

```
DigitalTwinsManager dtHelper = new DigitalTwinsManager();

string[] paths = new string[] {
    "Models/chapter5/airconditioningunit.json",
    "Models/chapter5/airconditioningcontroller.json",
    "Models/chapter5/airconditioningsystem.json"
};
dtHelper.CreateModels(paths);
Console.WriteLine("Models created");
```

Run the application in Visual Studio.

> **Tip**
> If you want to remove the models that have referring components, the order is important. First, start with the model that contains component references and then the components itself; otherwise, you will receive an error.

We will use the Azure Digital Twins Explorer to view the result. Refresh your browser and view the new models, as shown in the following screenshot:

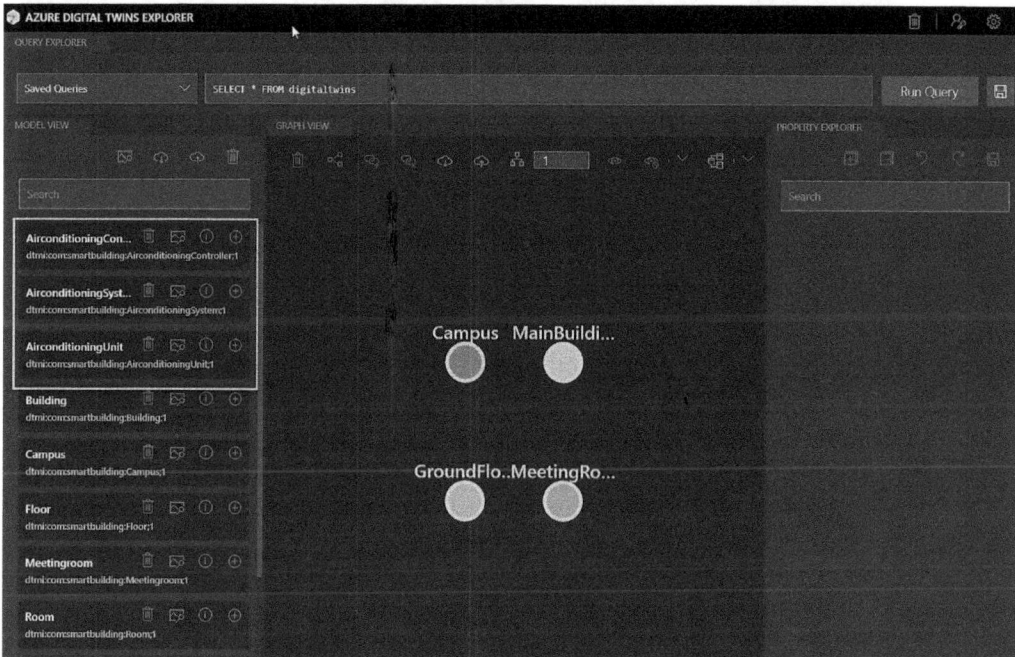

Figure 5.9 – Viewing the newly created models for our air conditioning system

The preceding screenshot shows how to create an air conditioning system based on the **AirconditioningSystem** model. Execute the following steps:

1. Press the + button of the **AirconditioningSystem** model. A pop - up dialog will appear, asking you to enter a new Digital Twin name. Enter `AirSystem` and press the **Save** button.

2. Press the **Run Query** button to refresh the **Graph View** area.

3. Select the Digital Twin node named **AirSystem**.

4.   View the properties of the selected Digital Twin, as shown in the following screenshot:

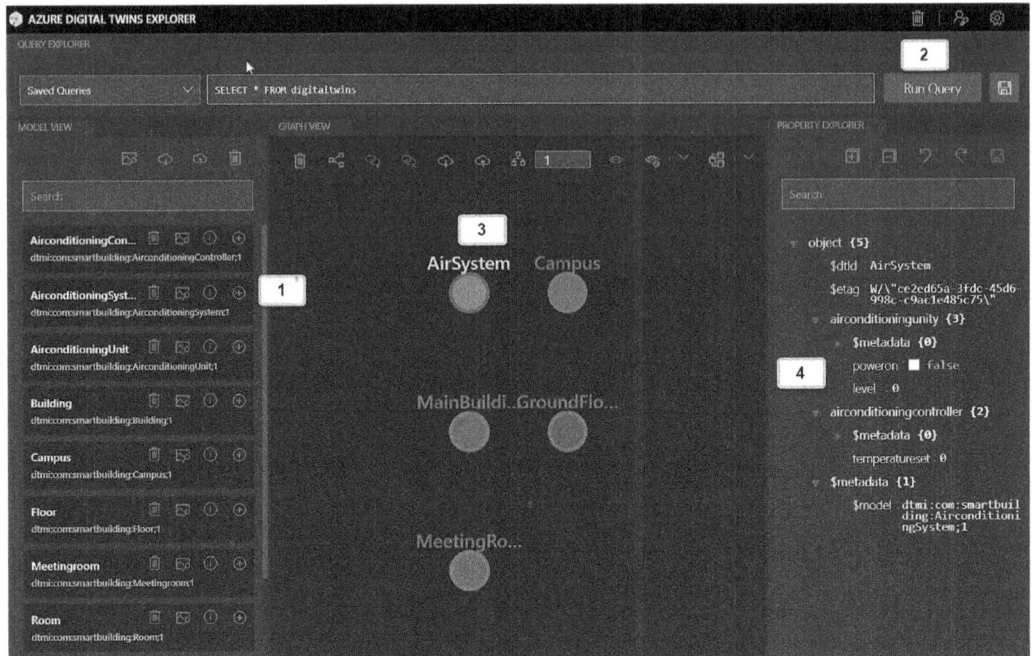

Figure 5.10 – Creating the air conditioning system and viewing its properties

Here, we can see that creating the air conditioning system has automatically instantiated a unit and a controller. The properties of the unit and controller can be found under the system.

> **Did you know?**
>
> When you remove a Digital Twin instance that contains components, it will remove all the referred components automatically.

Summary

In this chapter, you learned how to extend the models with primitive and complex properties, telemetry, and components. You learned that for telemetry, a property is used to store the actual data. Finally, you learned in what scenario you should use components, how they are created, and how they are instantiated.

The next chapter will explain how to define relationships in models and how to create, delete, and update relationships between Azure Digital Twin instances.

Questions

As we conclude, here is a list of questions for you to test your knowledge regarding this chapter's material. You will find the answers in the Assessments section of the Appendix:

1. Which of the following is typically used for storing a sensor value in a Digital Twin?

    a. Property

    b. Telemetry

    c. Component

2. Why would we model a real-time object as a component?

    a. The real-time object can exist on its own.

    b. This is a smart way of dividing the properties of a real-time object.

    c. The real-time object is part of another whole and cannot be seen as a self-sustained object.

3. What can you compare a Telemetry type to?

    a. C# event

    b. A property with `editable` set to `false`

    c. An interface to describe the relationship between a sensor and a Digital Twin

6
Creating Relationships between Azure Digital Twin Models

So far, we have been modeling and initiating Digital Twins and extending it with properties, telemetry, and components. But it is relationships that bring these Digital Twins together. This chapter will talk about how to define relationships between Azure Digital Twins. We will explain how to create a relationship between models, delete a relationship, and list all the relationships that have been defined.

The following topics will be covered in this chapter:

- Understanding relationships
- Creating relationships
- Getting relationships
- Deleting relationships

- Relationship properties
- Updating relationship properties

Let's get started!

Technical requirements

We will continue in the .NET console application using the .NET SDK to build up Azure Digital Twins instances using models. The Azure Digital Twins Explorer will be used to view the result of the .NET calls made by the console application.

Understanding relationships

A relationship is always between two Digital Twins called the *source* digital twin and the *target* digital twin. We can use descriptive language to describe the relationship between both Digital Twins. This makes it easier to understand what the relationship is about and in which direction the relationship has been set up. These descriptive names will also help us quickly identify the relationship when we use the Azure Digital Twins Explorer tool.

What is descriptive language and what does it look like? The following table contains several examples of descriptive language:

Relationship	Source	Descriptive name	Target
A building that has multiple floors.	Building	has	Floor
A sensor on a machine that generates alerts at a certain value.	Sensor	triggers	Alert
A document that has an approval process.	Document	approvedby	Manager
A factory floor where all kinds of machines are positioned.	Factory	contains	Machine

The descriptive name describes the relationship that the source digital twin has with the target digital twin, or vice versa. This will change the descriptive text:

Relationship	Source	Descriptive name	Target
A building that has multiple floors.	Floor	belongsto	Building
A sensor on a machine that generates alerts at a certain value.	Alert	generatedy	Sensor
A document that has an approval process.	Manager	approves	Document
A factory floor where all kinds of machines are positioned.	Machine	locatedat	Floor

It does not always make sense if you change the direction of the relationship. To make the right choice for a relationship, its direction, and description, we can take the following into consideration:

- Decide whether you really need a relationship. In some scenarios, we do not require one. This could be because the output data from a Digital Twin does not require the related digital twin data. You could replace the relationship with a *Component* if the metadata of both Digital Twins will inevitably be connected and you want to distinguish the different groups as separate parts.

- The relationship should make sense. We need to check whether the relationship between both Digital Twins is described properly and fits the real-world scenario. At some point, you will want to query the data from the Digital Twins. Having unclear relationships makes it more difficult to understand the results of your queries.

- We can retrieve data from the Digital Twin using queries. In some cases, by changing the direction, you can create less complex queries. The direction is how the relationship is defined. It can be defined from source to destination and vice versa. Relationships can help make queries easier to get certain data. In some situations, the direction can make it more difficult to get the right data. It can also influence the number of queries required to get the same data. This means that the structure and the data we need to have returned as output determines the way the relationship is implemented. Querying Digital Twins will be explained in more detail in *Chapter 8, Building Models Using Ontologies*.

- Determine whether the relationship is one-to-many or a one-to-one. This can influence the direction of the descriptive text. The direction is, again, best defined based on the impact of querying data from the Digital Twins. The descriptive text changes when you define the relationship from source to destination or vice versa.

> **Tip**
>
> Sometimes, it would be even better not to use a relationship but an Array property to store the relationship. You can do this by using a less complex query to get data from the Digital Twin. An example would be having a Floor containing Rooms where the Floor Digital Twin contains an array called Rooms that contains the list of rooms for that floor.

A relationship can also have properties. These properties allow us to specify metadata for a relationship, just like we do for a digital twin. An example is having the floor *level* specified in the relationship between a *Building* and the *Floor*.

With that, you have learned about the relationships between Digital Twins and how to use descriptive language to define the name of the relationship. In the next section, you will learn how to create a relationship using .NET code.

Creating relationships

We want to create a relationship between the Building digital twin and the Floor digital twin. The relationship is defined as *Building has Floor*. This means that we must add a property of the Relationship type to the source digital twin, which is the Building digital twin.

Create a new version of the model file for a Floor by executing the following steps:

- Create a new folder called `chapter6` under the `Models` folder of `SmartbuildingConsoleApp`. Right-click on the *Models* folder and select *Add → New folder*.

- Create a new `json` file that contains the following code:

```
{
    "@id": "dtmi:com:smartbuilding:Building;1",
    "@type": "Interface",
    "@context": "dtmi:dtdl:context;2",
    "displayName": "Building",
    "contents": [
        {
```

```
        "@type": "Property",
        "name": "rooms",
        ...
    },
    {
        "@type": "Relationship",
        "name": "has",
        "target": "dtmi:com:smartbuilding:Floor;1"

    }
  ]
}
```

The epsilon indication under the rooms property is to indicate that you need to leave that part intact. Only a new property of the *Relationship* type is added called has. target defines the digital twin that is related. In this case, it is related to the *Floor*. We can create this relationship using dtmi:com:smartbuilding:Floor;1.

To update the model, we must remove the model and create the model again based on the new file. Replace the code in the Program.Main method in the Program.cs file with the following code:

```
DigitalTwinsManager dtHelper = new DigitalTwinsManager();

string[] paths = new string[] {
    "Models/chapter6/building.json"
};

dtHelper.DeleteModel("dtmi:com:smartbuilding:Building;1");
dtHelper.CreateModels(paths);
Console.WriteLine("Model updated");
```

Now, let's create a new method for creating relationships. Add the following code to the DigitalTwinsManager class:

```
public void CreateRelationship(string twinSourceId, string
twinDestinationId, string description, Dictionary<string,
object> properties = null)
{
```

```
    string relationShipId = RelationshipId(twinSourceId,
twinDestinationId);

    BasicRelationship relationship = new BasicRelationship
    {
        Id = "buildingFloorRelationshipId",
        SourceId = twinSourceId,
        TargetId = twinDestinationId,
        Name = description,
        Properties = properties
    };

    try
    {
        client.CreateOrReplaceRelationship(twinSourceId,
relationShipId, relationship);
    }
    catch (RequestFailedException)
    {
    }
}
```

This method uses the BasicRelationship class to define the relationship. A relationship is based on the source digital twin and the target digital twin, its name, and its properties. Properties can be null. The client.CreateOrReplaceRelationship API method is used to create and/or replace the relationship.

Let's use a separate method called RelationshipId to create a relationship ID. Add the following method to the DigitalTwinsManager class:

```
public string RelationshipId(string twinSourceId, string
twinDestinationId)
{
    return string.Format("{0}-{1}", twinSourceId,
twinDestinationId);
}
```

This method will generate the `Id` property of the relationship by concatenating the `Id` property of the source digital twin and the `Id` property of the target digital twin.

Replace the code in the `Program.Main` method in the `Program.cs` file with the following code:

```
DigitalTwinsManager dtHelper = new DigitalTwinsManager();
dtHelper.CreateRelationship("MainBuilding", "GroundFloor",
"has");
```

Make sure that the description is equal to the name of the relationship definition in the model. It will fail to run otherwise. Run the application in Visual Studio.

Open Azure Digital Twins Explorer and execute the following steps, as shown in the following screenshot:

1. Refresh your browser if you have already started the Azure Digital Twins Explorer.

2. Run the query to refresh the **Graph View** area.

3. View the relationship in the **Graph View** area:

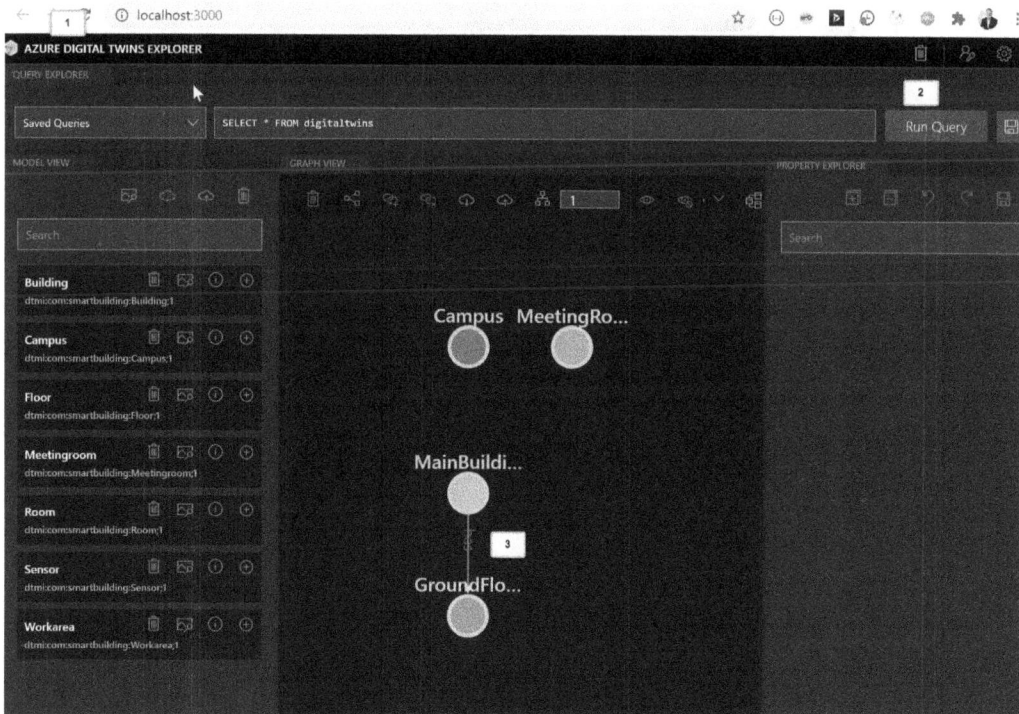

Figure 6.1 – Viewing the relationship between MainBuilding and GroundFloor

With that, you have learned how to create a relationship using .NET code. In the next section, you will learn how to retrieve relationships from the Azure Digital Twin.

Getting relationships

It is very handy to get relationships as we can get a related digital twin based on a relationship. An example of this is getting all the floors for a certain building. Relationships can be retrieved in two ways:

- **Getting a single relationship**: Get a relationship based on the source digital twin and the target digital twin.

- **Getting a list of relationships**: Get a list of relationships from a source digital twin.

Let's explain these in more detail. We will start by getting a single relationship.

Getting a single relationship

Let's start by creating a method that retrieves a relationship based on the source digital twin and the target digital twin. This allows us, for example, to access the properties specified for that relationship.

Create a new method for getting a relationship. Add the following code to the `DigitalTwinsManager` class:

```
public BasicRelationship GetRelationship(string twinSourceId,
string twinDestinationId)
{
    string relationshipId = RelationshipId(twinSourceId,
twinDestinationId);

    try
    {
        Response<BasicRelationship> relationship =
client.GetRelationship<BasicRelationship>(twinSourceId,
relationShipId);

        return relationship.Value;
    }
    catch (RequestFailedException)
    {
```

```
        }

        return null;
    }
```

This method is very simple and returns a `BasicRelationship` object based on the relationship that's found between two Digital Twins.

Replace the code in the `Program.Main` method in the `Program.cs` file with the following code:

```
DigitalTwinsManager dtHelper = new DigitalTwinsManager();

BasicRelationship relationship = dtHelper.
GetRelationship("MainBuilding", "GroundFloor");
if (relationship != null)
{
    Console.WriteLine(relationship.Id);
}
```

Run the application in Visual Studio. The code will output the `Id` property of the relationship to the console, which is built based on the IDs of the Digital Twins. The result can be seen in the following screenshot:

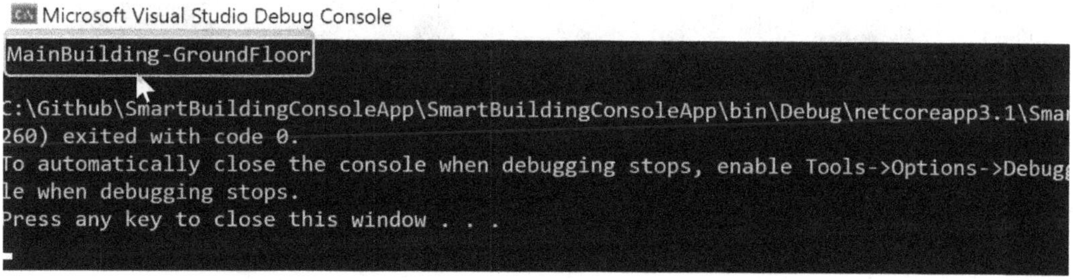

Figure 6.2 – The output of getting a single relationship

With that, you have learned how to retrieve a single relationship. In the next section, we will learn how to retrieve a list of relationships.

Getting a list of relationships

It is possible to get a list of relationships by only using the source digital twin. A source digital twin can have several relationships defined to one or more different target Digital Twins.

But before we start retrieving a list of Digital Twins, we want to create an additional relationship. We will be using the Azure Digital Twins Explorer to do this. Execute the following steps, as shown in the following screenshot:

1. Click on the + sign next to the *Floor* model to create a new digital twin. This will make a dialog box appear. Call it `1stFloor` and press the **Save** button.

2. Select the *source* digital twin named `MainBuilding`.

3. With the *Shift* key pressed, select the *target* digital twin named `1stFloor`.

4. Now, both Digital Twins will be selected. Click on the **+relationship** icon in the **Graph View** area to start creating a relationship:

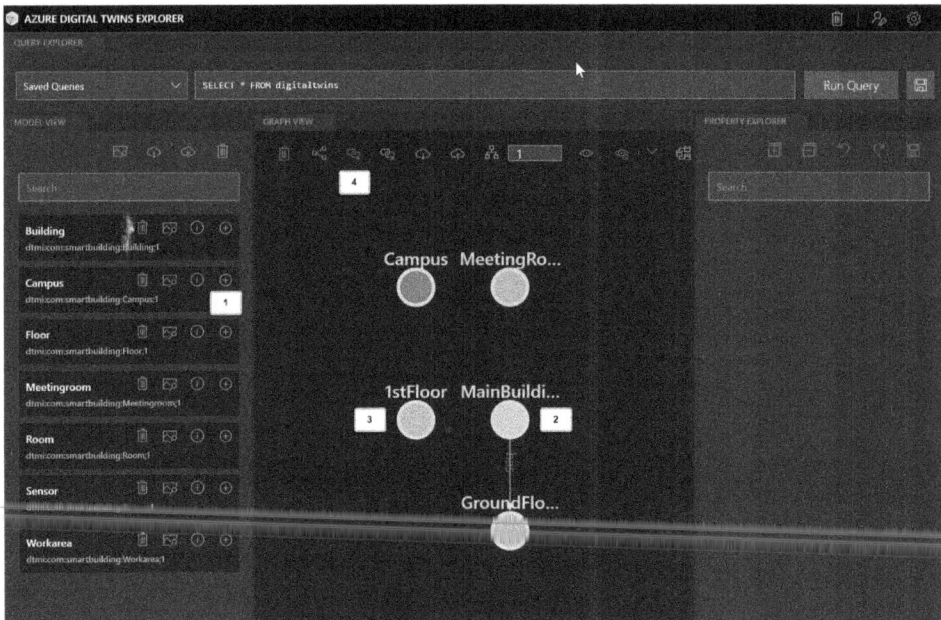

Figure 6.3 – Creating a new floor named 1stFloor

Once you have clicked on the *+relationship* icon in the **Graph View** area to start creating a relationship, the dialog shown in the following screenshot will appear. Execute the following steps:

1. Select the **has** relationship. It is possible to have multiple relationships of the same type. In that case, they will all appear in this dropdown.

2. Press the **Save** button to create the relationship:

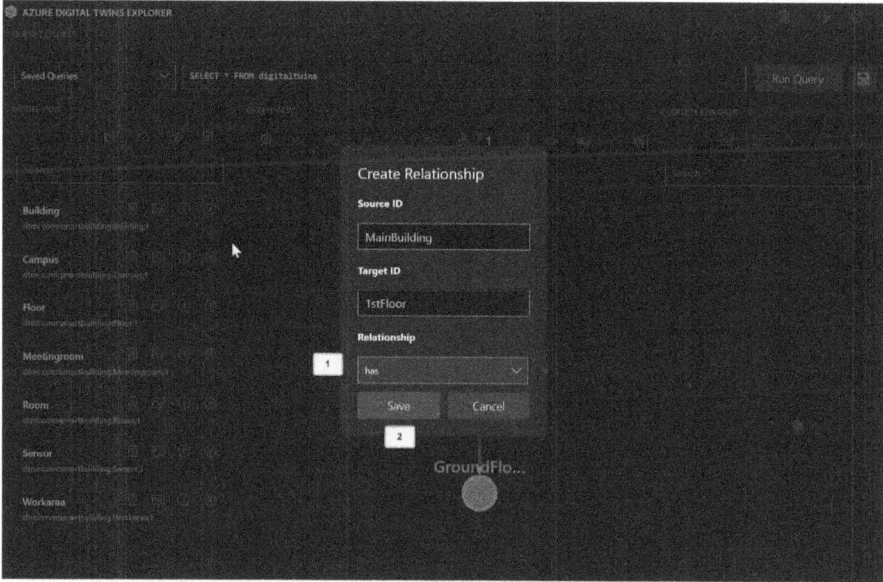

Figure 6.4 – Creating a relationship via Azure Digital Twins Explorer

The result of the relationship is shown in the following screenshot:

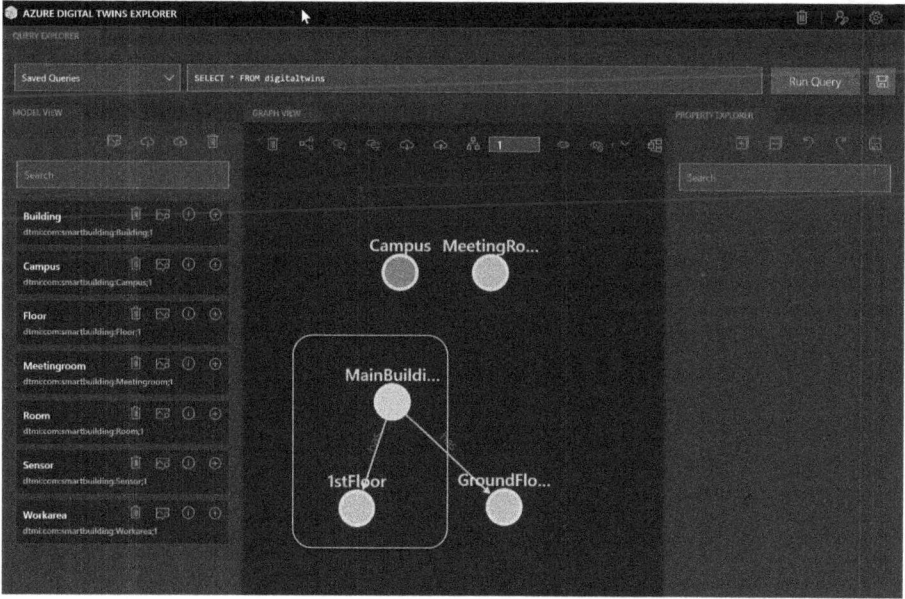

Figure 6.5 – The result of the created relationship

Create a new method for getting a list of relationships. Add the following code to the `DigitalTwinsManager` class:

```
public Pageable<BasicRelationship> ListRelationships(string
twinSourceId)
{
    try
    {
        Pageable<BasicRelationship> relationships = client.
GetRelationships<BasicRelationship>(twinSourceId);

        return relationships;
    }
    catch (RequestFailedException)
    {
    }

    return null;
}
```

This method returns a list of `BasicRelationship` objects. It is possible to use filters in the `GetRelationships<BasicRelationship>()` call. In this example, we won't use them, so all the relationships will be returned.

Replace the code in the `Program.Main` method in the `Program.cs` file with the following code:

```
DigitalTwinsManager dtHelper = new DigitalTwinsManager();

Pageable<BasicRelationship> relationships = dtHelper.
ListRelationships("MainBuilding");

foreach(BasicRelationship relationship in relationships)
{
    Console.WriteLine(relationship.Id);
}
```

Run the application in Visual Studio. The code will output the `Id` property of the relationship to the console, which is built based on the IDs of the Digital Twins. The result can be seen in the following screenshot:

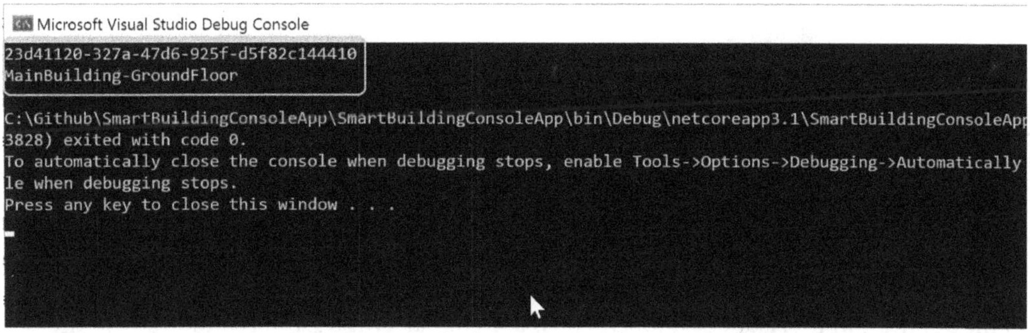

Figure 6.6 – The list of relationships for the MainBuilding Digital Twin

All the relationships of the *MainBuilding* digital twin are listed. As you can see, the first relationship is a `Guid`. This is the relationship that was created by using the Azure Digital Twins Explorer.

With that, you have learned how to get a single relationship and a list of relationships. In the next section, you will learn how to delete one of these relationships.

Deleting relationships

In this section, we will be deleting one of the relationships we created. For this, we must create a new method for deleting a relationship. Add the following code to the `DigitalTwinsManager` class:

```
public void DeleteRelationship(string twinSourceId, string
twinDestinationId)
{
    string relationShipId = RelationshipId(twinSourceId,
twinDestinationId);

    try
    {
        client.DeleteRelationship(twinSourceId,
relationShipId);
```

```
    }
    catch (RequestFailedException)
    {
    }
}
```

Replace the code in the `Program.Main` method in the `Program.cs` file with the following code:

```
DigitalTwinsManager dtHelper = new DigitalTwinsManager();
dtHelper.DeleteRelationship("MainBuilding", "GroundFloor");
```

Run the application in Visual Studio. Open Azure Digital Twins Explorer and refresh the browser. You will see that the relationship between *MainBuilding* and *GroundFloor* has been removed, as shown in the following screenshot:

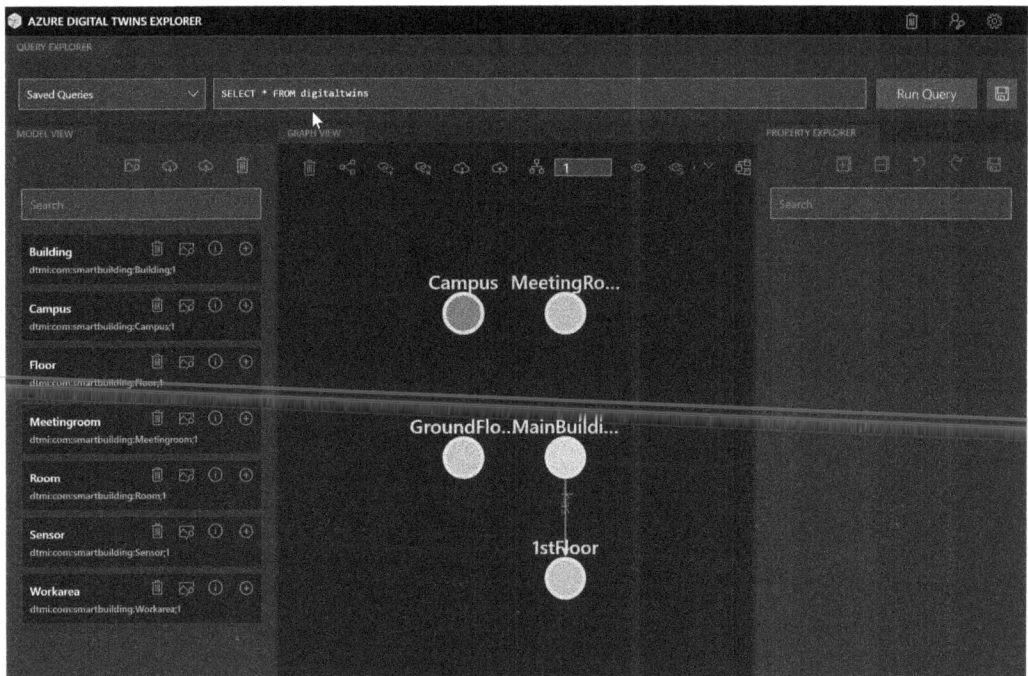

Figure 6.7 – An overview of the Digital Twin once the relationship has been removed

With that, you have learned how to remove a relationship using .NET code. We will explain how to use properties in a relationship in the next section.

Relationship properties

We want to create a relationship with a `level` property between *MainBuilding* and *GroundFloor*. Let's extend the relationship definition in the Building model. Open the `building.json` file in the `Models/Chapter6` folder. Replace the Relationship definition part with the following JSON payload, indicated in bold:

```
{
    "@id": "dtmi:com:smartbuilding:Building;1",
    "@type": "Interface",
    "@context": "dtmi:dtdl:context;2",
    "displayName": "Building",
    "contents": [
    ...
        {
            "@type": "Relationship",
            "name": "has",
            "target": "dtmi:com:smartbuilding:Floor;1",
            "properties": [
                {
                    "@type": "Property",
                    "name": "level",
                    "schema": "integer"
                }
            ]
        }
    ]
}
```

Replace the code in the `Program.Main` method in the `Program.cs` file with the following code:

```
DigitalTwinsManager dtHelper = new DigitalTwinsManager();

Dictionary<string, object> properties = new Dictionary<string,
object>();
properties["level"] = 0;
```

```
dtHelper.CreateRelationship("MainBuilding", "GroundFloor",
"has", properties);
```

Run the application in Visual Studio. This will create a relationship containing a property called `level` that's set to `0`.

Since we cannot see the properties of a relationship in the Azure Digital Twins Explorer, we will use some code to get the relationships and their properties.

Replace the code in the `Program.Main` method in the `Program.cs` file with the following code:

```
DigitalTwinsManager dtHelper = new DigitalTwinsManager();

Pageable<BasicRelationship> relationships = dtHelper.
ListRelationships("MainBuilding");

foreach(BasicRelationship relationship in relationships)
{
    Console.WriteLine(relationship.Id);

    foreach(string key in relationship.Properties.Keys)
    {
        Console.WriteLine(string.Format("{0}:{1}", key,
relationship.Properties[key]));
    }
}
```

Run the application in Visual Studio. You will see the output shown here:

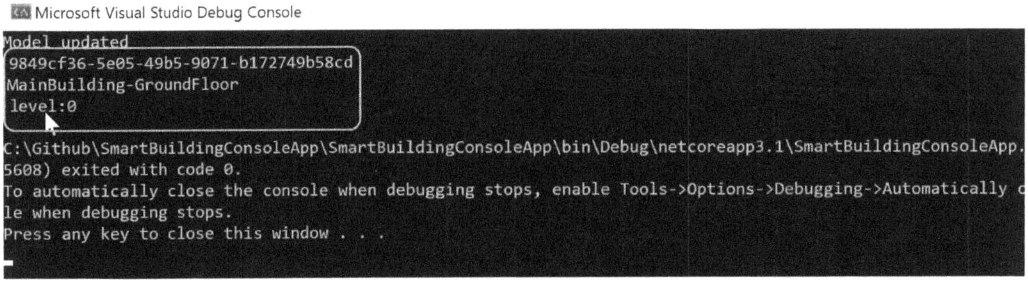

Figure 6.8 – The output of the code when adding a relationship with a single property

The preceding output shows two relationships. The second relationship – the one we created with the code – contains a `value` property set to `0`.

With that, we have learned how to create a relationship that has properties. In the final section of this chapter, we will learn how to update a property value.

Updating relationship properties

In some scenarios, the properties of relationships need to be updated. Updating properties can be done using the `JsonPatchDocument` class. This class allows us to execute several changes for properties at once. In this example, we will be updating one property at a time.

Create a new method for updating a relationship. Add the following code to the `DigitalTwinsManager` class:

```
public void UpdateRelationship(string twinSourceId, string
twinDestinationId, string property, object value)
{
    string relationShipId = RelationshipId(twinSourceId,
twinDestinationId);

    JsonPatchDocument patch = null;
    try
    {
        patch = new JsonPatchDocument();
        patch.AppendReplace("/" + property, value);

        client.UpdateRelationship(twinSourceId, relationShipId,
patch);
    }
    catch (RequestFailedException)
    {
    }
}
```

The preceding code uses the `AppendReplace` method of `JsonPatchDocument` to append or replace the value of a property. This method requires a path to the property. Since the property is defined at the highest level, we specified the path as `"/" + property`.

> **Important note**
> The property of a relationship can only be updated when the relationship was
> created with the property definition.

Replace the code in the `Program.Main` method in the `Program.cs` file with the
following code:

```
DigitalTwinsManager dtHelper = new DigitalTwinsManager();

dtHelper.UpdateRelationship("MainBuilding", "GroundFloor",
"level", 8);

Pageable<BasicRelationship> relationships = dtHelper.
ListRelationships("MainBuilding");

foreach(BasicRelationship relationship in relationships)
{
    Console.WriteLine(relationship.Id);

    foreach(string key in relationship.Properties.Keys)
    {
        Console.WriteLine(string.Format("{0}:{1}", key,
relationship.Properties[key]));
    }
}
```

Run the application in Visual Studio. The code will iterate through each relationship
and output the `Id` property of the relationship and each of its properties. The result is
shown here:

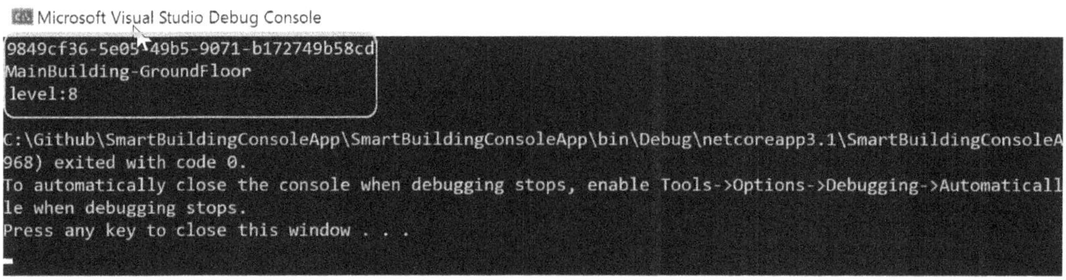

Figure 6.9 – The result of updating the level property to 8

While we have only updated a single property, the *JsonPatchDocument* allows us to add several changes to properties in a single call. This can be achieved by adding additional methods, such as *AppendReplace,* to the code.

Summary

In this chapter, you learned about the relationships between Digital Twins. We explained how descriptive language can be used to name relationships properly. Relationships can be extended with properties to provide additional information about the relationship.

In the next chapter, we will extend our application using asynchronous calls, which are normally used when applications are built around Azure Digital Twins. Several examples will show real-life examples mapped to Digital Twins.

Questions

As we conclude, here is a list of questions for you to test your knowledge regarding this chapter's material. You will find the answers in the Assessments section of the Appendix:

1. Which class allows you to update several properties of a relationship at the same time?

    a. `DigitalTwinClient`

    b. `JsonPatchDocument`

    c. `PropertiesDocument`

2. What is descriptive language used for?

    a. To create meaningful properties

    b. Digital Twin names

    c. Relationship names

3. Does the descriptive name of a relationship change when we change the direction of a relationship?

    a. Yes

    b. No

Further reading

- Managing a graph of Digital Twins using relationships: `https://docs.microsoft.com/en-us/azure/digital-twins/how-to-manage-graph`

7
Querying Digital Twins

From reading this book, you should now have mastered managing digital twin models and their relationships. In this chapter, we will explain the query language and how to query the data from these models and their relationships. We will start by setting up a usable set demo graph in an Azure Digital Twins instance. After that, we will execute several different types of queries using Azure Digital Twins Explorer and discuss the results. Finally, we will examine several examples of using the .NET SDK to query using C# code.

In this chapter, we will learn how to query and filter models and relationships of an Azure Digital Twins instance using Azure Digital Twins Explorer, and through .NET code.

The following sections can be found in this chapter:

- Setting up a demo graph
- Basic querying
- Querying by model
- Querying relationships
- Filtering results
- Querying using code
- Querying asynchronous calls using code

Technical requirements

We will continue in the .NET console application using the .NET SDK to further build up Azure Digital Twins instances using models. Azure Digital Twins Explorer will be used to view the results of the .NET calls made by the console application.

Setting up a demo graph

We require an example graph of several digital twins with relationships to understand how we can query digital twins. Therefore, we start by setting up a demo graph to support further steps.

Go to the following GitHub URL: `https://github.com/PacktPublishing/Hands-on-Azure-Digital-Twins`. Then, copy the contents of `/SmartBuildingConsoleApp/Models/chapter7` into your project. This contains several model files and an Excel file. The result should be similar to the following figure:

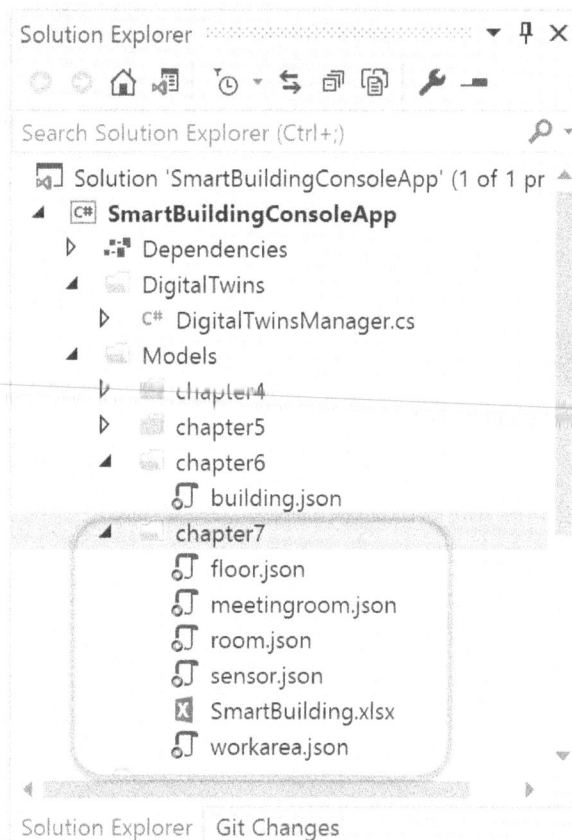

Figure 7.1 – The result after copying the contents of the chapter7 folder

We need to modify several existing models with additional properties in order to build our demo graph. In *Chapter 4, Understanding Models*, you learned how to manage models. Models can be removed and added via Azure Digital Twins Explorer. Execute the following steps to replace the models:

1. Open **Azure Digital Twins Explorer**.

2. In the model view of Azure Digital Twins Explorer, delete the models in the following order: Meetingroom, Workarea, Room, Sensor, Floor. The Meetingroom and Workarea models must be removed first before the Room model is removed since they are both derived from Room.

3. Add the models in the following order via the model view of Azure Digital Twins Explorer: Floor, Sensor, Room, Workarea, Meetingroom. Use the corresponding JSON files from the chapter7 folder to do this.

The next step is importing the demo graph into our Azure Digital Twins instance, using Azure Digital Twins Explorer. This is shown in *Figure 7.2*. Execute the following steps:

1. Click on the **Import Graph** button.

2. Select the Excel SmartBuilding.xlsx file.

3. Click the **Open** button to open the demo graph:

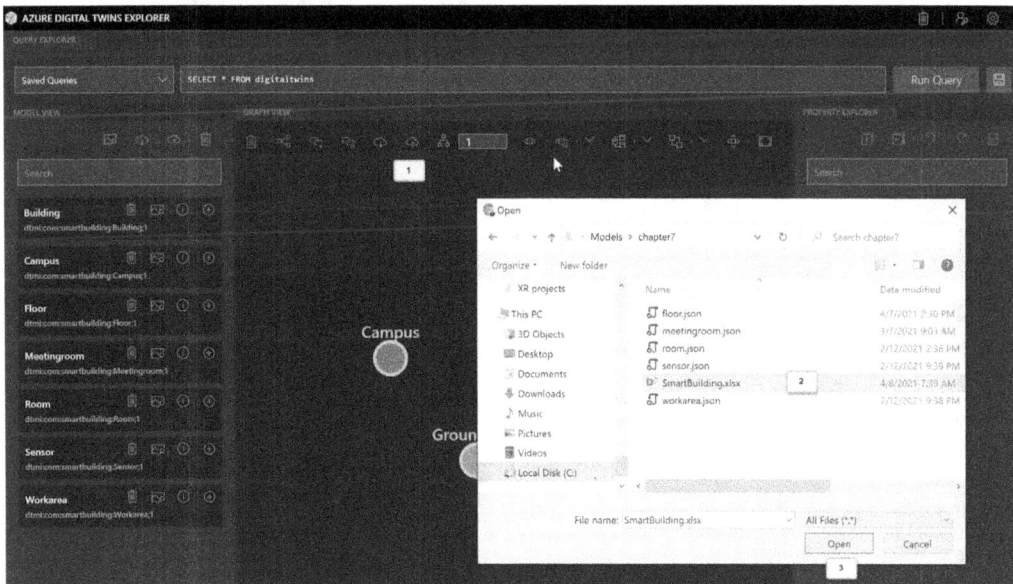

Figure 7.2 – Importing our demo graph data

The demo graph is loaded into the graph view as a preview, as shown in *Figure 7.3*. Click the **Save** button to store the demo graph into the Azure Digital Twins instance:

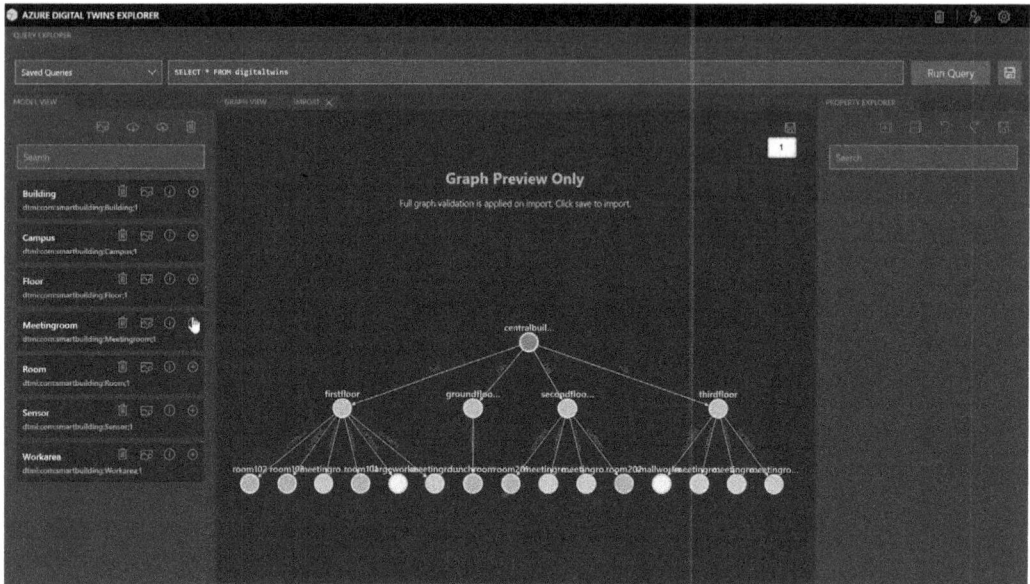

Figure 7.3 – Graph preview of the demo graph

Azure Digital Twins Explorer will show an **Import Successful** dialog after the demo graph is imported, as shown in *Figure 7.4*. It will also show how many digital twins and relationships are imported. Click the **Close** button to continue, as shown in the following screenshot:

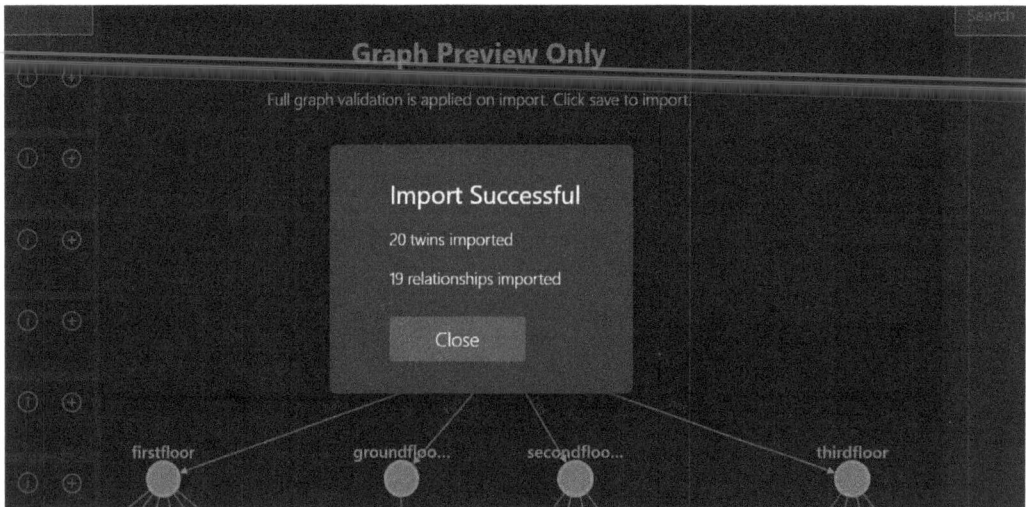

Figure 7.4 – The Import Successful dialog shows the result of the import

Azure Digital Twins Explorer still shows the result of our work from the previous chapters. An import does not replace the contents of the Azure Digital Twins instance. If we were to run the import again, we would get an additional structure with the same names, but with different underlying IDs:

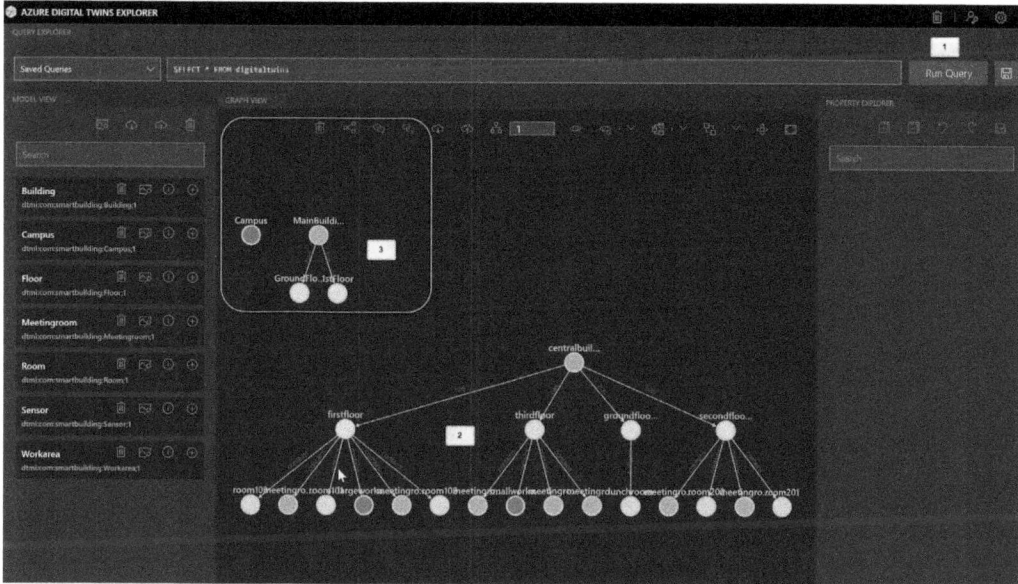

Figure 7.5 – Removing the structure of the work from the previous chapters

Let's clean up the structure of the previous chapter, as shown in *Figure 7.5*. Execute the following steps:

1. Click the **Run Query** button to refresh the graph view.

2. View the imported demo graph.

3.   Remove the structure of the previous chapters. Select each relationship one by
     one and click the **Delete Relationship** button. The digital twins can be removed
     all at once by multi-selecting them using the *Shift* button. Click the **Delete
     Selected Twins** button to remove them all at once. The following screenshot
     shows the results:

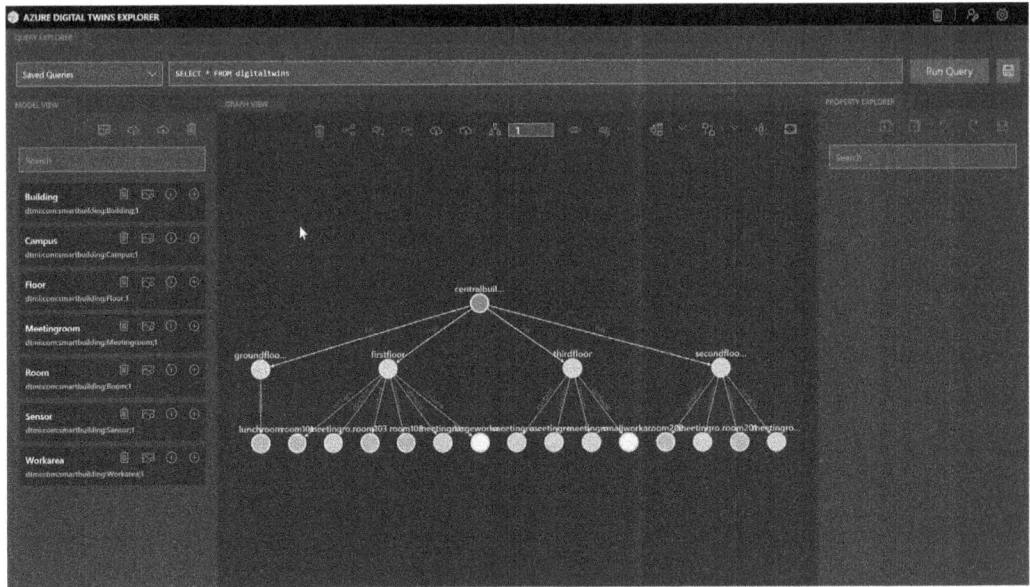

Figure 7.6 – The result after cleaning up the structure

The Azure Digital Twins instance now only contains the imported model.

In this section, we have set up a demo graph containing a large structure of several types
of digital twins. In the next section, we will begin to execute several basic queries on this
structure, in order to understand how they work.

Basic querying

Let's start with some basic queries that you can perform. We will be using Azure Digital
Twins Explorer to execute these queries. The most basic query is the one that we have
already been using. Enter the following query in the **Query** field and click the **Run Query**
button in Azure Digital Twins Explorer:

```
SELECT * FROM DIGITALTWINS
```

The result of the query can be viewed in *Figure 7.7*. We have set **Run Layout** to **fCoSE** to make it more readable. **fCoSE**, or **fast Compound Spring Embedder**, is a faster version of the CoSE algorithm. The algorithm combines several techniques to produce a more aesthetic version of a force-directed graph. It tries to create a visual model of all digital twins and their underlying relationships, where the relationships are drawn as short as possible while maintaining a readable layout:

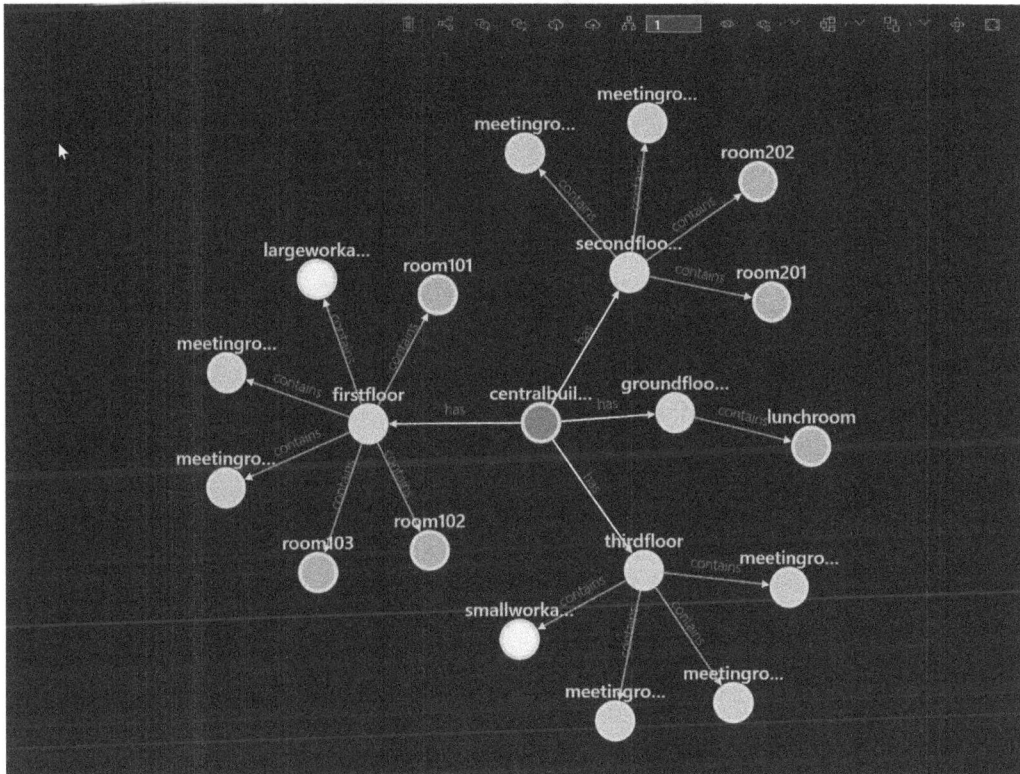

Figure 7.7 – The result of querying all digital twins

In the next query, we will add a filter. Enter the following query in the **Query** field and click the **Run Query** button in Azure Digital Twins Explorer:

```
SELECT * FROM DIGITALTWINS DT WHERE DT.lightson = TRUE
```

The filter in the query says to only look for digital twins that have a property called `lightson`, and that have the property set to the TRUE value. The result of the query is seen in *Figure 7.8*:

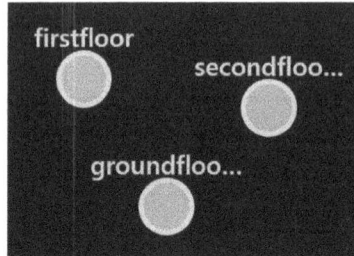

Figure 7.8 – Querying digital twins with a lights-on filter

We can see that only the digital twins that are based on the **Floor** model, and that also have the `lightson` property set to TRUE, are shown. All other related objects are not returned. To do that will require us to perform a JOIN. This will be explained later in this chapter.

The next query will be checking for a temperature between two values. Enter the following query in the **Query** field and click the **Run Query** button in Azure Digital Twins Explorer:

```
SELECT * FROM DIGITALTWINS DT WHERE DT.temperaturevalue>20.0
AND DT.temperaturevalue<21.0
```

The result of the query is seen in *Figure 7.9*:

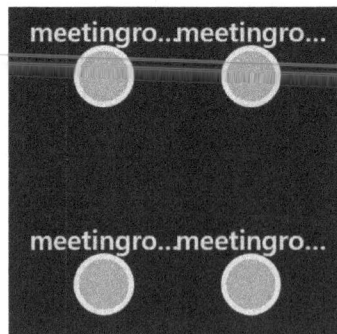

Figure 7.9 – A query returning rooms that have a temperature value between 20 and 21 degrees

Just like the previous query, only digital twins with the `temperaturevalue` property, and the value of that property set between both values, are returned.

> **Important note**
>
> A query with a property in the WHERE clause will return all digital twins that contain that property. This could mean that digital twins from different model types will be returned. We will later use IS_OF_MODEL to control the query result based on the model type.

The following query uses the IN operator. Querying using the IN operator can reduce the number of queries by specifying multiple outcome results. Enter the following query in the **Query** field, and click the **Run Query** button in Azure Digital Twins Explorer:

```
SELECT * FROM DIGITALTWINS DT WHERE DT.$dtId IN
['centralbuilding','thirdfloor','meetingroom301',
'meetingroom302']
```

This query checks if the ID of the digital twin is equal to one of the IDs named in the collection between the brackets after the IN operator. The result of the query is seen in *Figure 7.10*:

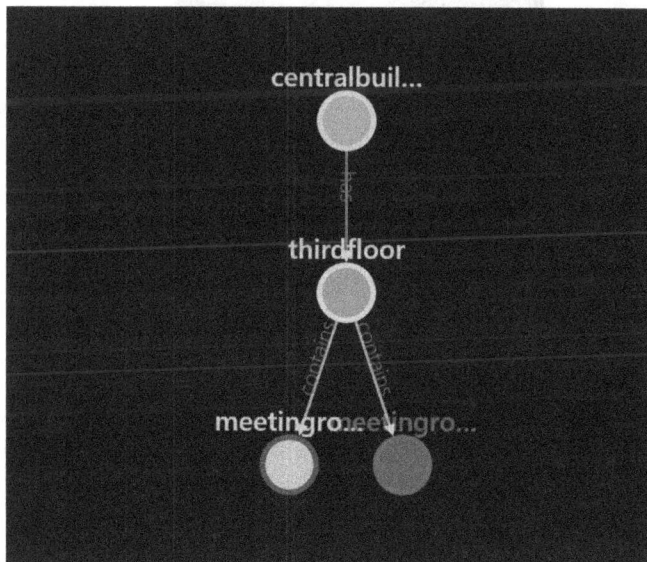

Figure 7.10 – Query result of an IN operator example

This query returns digital twins that have a relationship with each other. Azure Digital Twins Explorer is showing those relationships. While a digital twin contains the reference to another digital twin, the result does not contain the actual relationship object itself. Only digital twins are returned by the query, since the query output only specifies digital twins.

It is also possible to identify digital twins that have a certain property defined. Enter the following query in the **Query** field, and click the **Run Query** button in Azure Digital Twins Explorer:

```
SELECT * FROM DIGITALTWINS WHERE IS_DEFINED(occupied)
```

This query returns only digital twins that have a property called `occupied`. The result is shown in *Figure 7.11*:

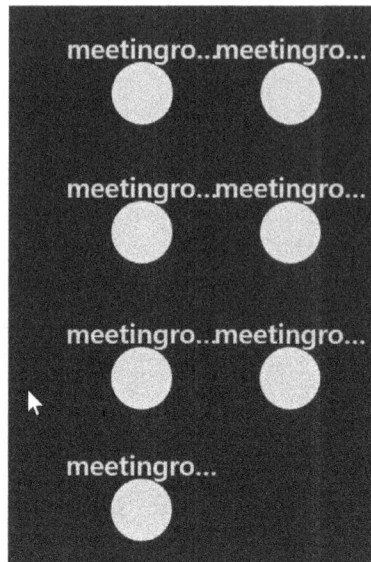

Figure 7.11 – A query returning only digital twins that have the occupied property

It is also possible to check if a property is of the NUMBER type. Enter the following query in the **Query** field, and click the **Run Query** button in Azure Digital Twins Explorer:

```
SELECT * FROM DIGITALTWINS DT WHERE IS_NUMBER(DT.
temperaturevalue)
```

This will return the same result as the previous query, since the same digital twins contain a property called `temperaturevalue` of the NUMBER type.

The next query allows us to get a certain amount of top results back. Enter the following query in the **Query** field, and click the **Run Query** button in Azure Digital Twins Explorer:

```
SELECT TOP(9) FROM digitaltwins
```

The query returns the top nine digital twins returned by the query. The result is shown in *Figure 7.12*:

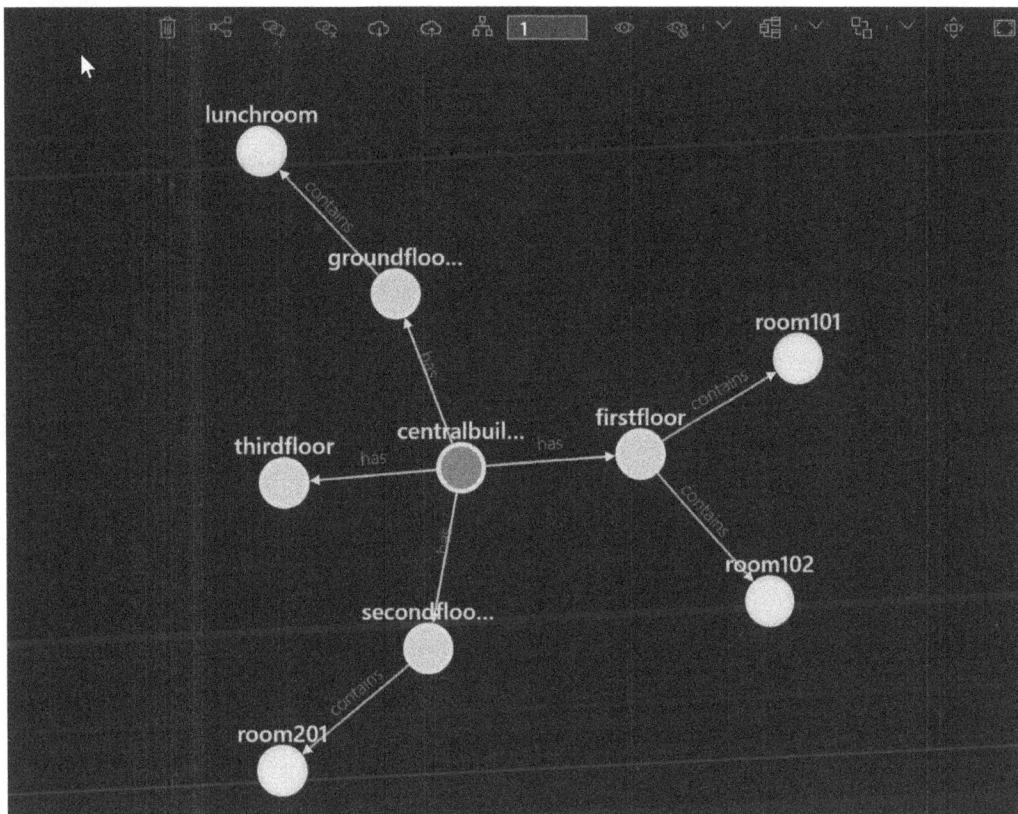

Figure 7.12 – The top nine results of a query

The order of digital twins is unspecified, causing this method to return the top rows based on the underlying database. Furthermore, this differs from the order in which digital twins and their relationships were created.

The last query helped us to determine the number of digital twins returned. Enter the following query in the **Query** field and click the **Run Query** button in Azure Digital Twins Explorer:

```
SELECT COUNT() FROM DIGITALTWINS
```

This query will not show any results in Azure Digital Twins Explorer, since it does not return any digital twins and/or relationships. It only returns the number of digital twins. It allows you to count the result set, and it can be used on queries with a WHERE clause, and even with the JOIN operator.

In this section, we have learned some of the basic ways to query digital twins. We have also learned about the IS_NUMBER, TOP, and COUNT operators. In the next section, we will learn how to query by model type.

Querying by model

In this section, we will explain how you can query by model. Each digital twin is derived from a model. The Query API allows us to use the IS_OF_MODEL operator. The definition of such a query is shown in the following:

```
SELECT * FROM DIGITALTWINS <Collection> WHERE IS_OF_MODEL(
<Collection>, <ModelId>, <Exact>)
```

The IS_OF_MODEL operator can have up to three parameters. Each of the parameters is explained here:

Parameters	Result set
<ModelId>	The result set contains all created digital twins that are based on this specific model and all models derived from this model. Only digital twins that have the same or higher version number as the model will be part of the result set.
<Collection>	The result set is based on the collection specified. This is mostly used when joining another result set in the query by using the JOIN keyword. This will be explained in a later section of this chapter.
<Exact>	The result set is based on the exact <ModelId> and version of the model by using the exact keyword.

We will explain the results of using these parameters in several examples using the IS_OF_MODEL operator. We start by defining the <ModelId>. Enter the following query in the **Query** field, and click the **Run Query** button in Azure Digital Twins Explorer:

```
SELECT * FROM DIGITALTWINS WHERE IS_OF_
MODEL('dtmi:com:smartbuilding:Room;1')
```

This query specifies to return all digital twins that are based on the Room model. The result is shown in *Figure 7.13*:

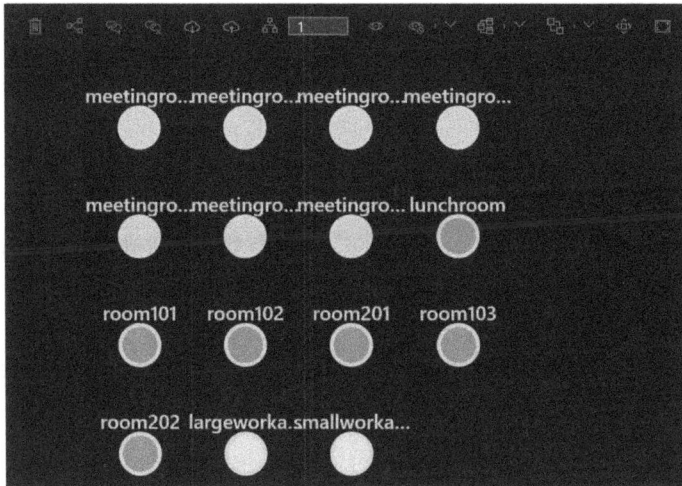

Figure 7.13 – All digital twins that are based on the Room model

We can see that it also returns the digital twins that are based on derived models, in this case, Meetingroom and ActivityArea.

We will add the exact keyword to the query. Enter the following query in the **Query** field, and click the **Run Query** button in Azure Digital Twins Explorer:

```
SELECT * FROM DIGITALTWINS WHERE IS_OF_
MODEL('dtmi:com:smartbuilding:Room;1', exact)
```

The exact keyword will make sure that only digital twins that are based on the exact model are returned. The result is shown in *Figure 7.14*:

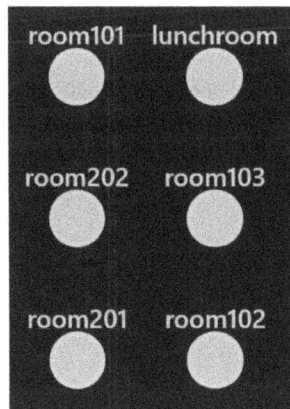

Figure 7.14 – All digital twins based on the Room model with the exact keyword specified

This section explained how to query based on models. This allows us to filter our results even more. In the next section, we will learn about querying relationships using the JOIN operator.

Querying relationships

This section will explain how we can query relationships by using the JOIN operator. The JOIN operator allows us to join (or combine, as we say) different result sets of digital twins. You can add up to five JOIN operators in one query.

There are some rules that we have to apply when using a JOIN operator. These rules are as follows:

- We need to specify a name for a result set. That means that we need to replace FROM DIGITALTWINS with, for example, FROM DIGITALTWINS DT. DT, is, in this case, the name of the result set. The name can be anything we want.
- The query is required to have a $dtId value in the WHERE clause.
- The SELECT * needs to be replaced with, for example, SELECT DT, RT. DT and RT, in this case, are both result sets. At least one result set or a specific field in a result set is required.

The following query does a JOIN operation between digital twins with a specified relationship called contains. In our current Azure Digital Twins instance, a contains relationship is only defined between Floor and Room. Enter the following query in the **Query** field and click the **Run Query** button in Azure Digital Twins Explorer:

```
SELECT DT,RT FROM DIGITALTWINS DT JOIN RT RELATED DT.contains
WHERE DT.$dtId = 'secondfloor'
```

The query contains a $dtId value in the WHERE clause. It is required when using the JOIN operator. Since we want to have all the rooms belonging to a specific floor, the $dtId value is referencing the ID of the floor called secondfloor. We will return DT, RT as the named result sets.

The result is shown in *Figure 7.15*:

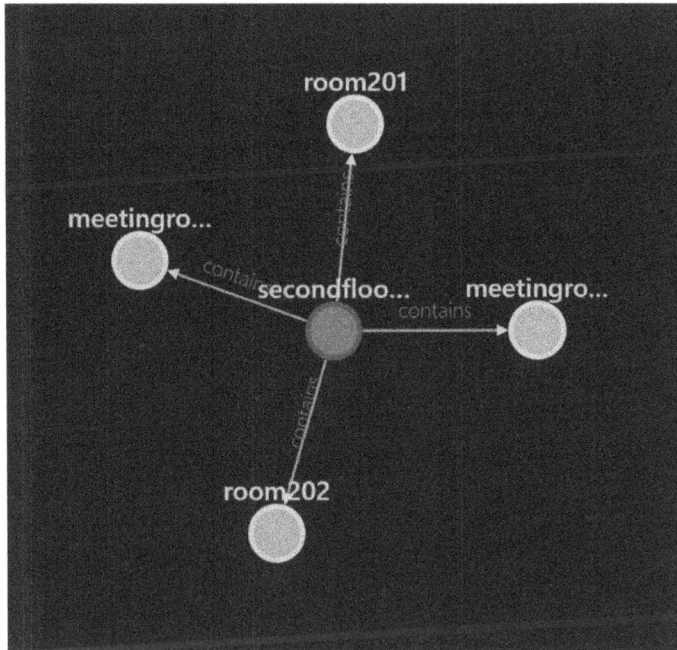

Figure 7.15 – The result of a query using a JOIN operator

The result contains all rooms available for the floor with the `secondfloor` name.

We want to have the building, a specific floor called `firstfloor`, and its meeting rooms returned. This query contains multiple `JOIN` operators. To explain the process of building such a query, we will specify each step. We start by getting the floors for buildings:

```
SELECT BU,FL FROM DIGITALTWINS BU JOIN FL RELATED BU.has
```

This query would return all buildings. Since we have a single building created in our Azure Digital Twins instance, only one building and its floors are returned.

We will extend the query by specifying a specific floor, as follows:

```
SELECT BU,FL FROM DIGITALTWINS BU JOIN FL RELATED BU.has WHERE
FL.$dtId='firstfloor'
```

We now extend the query by getting all the rooms of that floor, as follows:

```
SELECT BU,FL,RO FROM DIGITALTWINS BU JOIN FL RELATED BU.has
JOIN RO RELATED FL.contains WHERE FL.$dtId='firstfloor'
```

Finally, we want to specify that we only want to get meeting rooms returned. Enter this query in the **Query** field, and click the **Run Query** button in Azure Digital Twins Explorer:

```
SELECT BU,FL,RO FROM DIGITALTWINS BU JOIN FL RELATED
BU.has JOIN RO RELATED FL.contains WHERE IS_OF_
MODEL(RO,'dtmi:com:smartbuilding:Meetingroom;1') AND
FL.$dtId='firstfloor'
```

The result is shown in *Figure 7.16*:

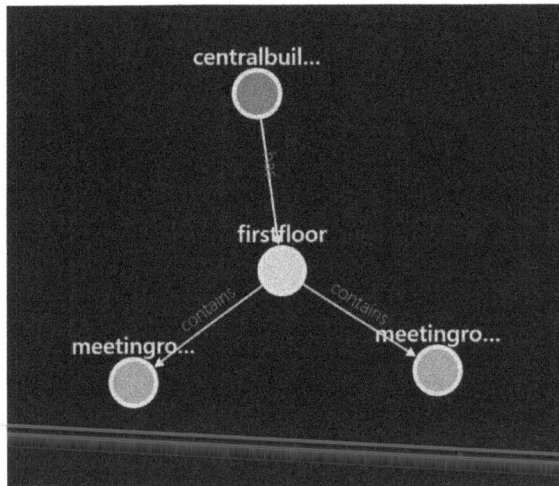

Figure 7.16 – A query returning digital twins using multiple JOIN operators and a WHERE clause

We have learned how to query relationships by using the JOIN operator. In the next section, you will learn how to filter the results of a query.

Filtering results

This section will explain how to filter your results more specifically by including relationships and single properties into the result of digital twins. The first example returns not only the digital twins, but also the relationships between the digital twins, by using something we call **projection**.

Projection means that we define a name for the relationship, and in case of the query, the JOIN operation. A small example will explain this more clearly. Normally, a JOIN operation would be specified as follows:

```
SELECT BU,FL FROM DIGITALTWINS BU JOIN FL RELATED BU.has WHERE
BU.$dtId='centralbuilding'
```

Now, we use projection to also return the relationship. Enter this query in the **Query** field, and click the **Run Query** button in Azure Digital Twins Explorer:

```
SELECT BU,FL,BF FROM DIGITALTWINS BU JOIN FL RELATED BU.has BF
WHERE BU.$dtId='centralbuilding'
```

The result is shown in *Figure 7.17*:

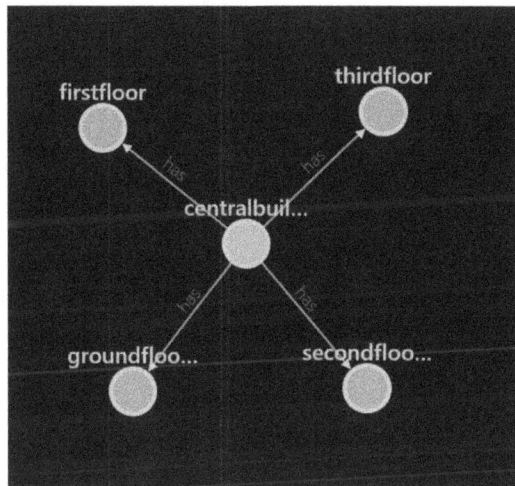

Figure 7.17 – A query result for getting digital twins and relationships

There is no difference in Azure Digital Twins Explorer with or without the relationships. The viewer will always show the result. However, the result of the query with **projection** will also contain the relationships next to the digital twins. This can be helpful if we have properties specified at a relationship.

In the following examples, we will show how to return properties of a digital twin or a relationship in the result of a query. When using properties, a primitive check is required. The primitive check can be done by adding the `IS_PRIMITVE(<property>)` to the `WHERE` clause of the query. This is shown in the following snippet:

```
SELECT temperaturevalue FROM DIGITALTWINS WHERE IS_OF_
MODEL('dtmi:com:smartbuilding:Meetingroom;1') AND IS_
PRIMITIVE(temperaturevalue)
```

It is of no use to execute the query in Azure Digital Twins Explorer, since it will not return any digital twins. The query only returns the `temperaturevalue` of all meeting rooms in the Azure Digital Twins instance.

The second example will return the `lightson` and `level` properties of different digital twins by using a `JOIN` operator in the query. All properties require a primitive check. Refer to the following code:

```
SELECT FL.lightson,BF.level FROM DIGITALTWINS BU JOIN FL
RELATED BU.has BF WHERE BU.$dtId='centralbuilding' AND IS_
PRIMITIVE(FL.lightson) AND IS_PRIMITIVE(BF.level)
```

No results can be shown in Azure Digital Twins Explorer, since it does not return digital twins.

In this section, you have learned about filtering using **projection** and the properties of digital twins. In the next section, you will learn how to execute all these queries using code.

Querying using code

Querying can be done through a call with the Query API using the .NET SDK. The result, however, can differ from the query request. It is mainly determined by the `SELECT` specification. Here, you will see several different return scenarios:

Type of query	Result item
`SELECT * FROM`	`BasicDigitalTwin`
`SELECT Building, Floor, Meetingroom FROM`	`Dictionary<string, BasicDigitalTwin>` where each result (`Building`, `Floor`, and `Meetingroom`) is a key in the dictionary
`SELECT Floor.level`	Integer

Since the result can differ per query, it is somewhat difficult to write a method that can be reused. It completely depends on what result you want the query to give back. However, it also allows you to write your own classes, supporting the serialization of results backward and forward.

A result set is always based on the `Pageable<>` class. It is a collection of values that support iteration over multiple service requests.

We will start by creating a method supporting the SELECT * type of query. Open the `DigitalTwinsManager` class and add the following code:

```
public Pageable<BasicDigitalTwin> QueryDigitalTwins (string
query)
{
    Pageable<BasicDigitalTwin> result = null;
    try
    {
        result = client.Query<BasicDigitalTwin>(query);
    }
    catch (RequestFailedException)
    {
    }

    return result;
}
```

Open the `Main` method in the `Program.cs` file and replace the code with the following:

```
Pageable<BasicDigitalTwin> result = dtHelper.
QueryDigitalTwins("SELECT * FROM DIGITALTWINS");

foreach (BasicDigitalTwin item in result)
{
    Console.WriteLine(item.Id);
}
```

Run the application. The result is shown in *Figure 7.18*. The code outputs the IDs of each digital twin returned in the result of the query:

Figure 7.18 – The result of executing a query to get all digital twins

The following example shows how to handle the result of a query where several digital twin names are specified in the SELECT statement. Open the DigitalTwinsManager class and add the following code:

```
public Pageable<Dictionary<string, BasicDigitalTwin>>
Query(string query)
{
    Pageable<Dictionary<string, BasicDigitalTwin>> result =
null;

    try
    {
        result = client.Query<Dictionary<string,
BasicDigitalTwin>>(query);
    }
    catch(RequestFailedException)
    {

    }

    return result;
}
```

This code does not differ greatly from the previous example. However, instead of returning a pageable collection of `BasicDigitalTwin` objects, it returns a pageable collection of `Dictionary<string, BasicDigitalTwin>` objects.

Open the `Main` method in the `Program.cs` file and replace the code with the following:

```
Pageable<Dictionary<string, BasicDigitalTwin>> result =
dtHelper.Query("SELECT BU,FL FROM DIGITALTWINS BU JOIN FL
RELATED BU.has WHERE BU.$dtId='centralbuilding'");

foreach (Dictionary<string, BasicDigitalTwin> item in result)
{
    BasicDigitalTwin BU = item["BU"] as BasicDigitalTwin;
    BasicDigitalTwin FL = item["FL"] as BasicDigitalTwin;

    Console.WriteLine(string.Format("{0}-{1}", BU.Id, FL.Id));
}
```

Each result of the pageable collection contains a dictionary. Since we have used `SELECT BU, FL` as a statement, the dictionary contains two keys that are similar to the specified digital twin sets in the `SELECT` statement. To get a part of the returned digital twins from the result item, we use the following call:

```
BasicDigitalTwin BU = item[<digital twin set >] as
BasicDigitalTwin
```

Run the application. The result is shown in *Figure 7.19*:

Figure 7.19 – Handling multiple result sets using a query

You have learned how to query code in order to return different forms of result sets. In the next section, you will learn to query asynchronous calls using code.

Querying asynchronous calls using code

The query calls of the Query API from Azure Digital Twins support paging. However, this requires another approach to be used in calling the methods. Until now, we have been calling methods synchronously. Since a call always depends on the availability of the service or the network connectivity, building production solutions requires using asynchronous calls.

Asynchronous calls also allow us to perform paging with the result of a query. Paging does immediately think about a fix number of results by page. In the case of executing queries, we talk more about handling each result from the query. The following example calls a method for each result in the query.

Open the `DigitalTwinsManager` class and add the following code inside the class definition, directly under the class properties:

```
public delegate void QueryResult(BasicDigitalTwin dt);
```

This code contains the definition of a delegate function. A **delegate function** describes the format of a function that could be, for example, given as a parameter to another method.

And that is exactly what we will do. Add the following two methods to the `DigitalTwinsManager` class:

```
public void QueryDigitalTwins(string query, QueryResult
onQueryResult)
{
    System.Threading.Tasks.Task task = System.Threading.Tasks.
Task.Run(
        () => QueryDigitalTwinsAsync(query, onQueryResult));
}

public async void QueryDigitalTwinsAsync(string query,
QueryResult onQueryResult)
{
    AsyncPageable<BasicDigitalTwin> result = client.
QueryAsync<BasicDigitalTwin>(query);
    try
    {
        await foreach (BasicDigitalTwin dt in result)
        {
            onQueryResult(dt);
```

```
            }
        }
    catch (RequestFailedException)
        {
        }
    }
```

The QueryDigitalTwinsAsync method contains an asynchronous call to the Query API. For each result of the query, the method specified by the onQueryResult parameter is called.

> **Tip**
>
> It is possible to replace the onQueryResult call with a call that has an array of a fixed number of results that allows you to specify a page size.

We want to have the QueryDigitalTwinsAsync method be executed in a separate thread. Therefore, we have the QueryDigitalTwins method that uses the System. Threading.Tasks.Task class to create a separate thread that executes the query.

Open the program.cs file and add the following code to the file:

```
public static void OnQueryResult(BasicDigitalTwin dt)
{
        Console.WriteLine(dt.Id);
}
```

This method will be called every time a search result is retrieved during the asynchronous call. Replace the contents of the Main method with the following code:

```
    DigitalTwinsManager dtHelper = new DigitalTwinsManager();

    dtHelper.QueryDigitalTwins("SELECT * FROM DIGITALTWINS",
OnQueryResult);

    while(true) { }
```

This method will call the QueryDigitalTwins method and give a reference to the static OnQueryResult method. The while(true) {} loop is required, since otherwise, the console application would already be closed before the second thread can start feeding back the results.

Run the application. The result of the application can be seen in *Figure 7.20*:

Figure 7.20 – Results are returned via an asynchronous call

Asynchronous calls will be used more often in the upcoming chapters.

Summary

In this chapter, you have learned how the query language provided by the Query API allows us to query digital twins and their relationships. We have learned how these queries are built up and how they return their results. We have also learned how to use the .NET SDK to execute these queries using C# code. In this chapter, we have mastered how to execute queries using Azure Digital Twins Explorer and through .NET code. Acquiring data from an Azure Digital Twins instance is done by executing queries almost every time.

In the next chapter, we will learn in several steps what is required to build our first model. Based on several scenarios and types of data, we will explain the best way to create your Azure Digital Twins instances.

Questions

As we conclude, here is a list of questions for you to test your knowledge regarding this chapter's material. You will find the answers in the *Assessments* section of the *Appendix*:

1. Which class is used when a result set is returned from a query?

    a. `Paging<>` class

    b. `Collection<>` class

    c. `Pageable<>` class

2. Which keyword is required in `IS_OF_MODEL` to make sure that the query only returns the specified model?

    a. No keyword is required.

    b. `exact`.

    c. The ID of the model.

3. What is the advantage of querying asynchronously?

    a. It is just another way of writing code.

    b. It allows you to implement paging and is more suitable for production.

Further reading

If you want to learn more on the subject, check out these resources:

* Query the Azure Digital Twins twin graph, available at the following URL: `https://docs.microsoft.com/en-us/azure/digital-twins/how-to-query-graph`.

8
Building Models Using Ontologies

This chapter will describe the strategy for modeling your digital twin using **ontologies**. Ontologies are predefined model sets for a certain industry. These ontologies can help us to create and define a model in order to obtain a faster solution. This chapter will offer a detailed explanation of ontologies, as well as what modeling strategies are available. We will examine in detail some of the industry-standard ontologies, and explain how to upload them into an Azure Digital Twins instance.

In this chapter, we will learn how to decide on the strategy that will work for us. We will also learn how to use some of the standard ontologies to help us get a quick start when building digital twin solutions. The following is a list of the section headings in this chapter:

- Understanding ontologies
- Modeling strategy
- Using industry-standard ontologies
- Uploading ontology models

Technical requirements

We will be downloading several ontologies via GitHub. We will download, build, and run a special Azure Digital Twins Model Uploader tool using *Visual Studio* in order to import one of the ontologies. The Azure Digital Twins Explorer will be used for viewing the results.

Understanding ontologies

Building a first model requires defining all the models and their underlying relationships with your digital twin. Starting a digital twin from scratch can be a very time-consuming undertaking. It requires the identification of the different entities that play a role within the digital twin. Additionally, it also requires thinking about what properties and relationships we need to consider to have our digital twin deliver what we want.

The building of digital twins is often bound to a certain industry, or a part of that industry. Industry-built digital twins are all about modeling complex systems with a lot of different assets, properties, and relationships.

Wouldn't it be great to start with a predefined set of models that suits your needs? That would make it far easier to design a digital twin. We could extend such a predefined set of models in order to accommodate our own needs. To do this, we need to use ontologies.

An ontology is a structure of entity classification and description in a specific area or domain describing its properties and relationships.

Ontologies are a set of models for a specific domain. Such a domain can be anything from a household to a large industry setup. An example is a building structure, as we have been using in our code samples in the previous chapters. Other examples are an IoT system for a specific factory setup, smart city, water supply system, or even a document library. There are several advantages to using an ontology:

- **Quick start**: Having a set of models already in place, and extending it where you want to build a digital twin more rapidly. Since you do not have to start from scratch, we can focus more on solving business problems and less on designing and modeling the digital twin.

- **Uniformity**: Using a similar set of models for different digital twin applications and services that act inside the same domain. Using the same ontology across your applications will make it far easier to maintain. It also allows us to integrate these products and services more easily.

- **Less cost-intensive**: Being able to conceptually model your digital twin more rapidly, instead of starting from scratch. This allows you to have the focus more on the application itself, rather than on the modeling process of the digital twin.

- **Best practices**: This especially applies to certain industries where there are ontologies available. Starting to build your own is like inventing the wheel again, and you may not be successful. You will require a lot of expertise and skills in the domain in question to design models, their properties, and their underlying relationships successfully.

- **Common vocabulary**: Ontologies will support a common use of terms and vocabulary across solutions. Due to this vocabulary, it will probably be easier to have a more seamless integration between different vendors and partners that provide solutions and services around digital twins.

As you can imagine, ontologies are a great starting point for building your digital twin solutions. They describe a set of models and relationships of entities that allow you to design, create, and build a digital twin graph with ease.

We have learned about ontologies and their advantages. In the next section, we will learn about modeling strategies. These modeling strategies can help us in designing a model.

Modeling strategy

Ontologies that are specifically written for digital twins are, in most cases, also written using the **Digital Twins Definition Language (DTDL)**. This allows us to use these ontologies with ease for Azure Digital Twins, since the models are also based on the DTDL.

There are several possible strategies we can follow for using ontologies with digital twins. They are given as follows:

Strategy	Description	Required skillset and knowledge
Adopt	Select an ontology that fits the domain we are looking for and use it. Minor changes, such as properties, are made to suit the requirements of the solution.	No (or minor) skillset or knowledge is required of the domain.
Conversion	Select an ontology that fits the domain for our solution as closely as possible. Make necessary changes to the models, their properties, and relationships to fit our solution. Additional models are added when needed.	An average skillset and knowledge are required of the domain.
Author	We are not able to find an ontology that suits our needs. We create a new ontology to fit our solution.	Expertise in your skillset and extensive knowledge of the domain is required.

Each strategy requires a certain skillset and knowledge of the domain. The following diagram explains the correct route to take when starting to use ontologies for building digital twins:

Figure 8.1 – The route for building digital twin solutions using ontologies

The research phase is all about determining which ontology strategy we want to use. We start by identifying the right domain for our solution. Research is used to identify available ontologies that meet our requirements. We need to determine the feasibility of using an ontology for our solution. It could be that the ontology is not yet written using the DTDL. In that case, we need to consider what we require in order to use a conversion strategy.

Based on the outcome of the research phase, we use one of the available ontology strategies. The strategy table shown previously will support us in making the right choice. We always try to keep the work to a minimum. That means that we always try to start by adopting an ontology. If that does not work for us, we need to see whether we can convert an existing ontology for our needs. And finally, if that is not possible, we can build our own ontology.

Finally, we need to validate the model. This validation process allows us to make sure that we can upload the models into the Azure Digital Twins instance and then start building our digital twin. We can make use of the validation tool to validate the model. This is explained in *Chapter 2, Requirements and Installation*. Keep in mind that, during validation, we will make minor changes, if required, in order to validate the model.

> **Tip**
> Make sure that you always validate your models before uploading them into the Azure Digital Twins instance. This can prevent a lot of additional work with regard to removing models and uploading them over and over.

In this section, we have learned what ontologies are and how we use these ontologies to build our digital twin solutions. We examined a best practice route to follow, as well as several other ontology strategies. The next section will describe industry-standard ontologies and how to use them with Azure Digital Twins.

Using industry-standard ontologies

There are several industry-standard ontologies available that are open source. These ontologies are, in most cases, easy to adopt and are extendable.

Microsoft offers, through GitHub, several industry-standard ontologies. Currently, the following are available:

Industry-standard ontologies	GitHub location
RealEstateCore for smart buildings	`https://github.com/Azure/` `opendigitaltwins-building`
Smart cities	`https://github.com/Azure/` `opendigitaltwins-smartcities`
Energy grid	`https://github.com/Azure/` `opendigitaltwins-energygrid/`

We will be focusing on two of them. These are specified in more detail in the following table:

	Smart building ontology	Smart city ontology
Description	To be used for modeling and controlling buildings, based on industry standards. It enables users to create a common ground for modeling.	This ontology enables city administrators and project planners to define representations of connected environments, which includes any element in the city, such as buildings, factories, energy networks, and much more.
Collaboration/ Partnership	RealEstateCore consortium	Open Agile Smart Cities (OASC) and Sirius
Target industry	Real estate	Cities
Open source	Yes	Yes
Other possible industry standards	BrickSchema, Project Haystack, W3C Building Topology Ontology	SAREF4CITY, CityGML, ISO
Location	`https://github.` `com/Azure/` `opendigitaltwins-` `building`	`https://github.` `com/Azure/` `opendigitaltwins-` `smartcities`

Most of the topologies you find online are not already converted to the DTDL language. Take, for example, the smart building ontology. It originates from an ontology that is represented by using the W3C **Web Ontology Language (OWL)**. OWL is a semantic language that allows us to design simple to complex ontologies by defining and storing the knowledge of entities, their properties, and their relationships. **Web-based Visualization of Ontologies** (**WebVOWL**) is a web-based application for interactive visualization of ontologies based on the OWL language. The following URL will open the WebVOWL for the smart building ontology, as shown in the following screenshot: `https://doc.realestatecore.io/3.3/full.html`:

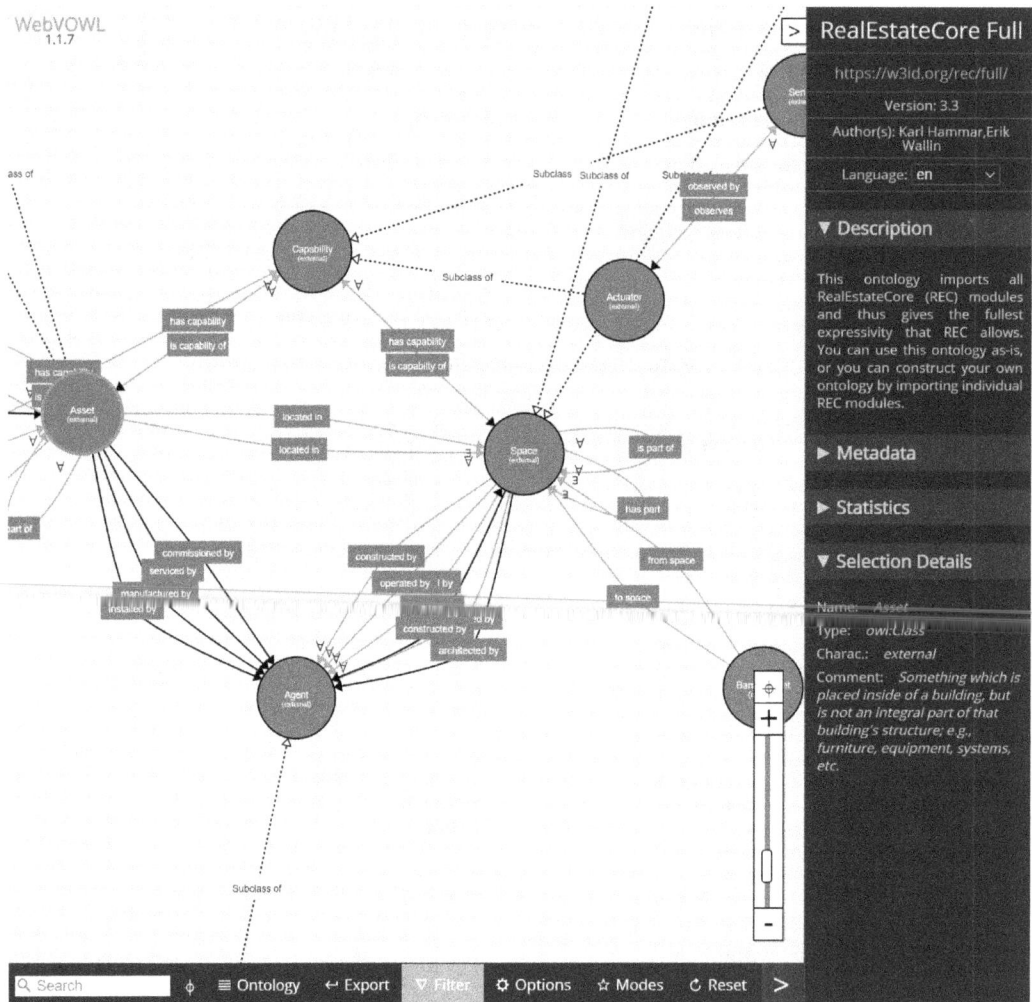

Figure 8.2 – The smart building ontology visualized using WebVOWL

Microsoft provides, as part of the Azure open Digital Twins tools, to convert OWL models to DTDL. We will not discuss and use this conversion tool further in this book. You can read more about the tool in question at the following location: `https://github.com/Azure/opendigitaltwins-tools`.

We want to use one of the ontologies in order to get its models into our Azure Digital Twins instance. As an example, we will download the smart building ontology. Execute the following steps:

1. Open a browser and go to the following URL: `https://github.com/Azure/opendigitaltwins-building`.

2. Press the **Code** button.

3. Select the **Download ZIP** option to download the smart building ontology to your local downloads folder:

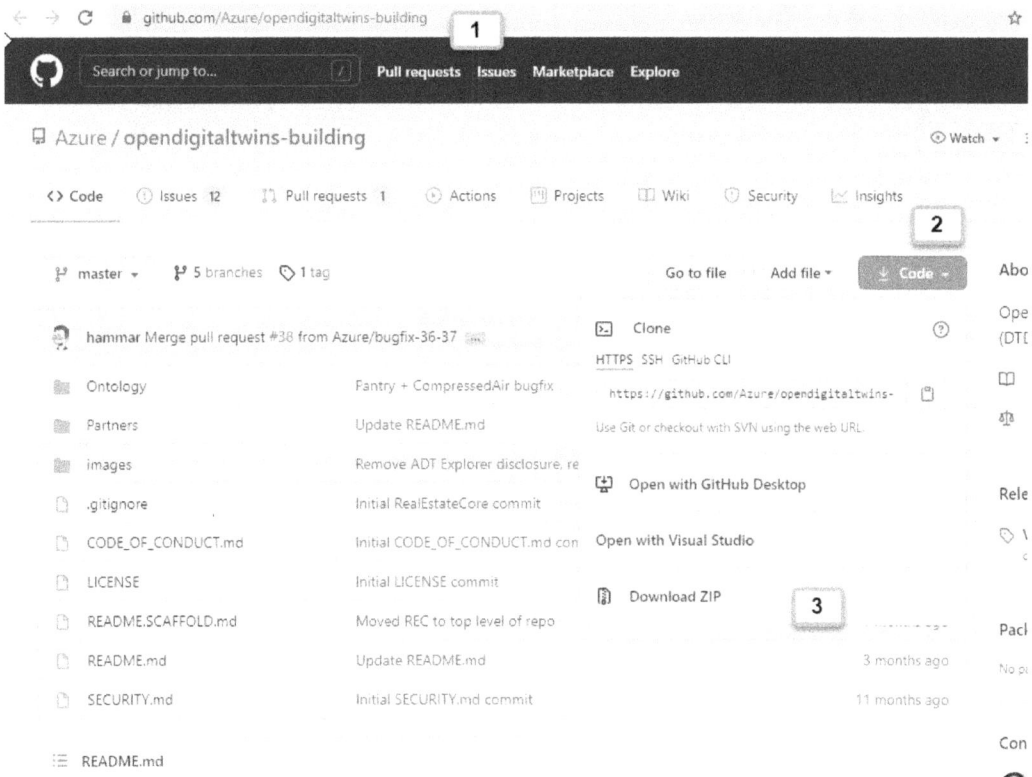

Figure 8.3 – Downloading the smart building ontology via GitHub

Unpack the content of the ZIP file to the `c:\github` location. Open Windows File Explorer and go to the `C:\Github\opendigitaltwins-building-master\Ontology` directory. We can see a hierarchical view of the ontology in the following screenshot:

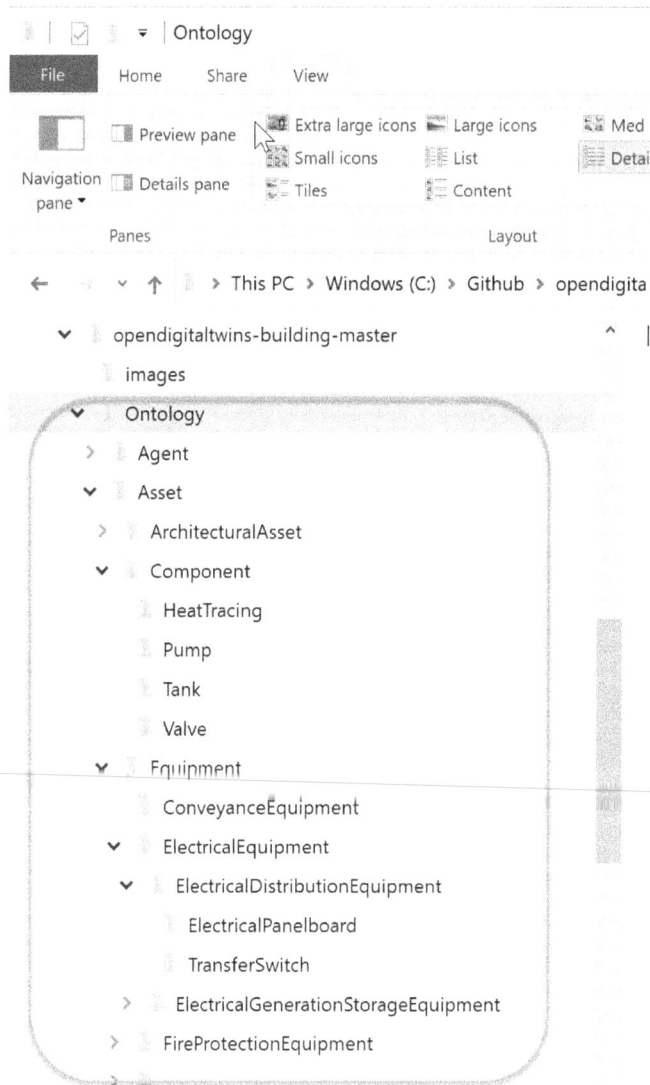

Figure 8.4 – The tree structure of the ontology

The `OWL2DTDL` tool creates a hierarchical folder structure of the models. Models that are inherited from another model are placed on a deeper level. If a model inherits from multiple models, it is in some cases placed even deeper, to indicate its inheritance.

We have learned about the available industry-standard ontologies, how they are structured, and how they can be used. In the next section, we will explain how to batch upload the smart building ontology structure to our Azure Digital Twins instance.

Uploading ontology models

Each of the discussed ontologies contains a large set of models. Adopting and uploading such models would be time-intensive. As we have learned in previous chapters, the order of uploading models is defined based on how models inherit from each other.

As you have noticed, these ontologies use a certain file structure containing the models. Models that inherit from another model are placed in folders deeper in the structure.

Microsoft provides an application, called the *Azure Digital Twins Model Uploader*, which is able to upload such a structure by traversing the content of the structure using the `Directory.EnumerateFiles` method. In previous examples, connecting to the Azure Digital Twins instance was handled by using the **identity and access management (IAM)** framework to add the **Azure Digital Twins Data Owner** role to a user. This application uses an app registration in **Azure Active Directory** (**AAD**) to give it access to the Azure Digital Twins instance.

This requires app registration. App registration can be carried out by executing the following steps:

1. Open the Azure portal using the following URL: `https://portal.azure.com`.
2. Log in with credentials that have at least administrator privileges.
3. Select the AAD service.
4. Select **App registrations** in the vertical menu on the left.
5. Select the **+ New registration** option in the top menu to create a new registration.

This will open the view as shown in *Figure 8.5*. Execute the following steps as given in *Figure 8.5*:

1. Enter `ADTUploaderApp` as the name for the registration.
2. Select the **Public client/native (mobile & desktop)** option, since we are connecting from a desktop application.
3. Enter a URI. This URI is returned when authentication has succeeded. The application will not use it for now. Fill in `http://localhost`.

4.   Press the **Register** button. Refer to the following figure:

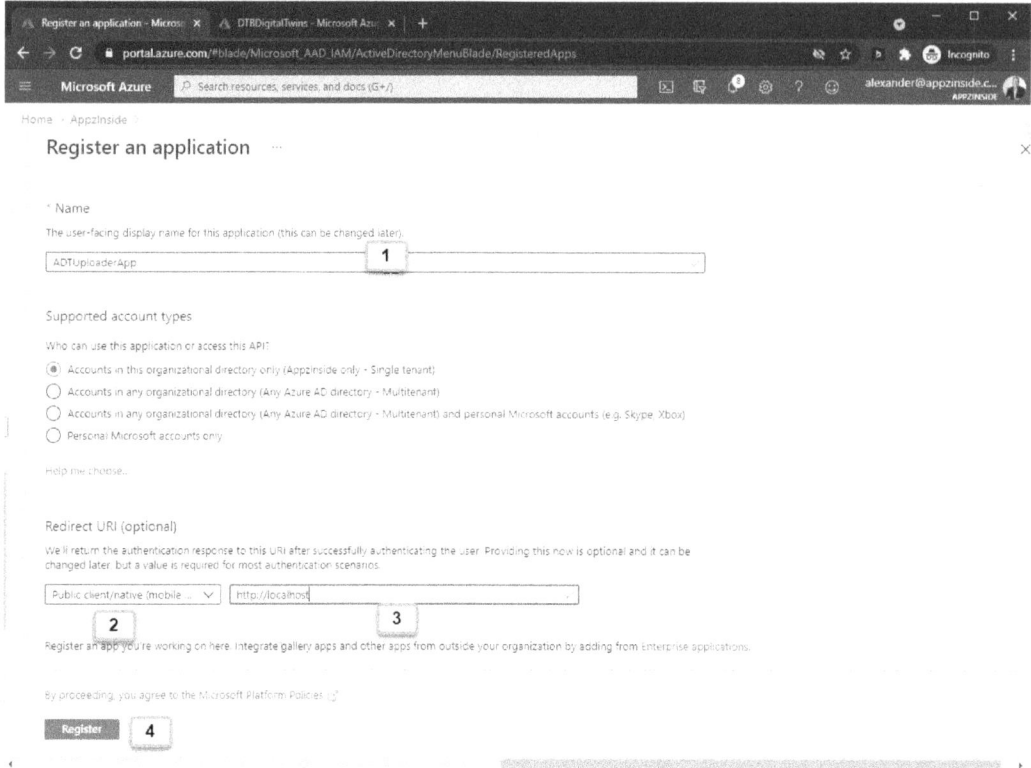

Figure 8.5 – Register an app registration for digital twin access

In the next step, we configure the required permissions for the app registration. These permissions are called **API permissions**, and define which service the app registration can gain access to.

> **Important note**
>
> An app uses app registration to get access to a service. It is important to understand that app registration only defines the permissions to that service. The account used to authenticate in your app will still require those permissions too, in order to get the defined permissions to that service.

We need to select the service for getting access to Azure Digital Twins instances. To do this, execute the following steps:

1.   Select **API permissions** in your created app registration.

2.   Select the **APIs my organization uses** tab.

3. Search for azure digital twins.

4. Select the **Azure Digital Twins** service from the search results. Refer to the following figure:

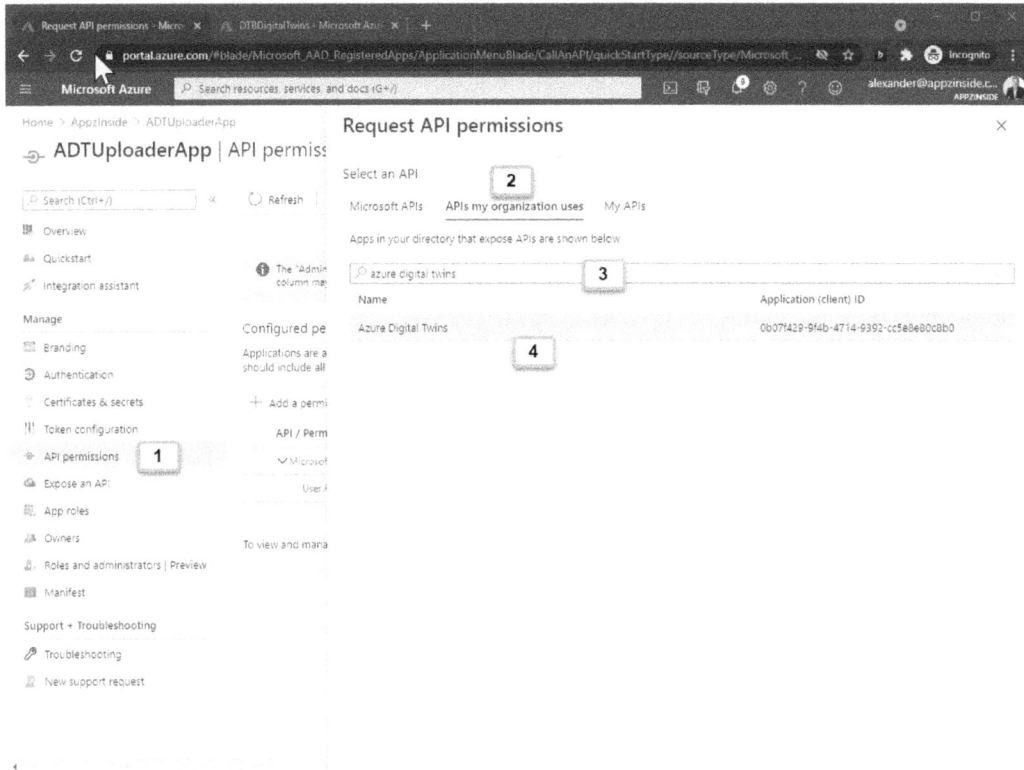

Figure 8.6 – Selecting the request API permissions for the app registration

Pressing the **Azure Digital Twins** service in the previous step will bring us to the screen shown in *Figure 8.7*. Execute the following steps to finalize adding the permission:

1. Check the permission type. It needs to specify **Azure Digital Twins**. We only have the option to delegate on behalf of a logged-in user. This means that the app registration will delegate the request made by the app on behalf of the logged-in user to the Azure Digital Twins instance.

2. Select the **Read/Write Access** option.

3.   Press the **Add permissions** button to add the permission to the app registration.

Refer to the following figure:

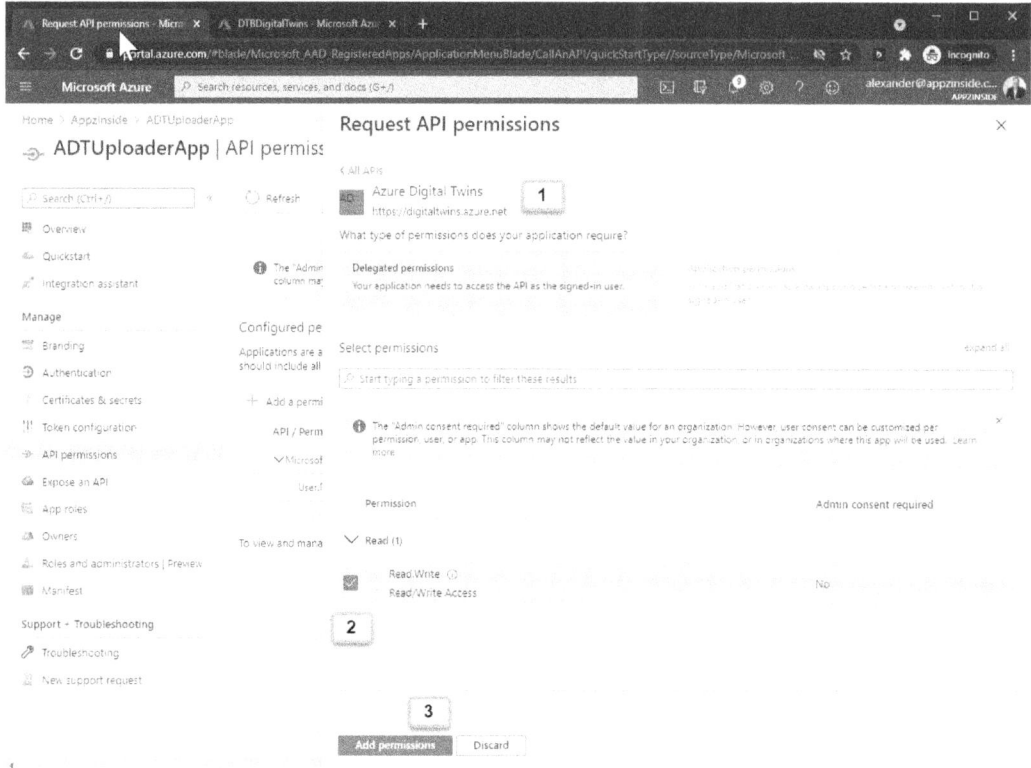

Figure 8.7 – Setting the request API permissions for the app registration

We have created an app registration with permission to read and write to Azure Digital Twins instances. In the next step, we need to download the Azure Digital Twins Model Uploader tool, configure its connection settings, and import an ontology.

Let's start by downloading the tool, as shown in *Figure 8.8*. Execute the following steps:

1.   Open a browser and enter the following URL: `https://github.com/Azure/opendigitaltwins-tools`.

2.   Press the **Code** button.

3. Select the **Download ZIP** option to download the code. Refer to the following figure:

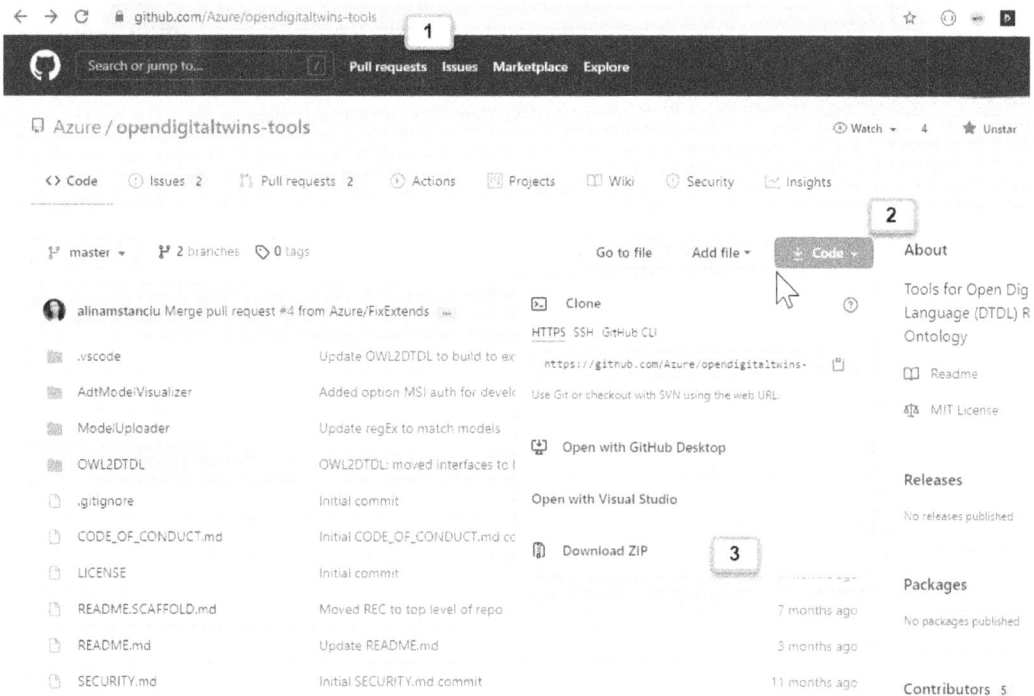

Figure 8.8 – Download the opendigitaltwins tool

We need to unpack the ZIP file to the c:\github directory. This will automatically create a folder called opendigitaltwins-tools-master under c:\github.

Open File Explorer and go to the C:\Github\opendigitaltwins-tools-master\ModelUploader directory. Double-click the ModelUploader.sln file. This will automatically start Visual Studio and open the Azure Digital Twins Model Uploader project.

We need to set several values in the serviceConfig.json file. Execute the following steps, as shown in *Figure 8.9*:

1. Double-click the serviceConfig.json file in Solution Explorer to open the file.

2. Open the app registration under AAD in the Azure portal and make sure the **Overview** option on the vertical menu is selected. The tenant ID is shown as **Directory (tenant) ID**.

3. We use the same **Overview** option to get the client ID. The client ID is shown as **Application (client) ID**.

4. The instance URL is the domain URL defined in the Azure Digital Twins instance in the Azure portal. This is the same URL used for accessing the Azure Digital Twins instance using the Azure Digital Twins Explorer:

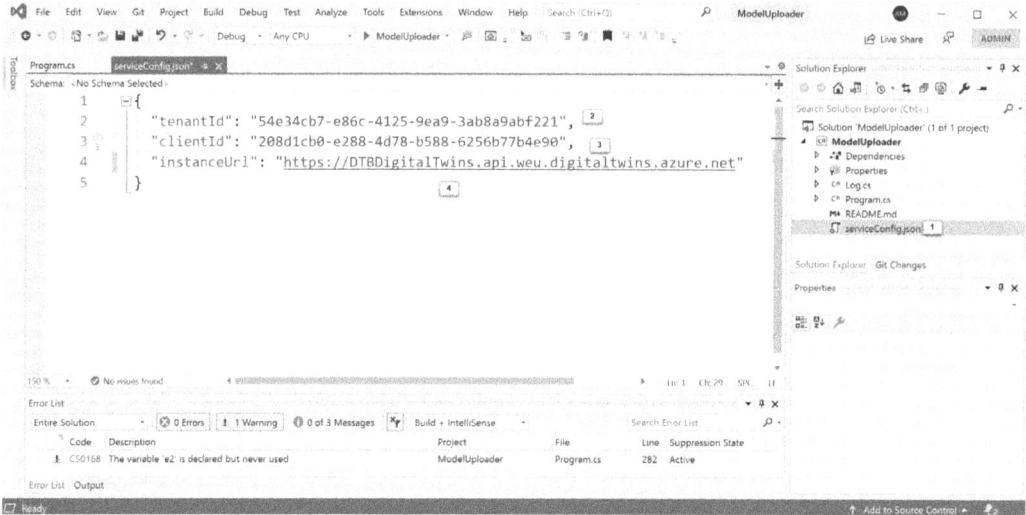

Figure 8.9 – Setting the properties in the serviceConfig.json file

Save the file and build the project. Open Windows PowerShell with administrative rights. Use the following command to move to the right folder:

```
cd C:\github\opendigitaltwins-tools-master\ModelUploader\bin\
debug\netcoreapp3.1
```

Check whether the `serviceConfig.json` file is copied to this build folder. If that is not the case, make sure to copy the file manually to this build folder.

We will take the smart building ontology as an example. We will be using the Azure Digital Twins Model Uploader tool to remove all existing models and replace them with the models from the smart building ontology. Enter the following command:

```
./ModelUploader -p C:/Github/opendigitaltwins-building-master/
Ontology -d
```

The -d option will remove all current models from the Azure Digital Twins instance. The -p option specifies the path of the ontology structure on the disk. Make sure that all backward slashes are replaced by forward slashes.

As soon as we run the command, the tool will open a browser asking for credentials. Use the credentials that have access to the Azure Digital Twins instance. You will get a window asking for your consent, as shown here:

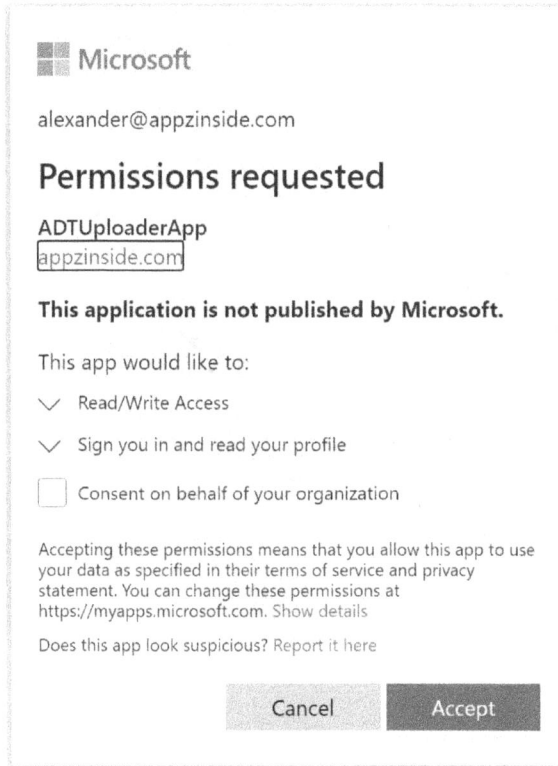

Figure 8.10 – Giving consent for the app

The dialog contains the list of permissions requested by the application. These permission requests are on behalf of the currently logged-in user. By pressing the **Accept** button, we give applications the ability to perform actions given by these permissions.

This will allow the tool to access the Azure Digital Twins instances using both read and write access by using the account of the logged-on user. The browser will show the following message: **When logged in successfully, close the browser**. This indicates that the application is logged in.

The tool will then remove all the current models and start uploading models one by one, maintaining the directory structure as shown in this screenshot:

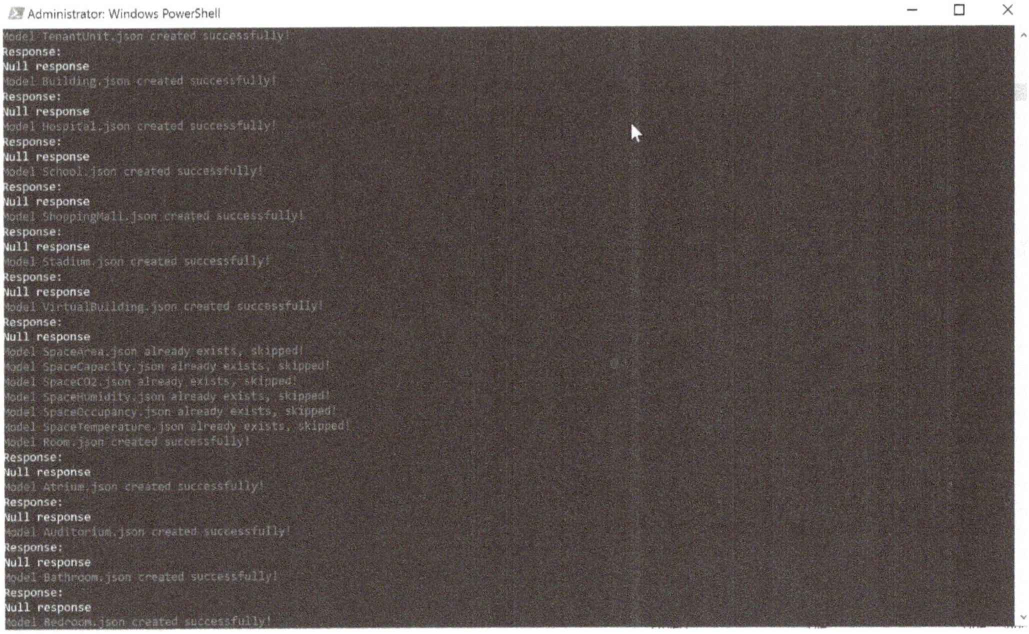

Figure 8.11   The output result of the Model Uploader application

We can use the Azure Digital Twins Explorer to view the result, as shown in the following screenshot:

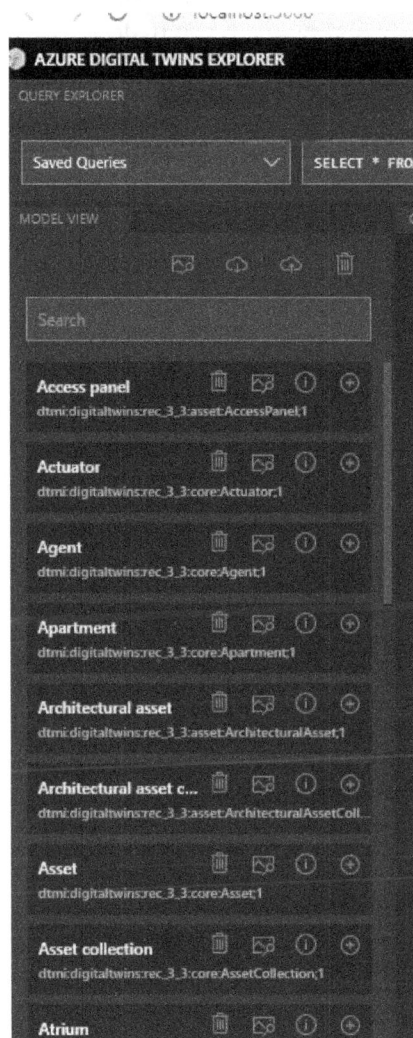

Figure 8.12 – The result of the new uploaded models in the Azure Digital Twins Explorer

When we run the query again in Azure Digital Twins Explorer, the existing digital twins we created earlier are still there. They can live without the models that we removed using the previous command. The next steps would be to remove these digital twins, and then start creating digital twins based on the smart building ontology models.

Summary

We have learned how ontologies can contribute to modeling a digital twin solution. We provided a modeling strategy that can help you to decide how to start with modeling. We also showed how to upload an ontology into our Azure Digital Twins instance. We should now be able to choose the right strategy when creating a digital twin, and also understand how we can upload one of the available standard ontologies into the Azure Digital Twins instance.

In the next chapter, we will learn more about the different SDKs and APIs available for Azure Digital Twins.

Questions

As we conclude, here is a list of questions for you to test your knowledge regarding this chapter's material. You will find the answers in the *Assessments* section of the *Appendix*:

1. In which order can we best use ontology strategies?

    A. Adopt, Conversion, Author

    B. Author, Adopt, Conversion

    C. Conversion, Adopt, Author

2. Which tool allows us to batch upload ontology models?

    A. Azure Digital Twins Explorer

    B. Azure Digital Twins Model Uploader

    C. Azure Portal

3. What tool can be used to visualize an OWL?

    A. WebVOWL

    B. OWL language

    C. Azure Digital Twins Explorer

Further reading

To learn more on the subject, refer to the following links:

- *What is an ontology?*:

  https://docs.microsoft.com/en-us/azure/digital-twins/
  concepts-ontologies

- *Adopting an industry ontology*:

  https://docs.microsoft.com/en-us/azure/digital-twins/
  concepts-ontologies-adopt

- Web Ontology Language (OWL):

  https://www.w3.org/2001/sw/wiki/OWL

- Web-based Visualization of Ontologies:

  http://vowl.visualdataweb.org/webvowl.html

Section 3: Digital Twins Advanced Techniques

This section of the book will focus on more advanced techniques of using Azure Digital Twins. We will go in depth to understand the various APIs and SDKs available. We will start building our first digital twin pipeline and route event messages using event routing. We will set up and integrate a digital twin solution using Azure Maps. We will conclude this section by explaining how to monitor and troubleshoot our Azure Digital Twins instance.

This part of the book comprises the following chapters:

- *Chapter 9, APIs and SDKs*
- *Chapter 10, Building a Digital Twins Pipeline*
- *Chapter 11, Updating the Model*
- *Chapter 12, Event Routing*
- *Chapter 13, Setting Up Azure Maps*
- *Chapter 14, Integrating Azure Maps*
- *Chapter 15, Monitoring and Troubleshooting*

9
APIs and SDKs

There are several APIs and SDKs available to access the Azure Digital Twins services. In this chapter, we will learn the difference between. We will learn how to control and manage the Azure Digital Twins instances using the REST API together with the Postman tool. It is explained how to use the metrics on the service to monitor and take control of the service. We will also explain how to directly use the Azure CLI to perform the same tasks using Windows PowerShell. The chapter concludes by explaining the service limits and what you can do to prevent them or to resolve them.

In this chapter, we'll cover the following topics:

- Understanding the developer landscape
- Understanding the REST API
- Monitoring API metrics
- Using the Azure CLI to manage Azure Digital Twins
- Understanding service limits

Technical requirements

We will be downloading several ontologies via GitHub. We will download, build, and run a special Azure Digital Twins Model Uploader tool using Visual Studio to import one of the ontologies. The Azure Digital Twins Explorer is used to view the result.

Understanding the developer landscape

Building applications requires understanding the developer landscape. The developer landscape is everything that is available to build our solutions. There are APIs and SDKs available to access and manage Azure Digital Twins instances. But what is the difference between an API and an SDK? Let's start explaining each of them.

API stands for **Application Programming Interface**. As the name suggests, it's an interface. An interface allows software to interact with other software even if they use a different type of language and/or definition. The interface is an intermediate between the software and uses a uniform and consistent language. There are different forms of APIs. An example that we will discuss later in this chapter is the REST API.

SDK stands for **Software Development Kit**. As the name indicates, it is a kit. A kit indicates a set of tools to make our lives easier. Tools contain software libraries, documentation, code samples, and much more to support us in building applications. In most situations, the SDK includes an API to perform its functionality.

A great example is the .NET SDK we have been using in the past chapters when building .NET code solutions against the Azure Digital Twins instance. In fact, we have been using the Azure.DigitalTwins.Core NuGet package in our projects. This is, in fact, a .NET SDK for accessing and managing Azure Digital Twins instances, and it includes the REST API, which is the underlying API for performing all the requests made to and from the Azure Digital Twins instance.

Another advantage is that in most cases, an SDK has an easy implementation for authenticating a user. Normally, you would require multiple HTTP requests using tokens to get a user authenticated via the API.

We have learned the differences between APIs and SDKs. Both can be used when working with Azure Digital Twins. In the next section, we will understand and learn about using the REST API, which is the underlying API for most SDKs.

Understanding the REST API

REST is the abbreviation for **Representational State Transfer**. It is an API that uses HTTP requests to access and use data based on data types. REST uses HTTP request commands such as GET, PUT, POST, and DELETE, which refer to *read*, *update*, *create*, and *delete* operations.

A REST API call is an HTTP call to a REST API endpoint using a URL. Each URL is built up from a domain, port, path, and query string. A call can return a single result or even a complete data structure.

The Azure Digital Twins REST API has its API divided between two separate planes. Each plane manages a specific part of the Azure Digital Twins service. The following planes are defined.

- **Control Plane**: This plane is used to manage the Azure Digital Twins instance.
- **Data Plane**: This plane is used to manage the elements within an Azure Digital Twins instance.

Each of these planes has numerous REST API calls available that are used by SDKs to perform actions and to manage the Azure Digital Twins instance. To experience how this works, we will be accessing the REST API by using a tool called Postman. **Postman** is a tool that allows us to execute a REST API call and view the results accordingly.

Download the Postman tool from the Postman website via https://www.postman. com. There is a free plan available that has sufficient functionality to perform the different tasks in this chapter. You will need to sign up first when you access the website. You can sign in when you have an account with Postman.

The tool will start in the browser, but can also be downloaded as an application running locally. Create a new workspace. This will bring you to the **Create New Workspace** page, as shown in *Figure 9.1*. Perform the following steps:

1. Enter the name `DigitalTwinsBook` as the name of the workspace.

2. Press the **Create workspace and team** button:

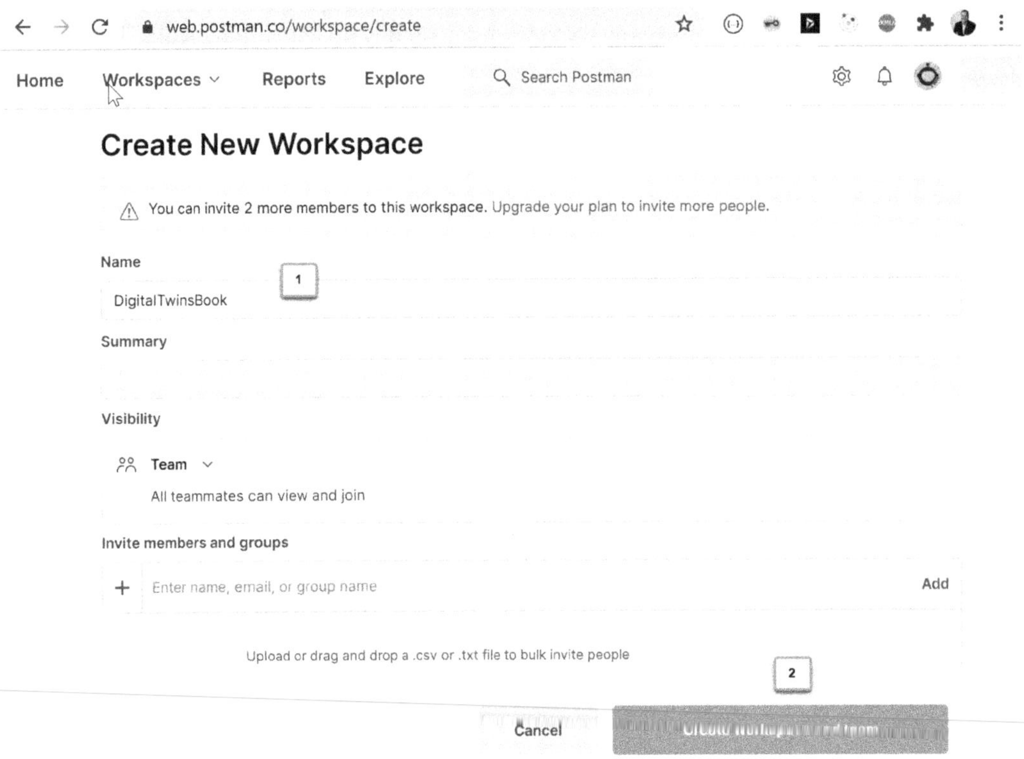

Figure 9.1 – Creating a new workspace in Postman

3. You will be redirected to the overview page as shown in *Figure 9.2* after the workspace has been created. Select your workspace if you have already created one previously:

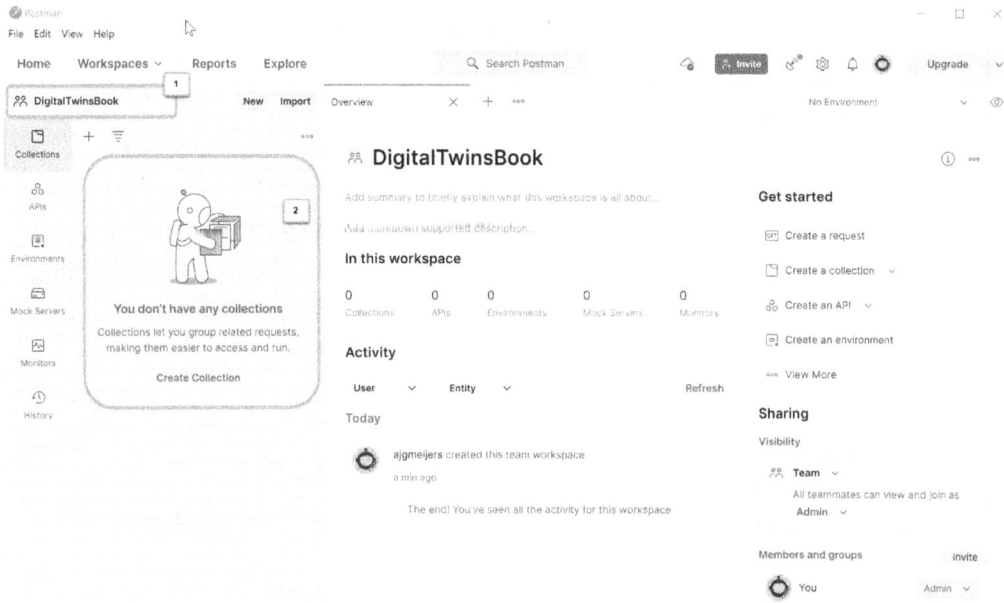

Figure 9.2 – The workspace overview in the Postman tool

The name of the workspace is visible in the top-left corner of *Figure 9.2*. We will not have any collections yet. **Collections** allow us to have a set of predefined API calls to help us out. We can also set the access token, required for authentication, in the properties of a collection. In the following control plane and data plane sections, we will be importing a collection to execute REST API calls.

Control plane

The control plane is used to control the Azure Digital Twins instance. This allows us to create and delete a complete instance. The following is available:

Azure Digital Twins	REST API
Instances	Check name availability, `CreateOrUpdate`, `Delete`, `Get`, `List`, `ListByResourceGroup`, and `Update`
Endpoints	`CreateOrUpdate`, `Delete`, `Get`, and `List`
Private endpoints	`CreateOrUpdate`, `Delete`, `Get`, and `List` (both for connections and linking resources)
Available REST API operations	`List`

Let's use the Postman tool to execute a REST API call for the control plane. We begin by importing a collection into Postman, which will give us all the available REST API calls. This file can be found at the following URL:

```
https://github.com/Azure/azure-rest-api-specs/tree/master/
specification/digitaltwins/resource-manager/Microsoft.
DigitalTwins.
```

We will need to go a few levels deeper based on the version we want to use. The best thing is looking for a stable version. The file we are looking for is called `digitaltwins. json`. While writing the book, we used the following file:

```
https://raw.githubusercontent.com/Azure/azure-rest-api-specs/
master/specification/digitaltwins/resource-manager/Microsoft.
DigitalTwins/stable/2020-12-01/digitaltwins.json.
```

Download and save the file at a location you will remember. Import the file as shown in *Figure 9.3*. Perform the following steps:

1. Click on the **Import** button.
2. Select the **File** tab.
3. Click on the **Upload files** button.
4. Select the file in the pop-up explorer.
5. Click on the **Open** button:

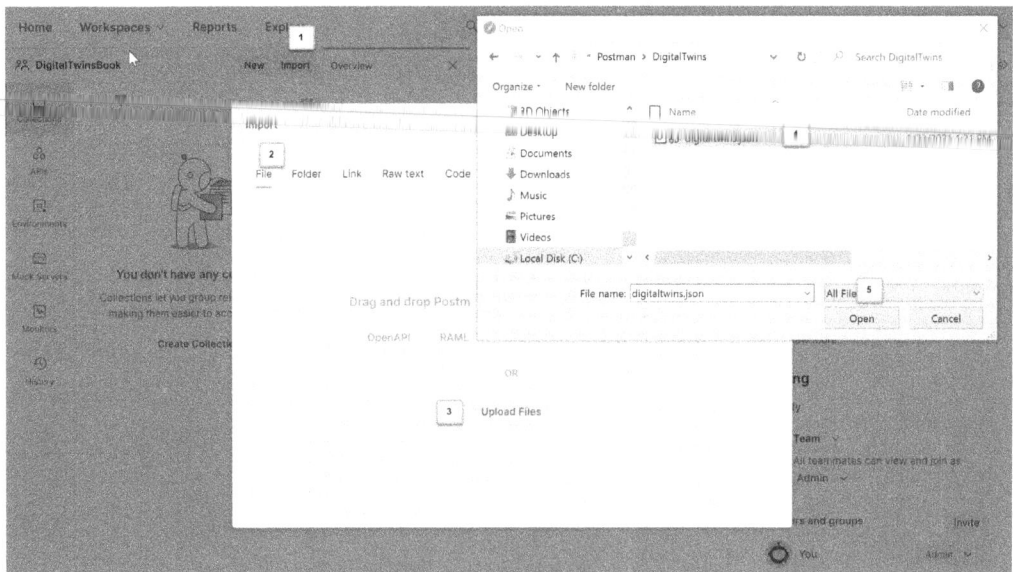

Figure 9.3 – Importing a collection into the Postman tool

The dialog will change when we have selected the file to upload. The dialog will look like the one in *Figure 9.4*:

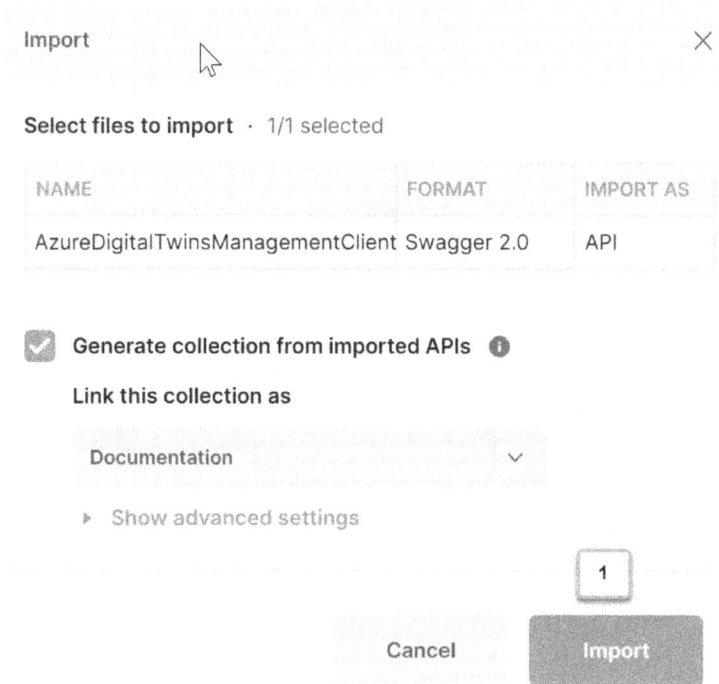

Figure 9.4 – Finalizing the importing of the collection

Press the **Import** button. The dialog will refresh, and a final **Confirm and close** button will appear. Press that one as well.

Before we can start using REST API calls, we need to have an access token set up. Perform the following steps and view the result as shown in *Figure 9.5*:

1. Open a Windows PowerShell.

2. Log in with `az login` and enter your credentials in the browser.

    Use the following command to get an access token with the resource set to `https://management.azure.com`:

    ```
    az account get-access-token --resource https://
    management.azure.com/
    ```

3. Copy the result of `accessToken` into a temporary notepad:

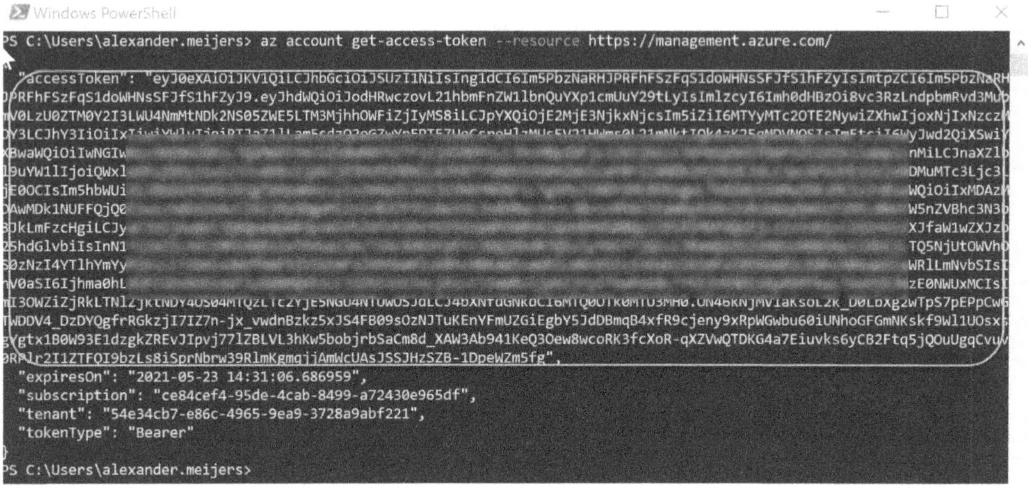

Figure 9.5 – Obtaining the access token for the control plane using Windows PowerShell

The final step is to configure the access token in the collection. Go to the Postman tool and perform the following steps, as shown in *Figure 9.6*:

1. Select the **AzureDigitalTwinsManagementClient** collection in the **Collections** view.

2. Make sure that the **Auth** tab is selected.

3. Copy the access token from the notepad into the **Access Token** field:

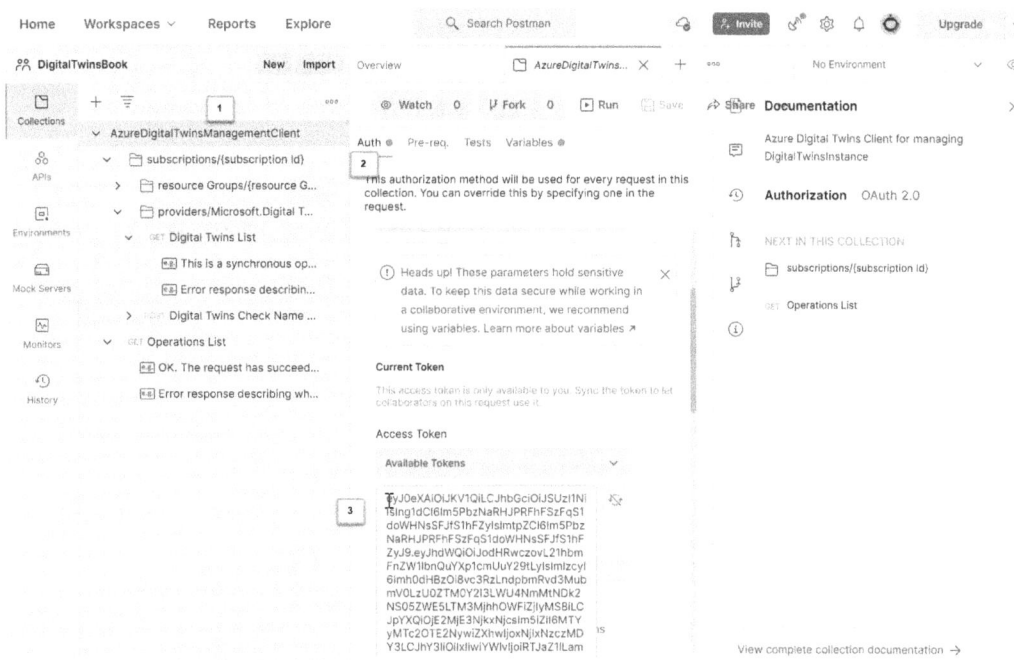

Figure 9.6 – Configuring the access token of the collection

We will be using the Postman tool to retrieve all Azure Digital Twin instances. Perform the following steps, as shown in *Figure 9.7*:

1. Expand the nodes in the collection view until the **Digital Twins List** node is visible. Make sure to select the **Digital Twins List** node.

2. The URL of the REST API is shown in the URL field of the GET method. The **{{baseURL}}** part will be replaced by the value set under the **Variables** tab when the collection is selected. This should be set to `https://management.azure.com`.

3. Open one of the resources of the *DigitalTwinsBook* resource group in the Azure portal via `https://portal.azure.com`. Copy the subscription ID and paste it into the value field of the `subscriptionId` key.

4. Click on the **Send** button.

5. View the result of the REST API call. In my case, it shows several Azure Digital Twins instances and their metadata. In your case, it will probably only show the instance you have created while reading this book:

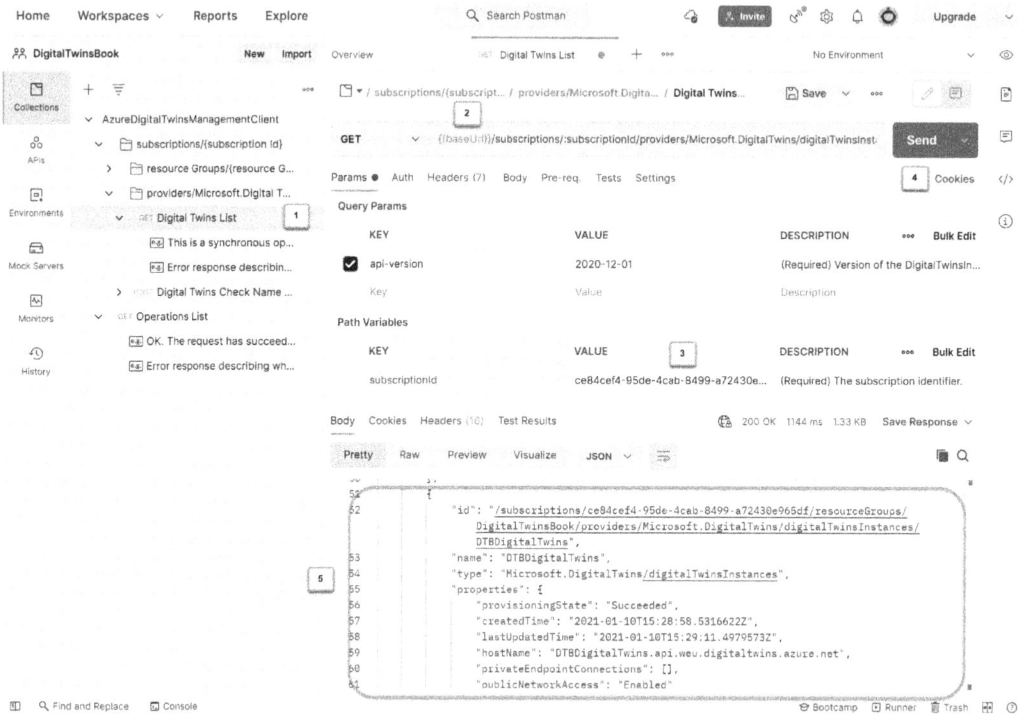

Figure 9.7 – Executing a REST API call to get all Azure Digital Twins instances

We have now executed one of the REST API calls for the control plane. In the next section, we will be using REST API calls for the data plane.

Data plane

The date plane is used to access and manage any data stored in an Azure Digital Twins instance. This allows us, for example, to create models and instantiate a digital twin based on a model. The following is available:

Azure Digital Twins	REST API
Event routes	`Add`, `Delete`, `GetById`, and `List`
Models	`Add`, `Delete`, `GetById`, `List`, and `Update`
Query	`Query`
Digital Twins	`Add`, `Delete`, `GetById`, `Update`, and `SendTelemetry`
Relationships	`AddRelationship`, `DeleteRelationship`, `GetRelationshipById`, `ListRelationships`, and `UpdateRelationship`
Components	`GetComponent`, `UpdateComponent`, and `SendComponentTelemetry`

Let's use the Postman tool to execute a REST API call for the data plane. We begin by importing a collection into Postman, which will give us all the available REST API calls. This file can be found at the following URL:

`https://github.com/Azure/azure-rest-api-specs/tree/master/specification/digitaltwins/data-plane/Microsoft.DigitalTwins.`

We will need to go a few levels deeper based on the version we want to use. The best thing is looking for a stable version. The file we are looking for is called `digitaltwins.json`. While writing the book, we used the following file:

`https://github.com/Azure/azure-rest-api-specs/blob/master/specification/digitaltwins/data-plane/Microsoft.DigitalTwins/stable/2020-10-31/digitaltwins.json.`

Download and save the file at a location you will remember. Import the file as shown in *Figure 9.8*. Perform the following two steps:

1. Click on the **Import** button. Follow the exact same steps as we did before while importing the file for the control plane. The file will have the same name.

2. The collection is imported and visible in the collection view:

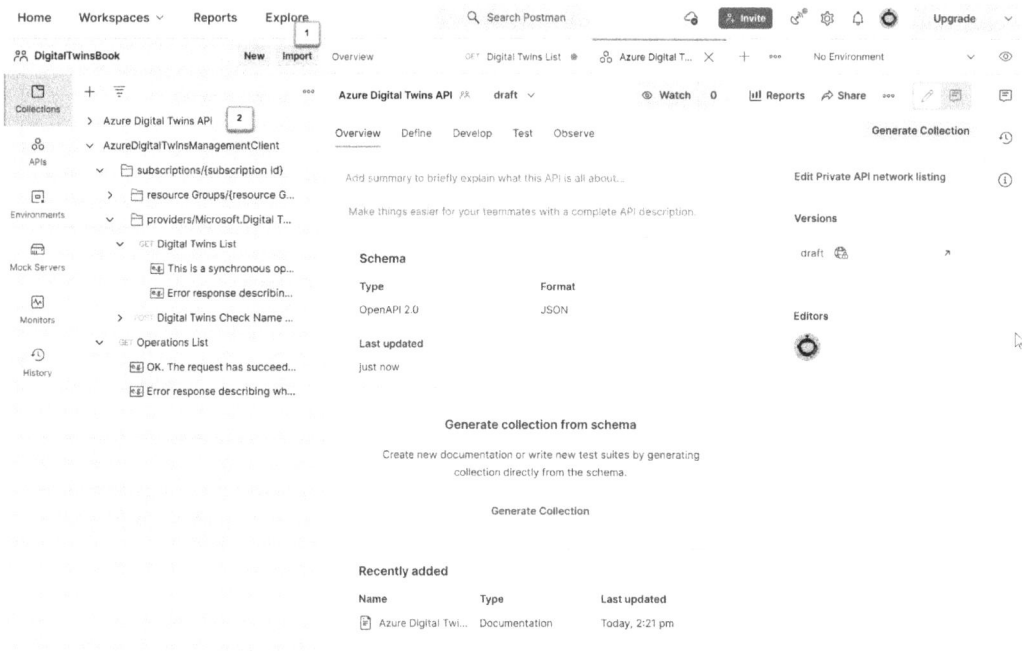

Figure 9.8 – The imported collection for the data plane

Before we can start using REST API calls, we need to have an access token set up. Perform the following steps and view the result, as shown in *Figure 9.9*:

1. Open a Windows PowerShell.

2. Log in with `az login` and enter your credentials in the browser.

3. Use the following ID to get an access token with the resource set to `0b07f429-9f4b-4714-9392-cc5e8e80c8b0`. This is the resource ID for the Azure Digital Twins service endpoint:

```
az account get-access-token --resource 0b07f429-9f4b-
4714-9392-cc5e8e80c8b0
```

4. Copy the result of `accessToken` into a temporary notepad:

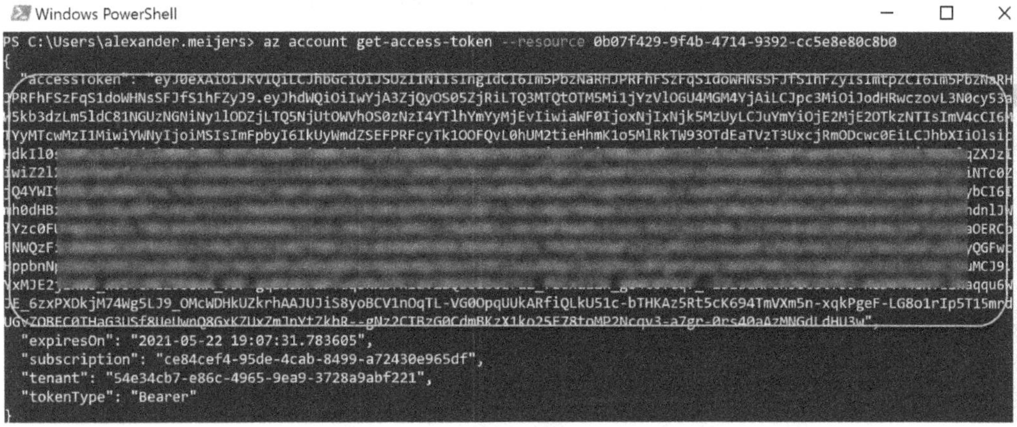

Figure 9.9 – Getting the access token for the data plane

The final step is to configure the access token in the collection. Go to the Postman tool and perform the following steps, as shown in *Figure 9.10*:

1. Select the **Azure Digital Twins API** collection in the collection view.

2. Make sure that the **Auth** tab is selected.

3. Copy the access token from the notepad into the **Access Token** field:

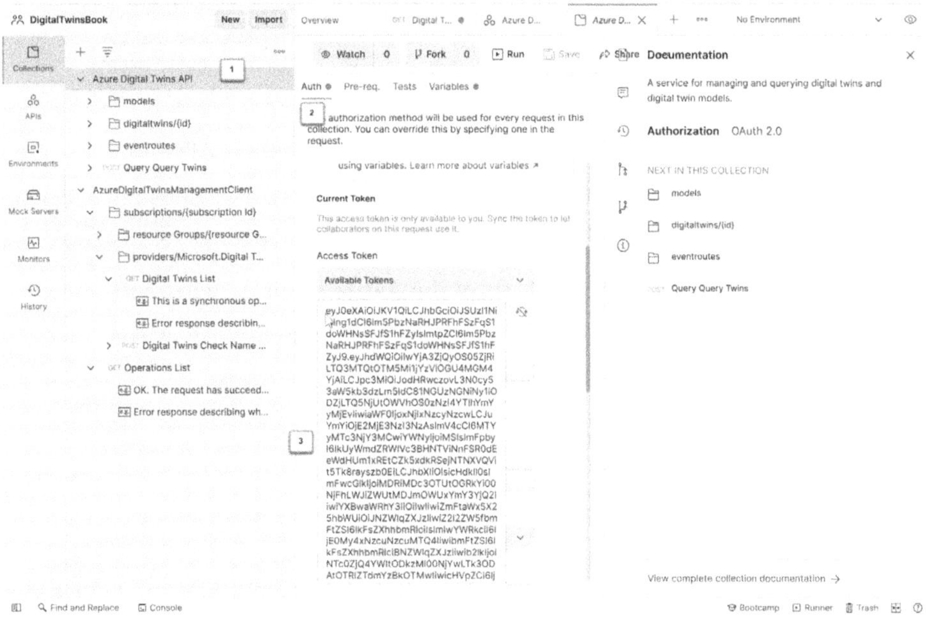

Figure 9.10 – Configuring the access token for the data plane collection

Data plane calls are specific to an Azure Digital Twins instance. Therefore, we need to configure the base URL of the collection. Perform the following steps, as shown in *Figure 9.11*:

1. Select the **Azure Digital Twins API** collection.

2. Click on the **Variables** tab.

3. Set the value of the `baseUrl` to the hostname of our Azure Digital Twins instance. The hostname can be found by opening the Azure Digital Twins resource in the Azure portal at `https://portal.azure.com`:

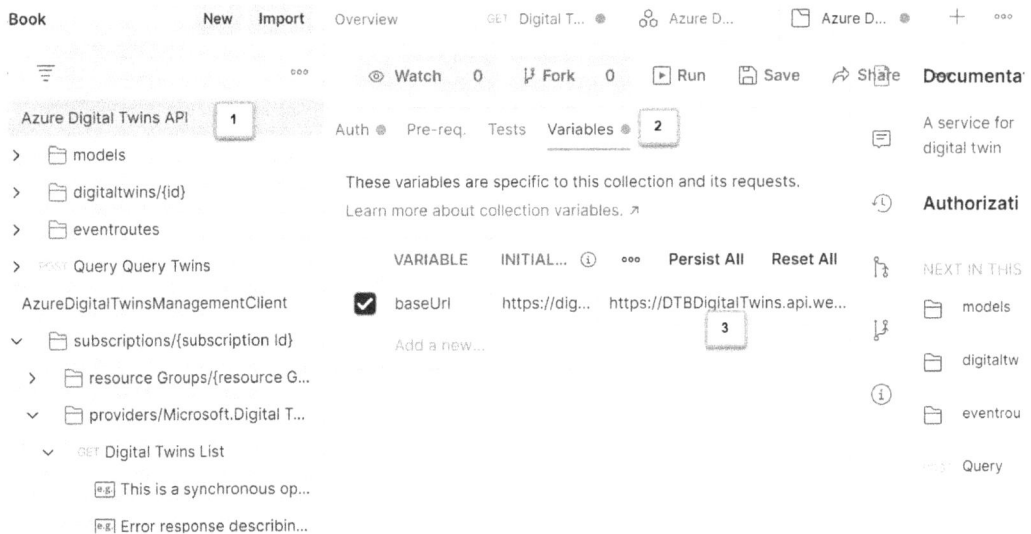

Figure 9.11 – Configuring the base URL for the data plane REST API calls

We will be querying all the digital twins within our Azure Digital Twins instance. Perform the following steps, as shown in *Figure 9.12*:

1. Select the **Query Query Twins** node from the **Azure Digital Twins API** node.

2. Select the **Header** tab. Make sure you have everything unchecked except `Content-Type`.

3. Select the **Body** tab.

4. Enter the following JSON payload into the body field:

```
{
    "query": "SELECT * FROM DIGITALTWINS"
}
```

5. Click on the **Send** button.

6. View the result. It will contain all the digital twins that we have created in the previous chapters:

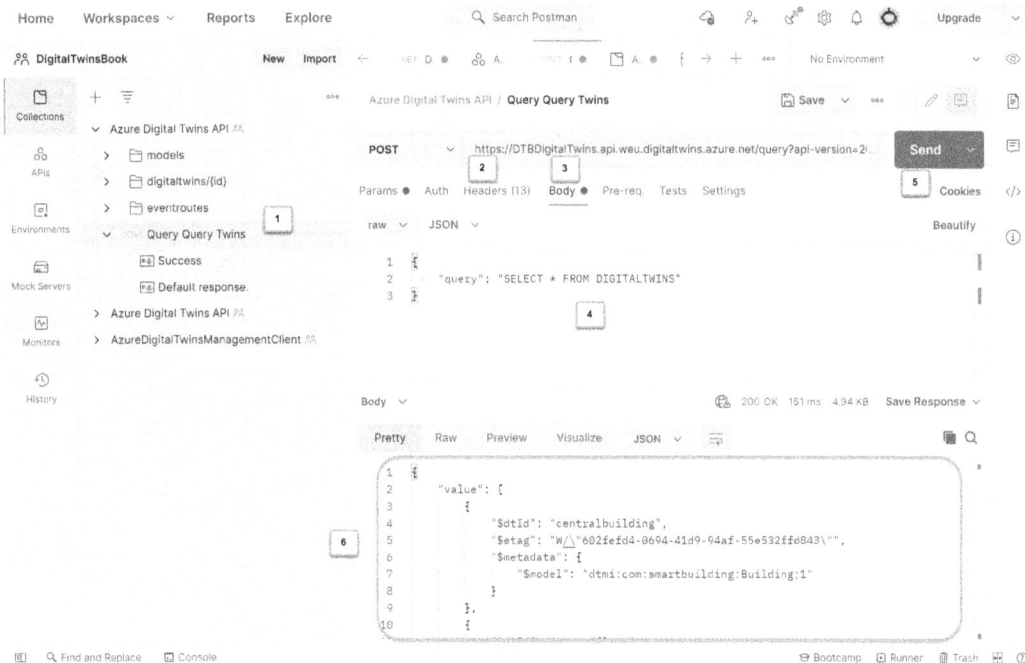

Figure 9.12 – Using the REST API to query the Azure Digital Twins instance

We have learned how to use one of the REST API calls of the data plane to perform queries on the current Azure Digital Twins instance. In the next section, you will learn about the different SDKs available.

SDKs

There are several SDKs available based on this REST API for Azure Digital Twins. The following list includes an explanation of the different SDKs and whether it is available for the control plane and the data plane:

SDK	Description	Control Plane	Data Plane
.NET SDK	The .NET SDK is specifically for .NET projects based on C#. This SDK manifests itself via NuGet packages. We have the `Microsoft.Azure.Management.DigitalTwins` and `Azure.DigitalTwins.Core` packages. We have been using the last one in our code examples in the previous chapters.	Yes	Yes
Java SDK	This is an SDK created for Maven. Maven is an open source build tool for enterprise Java projects. This SDK provides a means of accessing the Azure Digital Twins instances from those projects.	Yes	Yes
GO SDK	GO is also an open source programming language for building software. This SDK allows GO programs to access the Azure Digital Twins instances. Unfortunately, only the package for the control plane is available.	Yes	No
JavaScript SDK	This SDK is provided through the NPM package manager for the Node JavaScript platform. We have discussed and used NPM in one of the previous chapters. This allows us to access the Azure Digital Twins instances from node-based programs.	Yes	Yes
Python SDK	Python is an object-oriented programming language focusing on high-level programming. It is used a lot when building solutions around machine learning. This SDK allows us to access the Azure Digital Twins instances from Python.	Yes	Yes
PowerShell SDK	While we could debate whether this is actually an SDK, it is nevertheless important to mention this one. It allows you to execute calls for both control and data planes. PowerShell is explained in more detail in one of the next sections of this chapter.	Yes	Yes

Since this book focuses only on the .NET SDK, other SDKs will not be discussed in any further detail. In the next section, we will talk about how to monitor and view metrics of API calls.

Monitoring API metrics

The Azure portal provides us with the means to monitor API metrics. API metrics are measurements performed around the execution of the API. There are many metrics available. We can divide the metrics into several categories, as shown here:

Category	Description	Metrics
API requests	Metrics regarding API requests in relation to read, write, delete, and query operations	Requests, failure rates, and latency
Billing	Metrics regarding the count within the Azure Digital Twins instance in relation to billing costs	API operations, processed messages, and querying units
Ingress	Metrics regarding incoming events into the Azure Digital Twins instance	Events, event failure rates, and event latency
Messages	Message count to endpoints such as Event Hub, Event Grid, and Service Bus	Routed
Routing	Failure count and the time that has elapsed between messages via routing	Failure rate, latency
Other	Approaching service limits for maximum counts within the Azure Digital Twins instance	Model count, twin count

Perform the following steps to get access to the metrics of the Azure Digital Twins instance, as shown in *Figure 9.13*:

1. Open the Azure portal using `https://portal.azure.com` and log in with your credentials. Select the **DTBDigitalTwins** resource. This is our Azure Digital Twins instance.
2. Select **Metrics** in the left menu.
3. Select one of the metrics within the chart.

4.   The metric selected at *step 3* is permanently added to the chart:

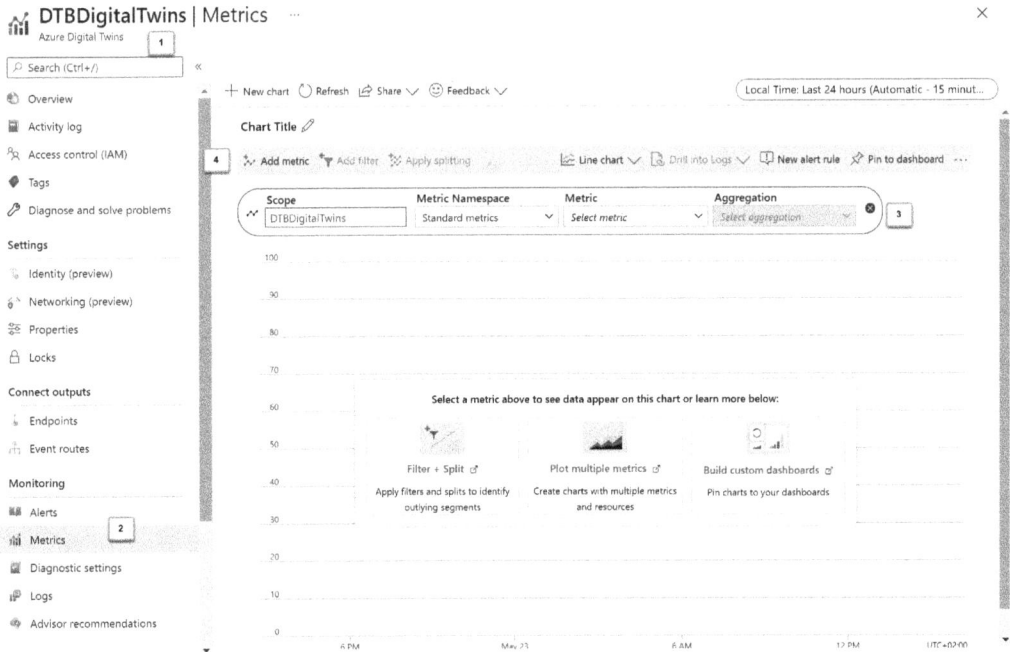

Figure 9.13 – Viewing the metrics of the Azure Digital Twins instance

You can repeat *steps 3* and *4* to add more metrics to the same chart. The result of showing the API Requests and API Request Latency metrics is indicated in *Figure 9.14*:

Figure 9.14 – Working with metrics of the Azure Digital Twins instance

The metrics view allows us to change a lot of different parts in relation to metrics. Each of those parts, as shown in *Figure 9.14*, is described in more detail here:

Part	Description
A	This area allows us to view all the currently visible metrics. In this example, we have the API Requests and API Request Latency visible. We can remove metrics and add metrics again.
B	There are several choices for visualizing metrics. We have Line chart, Area chart, Bar chart, Scatter chart, and Grid. By default, it is set to Line chart.
C	We can create multiple alert rules. Alerts can be generated based on a condition regarding metrics. When the condition is met, the corresponding action is executed.
D	It is possible to create multiple charts. Each created chart can be deleted. Charts can also be pinned to the dashboard.

We have learned how to access metrics to get more insights into the REST API usage of an Azure Digital Twins instance. In the next section, we will learn how to control and manage the Azure Digital Twins instance using PowerShell.

Using the Azure CLI to manage Azure Digital Twins

It is also possible to manage Azure Digital Twins instances using Windows PowerShell. We will require the `azure-iot` extension, which contains all the cmdlets. While the extension is automatically installed when running command lets, we will install the extension by ourselves.

Open a Windows PowerShell and use the `az login` command to log in with your credentials. Use the following command to install the extension:

```
az extension add --upgrade --name azure-iot
```

The extension is updated if it already exists. Just like the REST API, we have control plane and data plane command lets. An extensive list of all available command lets can be found here:

`https://docs.microsoft.com/en-us/cli/azure/dt?view=azure-cli-latest&preserve-view=true#az_dt_list.`

Let's start with an example of the control plane. Run the following command let:

```
az dt list --output table
```

This command let will generate a list of all available Azure Digital Twins instances inside the environment, as shown in *Figure 9.15*. As you can see, there is more than one instance in my environment:

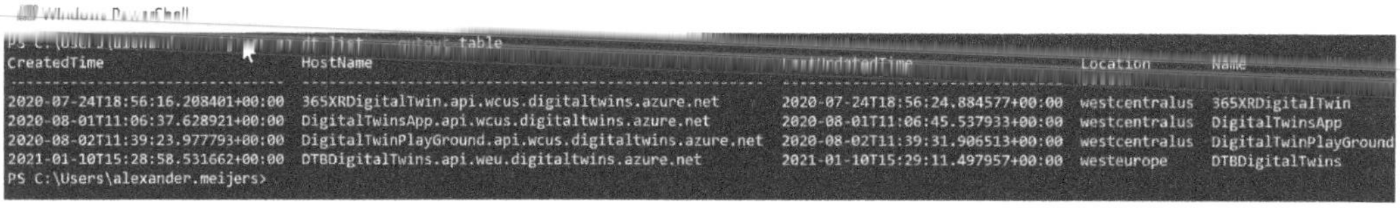

Figure 9.15 – An overview of all Azure Digital Twins instances using PowerShell

The following is an example of a command let for the data plane. The following command let will execute a query requesting all digital twins in the Azure Digital Twins instance with the name `DTBDigitalTwins`:

```
az dt twin query --dt-name DTBDigitalTwins -q "SELECT * FROM DIGITALTWINS"
```

Run the command. It will generate a JSON payload containing all digital twins, as shown in *Figure 9.16*:

```
Windows PowerShell                                                        —    □    ×
C:\Users\alexander.meijers> az dt twin query --dt-name DTBDigitalTwins -q "SELECT * FROM DIGITALTWINS"

"result": [
    {
        "$dtId": "centralbuilding",
        "$etag": "W/\"602fefd4-0694-41d9-94af-55e532ffd843\"",
        "$metadata": {
            "$model": "dtmi:com:smartbuilding:Building;1"
        }
    },
    {
        "$dtId": "secondfloor",
        "$etag": "W/\"1dae57be-7dd7-4a42-a074-e93957f963c8\"",
        "$metadata": {
            "$model": "dtmi:com:smartbuilding:Floor;1",
            "floornumber": {
                "lastUpdateTime": "2021-04-08T06:15:23.4237874Z"
            },
            "lightson": {
                "lastUpdateTime": "2021-04-08T06:15:23.4237874Z"
            }
        },
        "floornumber": 2,
        "lightson": true
    },
    {
        "$dtId": "groundfloor",
        "$etag": "W/\"bfaa3d12-d9c8-4830-bcc8-29d036b84b5f\"",
        "$metadata": {
            "$model": "dtmi:com:smartbuilding:Floor;1",
            "floornumber": {
                "lastUpdateTime": "2021-04-08T06:15:23.5387306Z"
            },
            "lightson": {
```

Figure 9.16 – The query result of a query to retrieve all digital twins using PowerShell

We have learned how to use Windows PowerShell to control and manage Azure Digital Twins. In the next section, we will learn about the service limits when using APIs and SDKs.

Understanding service limits

It is important to understand that using the services, via APIs and SDKs, is not unlimited. The Azure Digital Twins instance has several types of limits. These types are explained here:

- **Functional Limits**: Functional limits are limits that are defined by the ability of the service. Think of the number of instances, models, digital twins, and joins inside a query or the size of a certain message.

- **Rate Limits**: These limits are limits of the requests we make to the service. Think of the number of read requests per second. When we try to read more than these limits, we are throttled on the result of the request. One such example is executing queries. When you create a solution, which is used by a lot of people, and the solution is continuously requesting data by querying, you could easily reach the query limit.

Service throttling takes place on additional requests when a certain limit is reached. We will receive the `429` error as a response when limits are reached. Such errors could interrupt the application or service. There are several ways to prevent these limits from being reached:

- **Solution deployment across multiple instances**: It is always best to not have a single point of failure. Deploying to multiple instances will divide the number of API requests across the different instances and hopefully keep you within the service limits. However, this scenario will also require having the digital twin synchronized across all instances.

- **Caching mechanisms**: Another way is to cache the results of requests to the digital twin instance. This will downsize the number of requests to the service and hopefully keep you far away from reaching the limits. The downside is that your application will not work the entire time with the latest values from the digital twin. It depends entirely on how often your digital twin is refreshed with new data.

There are several ways to handle situations when these limits are (almost) reached. In those scenarios, you need some business or functional logic to prevent or to solve the issue:

- **Alerts on metrics**: Alerts on metrics can be used to warn your end users or to take preventative actions inside your application to make sure that the limits are not reached.

- **Retry logic for failed requests**: If there is no way to prevent limits from being reached, for example, due to the large number of users, we could incorporate retry logic. Retry logic is taking care of attempting to access the service again to get new data. Retry logic can be written in several ways to support the scenario where you need it.

The list of limits is extensive. It does not make sense to include one in this book. Check out the limits at the following location:

```
https://docs.microsoft.com/en-us/azure/digital-twins/
reference-service-limits.
```

You will notice that the table contains something called `Adjustable?`. If a certain limit is adjustable, this means that you can discuss your business requirements with Microsoft and request additional resources via a support ticket.

This chapter has guided us through the developer landscape of available SDKs and APIs. We have learned how the underlying REST API provides us with the means to control the Azure Digital Twins instances. We have learned about API metrics and how to monitor them. We have even used the Azure CLI to control our Azure Digital Twins instance. And finally, we have learned and understood the service limits of using APIs and SDKs. We have also learned what to do with service throttling and how to prevent these limits from being reached.

Summary

We have learned in this chapter to understand the differences between APIs and SDKs. We have learned that the underlying REST API for Azure Digital Twins is divided into a control plane and a data plane. The chapter has demonstrated several examples of using the Azure CLI, which allows us to provision and control Azure Digital Twins instances via PowerShell. Finally, we have learned how to monitor the API requests and what the impact is of the service limits. As a reader, you now have the skills to use the REST API and the Azure CLI to access Azure Digital Twins.

In the next chapter, we will learn how to set up a pipeline in Microsoft Azure using several Azure services to get data from a demo sensor onto an Azure Service Bus namespace.

Questions

As we conclude, here is a list of questions for you to test your knowledge regarding this chapter's material. You will find the answers in the *Assessments* section of the *Appendix*:

1. What is the main difference between an API and an SDK?

    a. An API is used by an SDK to use a service.

    b. There is no difference.

    c. SDKs do not use an API to access a service.

2. What tool is best used to execute REST API calls?

    a. Postman

    b. Chrome browser

    c. Visual Studio

3. Why are service limits so important to consider when building digital twins' solutions?

    a. They are not that important since the service is always running in the cloud.

    b. Limits will prevent content from being saved to the digital twin.

    c. To prevent getting results back from the digital twin.

Further reading

To learn more on the subject, refer to the following links:

- Azure Digital Twins APIs and SDKs: `https://docs.microsoft.com/en-us/azure/digital-twins/concepts-apis-sdks`

- Azure Digital Twins REST API reference: `https://docs.microsoft.com/en-us/rest/api/azure-digitaltwins/`

- Azure Digital Twins service limits: `https://docs.microsoft.com/en-us/azure/digital-twins/reference-service-limits`

- Azure Digital Twins CLI command set: `https://docs.microsoft.com/en-us/azure/digital-twins/concepts-cli`

- Az dt: `https://docs.microsoft.com/en-us/cli/azure/dt?view=azure-cli-latest&preserve-view=true`

- How to use Postman to send requests to the Azure Digital Twins APIs: `https://github.com/MicrosoftDocs/azure-docs/blob/master/articles/digital-twins/how-to-use-postman.md`

10
Building a Digital Twin Pipeline

An important aspect of having a digital twin is getting data into your model. This data needs to be bound to a digital twin that is built from different models and properties. In this chapter, we will learn how to get the data in the digital twin property by building a pipeline, using several Azure services to transport data from an actual sensor as a demo sensor into the digital twin of a sensor.

In this chapter, we'll cover the following topics:

- Understanding application architecture
- Setting up a demo sensor using Azure IoT Central
- Getting sensor messages on Azure Service Bus

Technical requirements

We will use Visual Studio to create a new project to connect to an Azure service. We will also use the Azure portal extensively to create Azure services and connect everything together. **Azure IoT Central** will be used to set up and configure a demo sensor.

Understanding application architecture

Up till now, we have been creating a simple console application that runs locally, in order to manage our Azure Digital Twins instance. A new approach is required when we need to connect an actual sensor to a digital twin. We will start with the simplest, most production-ready solution. This will involve leveraging several Azure services, as shown in *Figure 10.1*.

Azure IoT Central is an application that allows us to connect IoT devices to cloud services. It is a secure and scalable application that allows us to build rich IoT solutions. The platform has a rich user interface that allows us to connect devices and monitor their conditions. Azure IoT Central allows us to specify rules to support different business scenarios. It also supports Azure IoT Edge. This is a managed service built on top of Azure IoT Hub, allowing us to run IoT devices via standard containers. Furthermore, it integrates with a variety of other services, such as email, Power Automate, Azure Blob Storage, Azure Event Hubs, and Azure Service Bus. Azure IoT Central allows us to create demo instances of actual sensor devices. This allows us to mimic an actual sensor. This demo sensor can always be replaced by an actual sensor to generate real-world IoT data. The sensor (or the demo variant of it) is registered in the Azure IoT Central application.

The Azure IoT Central application allows you to export data to a variety of cloud services, such as Azure Blob Storage, Azure Event Hubs, Azure Service Bus, and even webhooks. In our example, we will be using an Azure Service Bus queue to transport our IoT data as messages to the Azure Digital Twins instance:

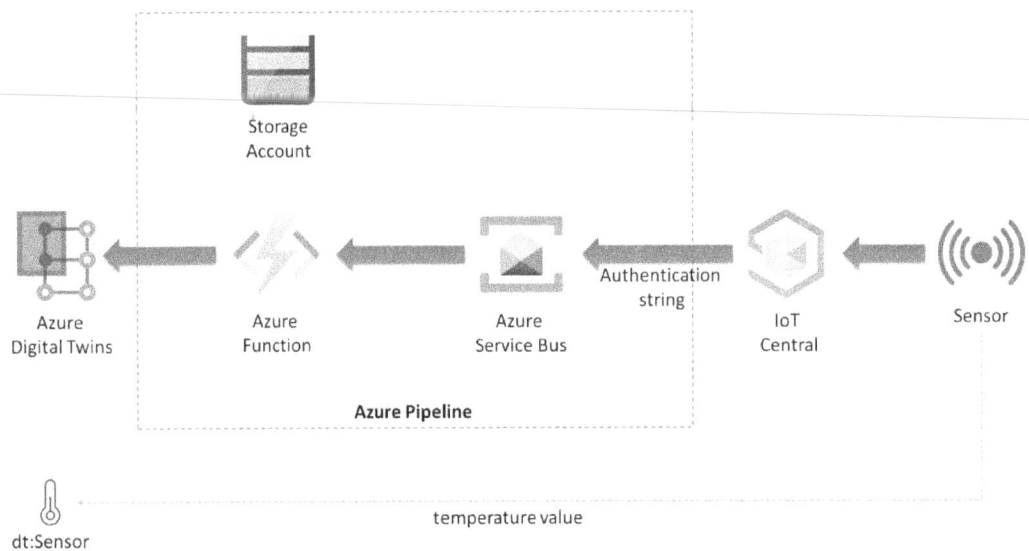

Figure 10.1 – Application architecture of a sensor connected to an Azure digital twin

The connection between Azure Service Bus and the Azure IoT Central application is based on a connection string generated by a **shared access policy**. A shared access policy is a shared access signature that allows an application to connect to a specific Azure service for a predefined time and with predefined permissions. The export of the Azure IoT Central application puts messages on the **Azure Service Bus queue**, with the telemetry selected by the configuration of the export. Each message contains the unique identification name of the sensor. This allows us to relate the telemetry data inside the message to a specific sensor.

We will be using **Azure Functions** to read the messages from the Azure Service Bus queue. Azure Functions is an event-driven application that is highly scalable. Azure Functions provides on-demand computing power. This means that Azure Functions only requires a very small amount of computing power to run. It only runs when an event is fired or when an Azure function is called. There are several events on which a function can fire. One of them is when a message is placed in an Azure Service Bus queue.

Azure Functions connects to an Azure Digital Twins instance by using a registered principal ID. This has the role of a digital twins owner assigned to it. The digital twin of the sensor is found via its name. This name is provided through the unique name in the message. The property – in this case, temperature – is updated accordingly.

We have now learned about the architecture required to get data from a sensor into a digital twin.

In the next section, we will focus on setting up a demo sensor and getting its data onto Azure Service Bus. In the next chapter, we will be finishing the pipeline by bringing the data into Azure Digital Twins using Azure Functions.

Setting up a demo sensor using Azure IoT Central

We will start by setting up a demo sensor using the Azure IoT Central application. To do this, we will need to use the Azure portal. Go to the Azure portal at `https://portal.azure.com` and log in with your credentials as before. Execute the following steps, as shown in *Figure 10.2*:

1. Open the hamburger at the top left and select **+ Create resource**. Enter `IoT Central Application` in the search box and press *Enter*.

2. Click the **Create** button in the IoT Central application box and select the **IoT Central application** option:

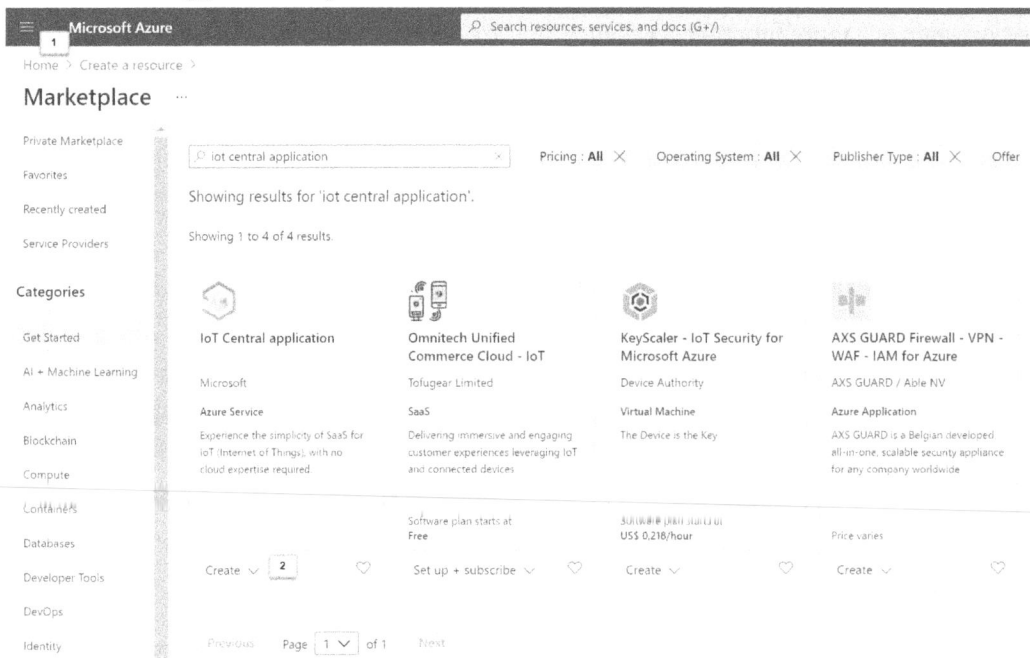

Figure 10.2 – The marketplace in the Azure portal for selecting the IoT Central application

Next, execute the following steps, as shown in *Figure 10.3*:

1. Use `dtbdemosensors` as the resource name. This will fill in the application URL accordingly. If, for some reason, the name is already taken, try using another unique name.

2. Select the subscription you have been using throughout this book.

3. Select our resource group, called **DigitalTwinsBook**.

4. Choose the Standard Tier 0 pricing plan. A free plan is not available. However, the cost of a single IoT device per month with some usage is very low, and will add up to no more than a dollar using these examples.

5. Select **Custom application** as a template. This allows us to create our own destinations, exports, devices, and more. The other options are templates for specific industry-related situations with predefined components. Further explanation of these templates can be found at the following URL: `https://docs.microsoft.com/en-us/azure/iot-central/core/concepts-app-templates`.

6. Select the location. Choose the location that is nearest to you.

7. Click the **Create** button to continue:

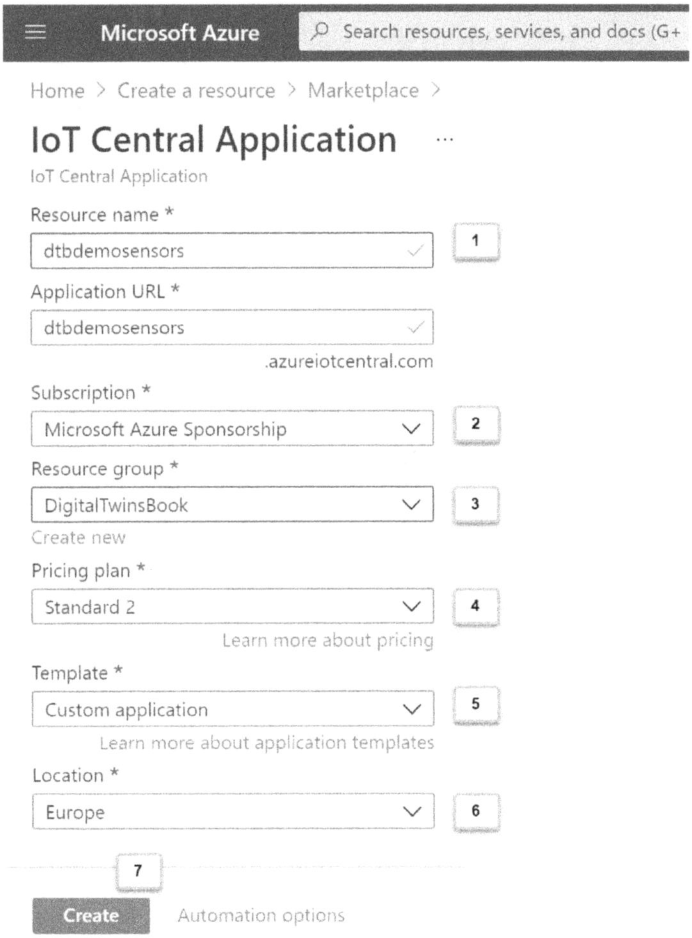

Figure 10.3 – Setting up the Azure IoT Central application

The deployment of this resource will begin. When it is finished, click the **Go to resource** button to access the resource overview of the Azure IoT Central application resource.

Open the Azure IoT Central application URL by clicking on it, as shown in *Figure 10.4*. This will open another tab with the Azure IoT Central dashboard. In some situations, you will need to log in again. Use the same account that was used to create the resource:

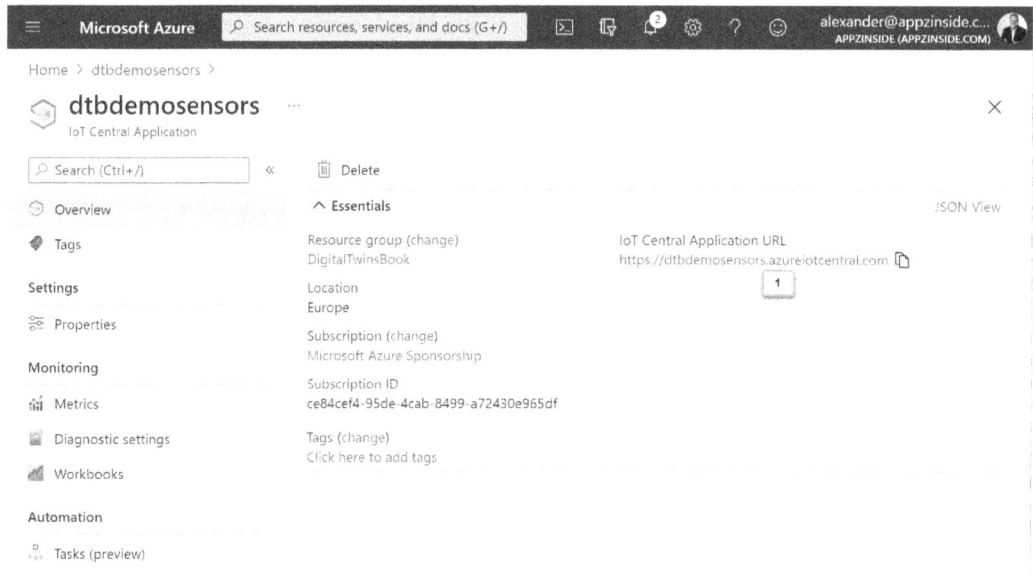

Figure 10.4 – The view of the Azure IoT Central application resource

We now need to create a device template before we can register or create a device. Execute the following steps, as shown in *Figure 10.5*.

1. Click on **Device templates** in the left menu.

2. Click on the **+ New** or **Create a device template** button:

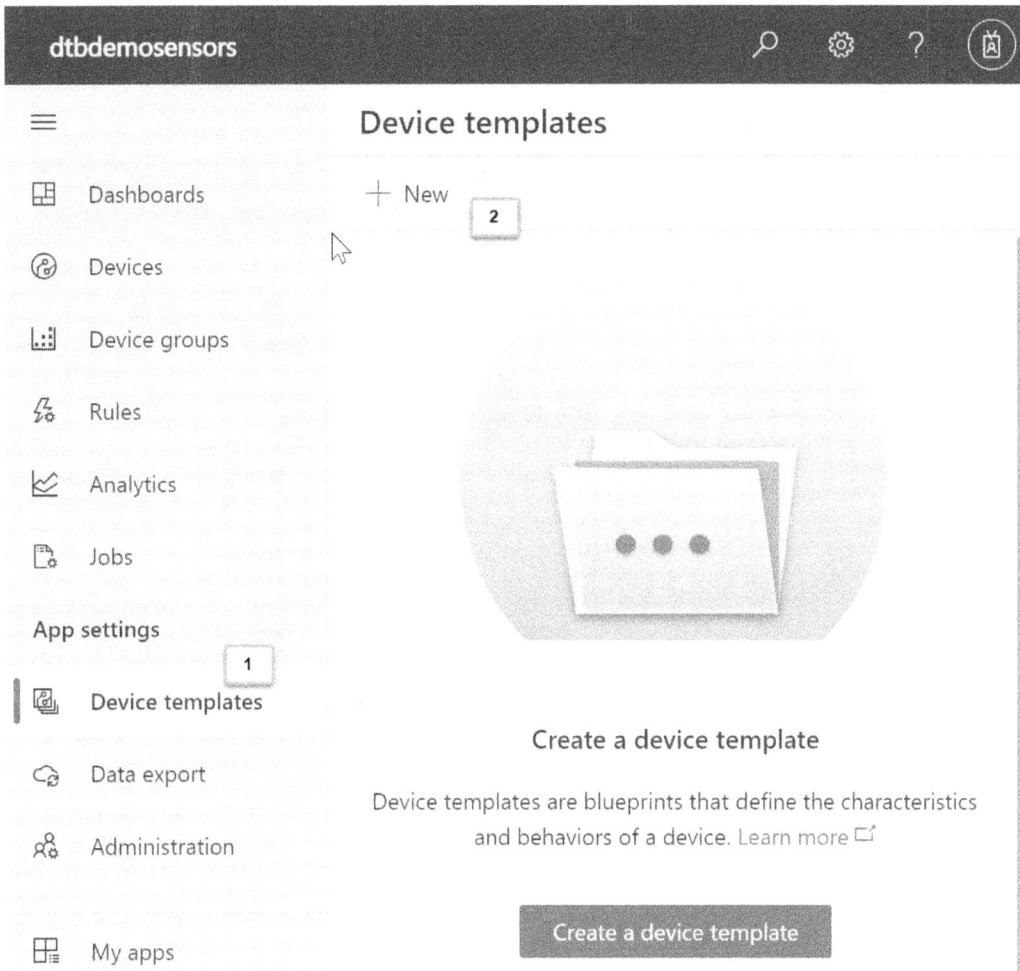

Figure 10.5 – The device templates overview in the Azure IoT Central application

This will open a very long list of several IoT devices. Open the search box in your browser and look for **MXCHIP**. The MXCHIP is a device that is part of a less expensive IoT development kit that you can buy online. The device contains several sensors for temperature, motion, and magnetic and atmospheric measurements. It is a fun IoT development kit to start with and to learn more about IoT devices. More information about this device can be found at the following URL: `https://en.mxchip.com/az3166`.

While in this example we use a demo variant of this device, it can be replaced by such a device and let it generate values based on actual sensors.

Execute the following steps, as shown in *Figure 10.6*:

1. Select the device.
2. Click on the **Next: Review** button:

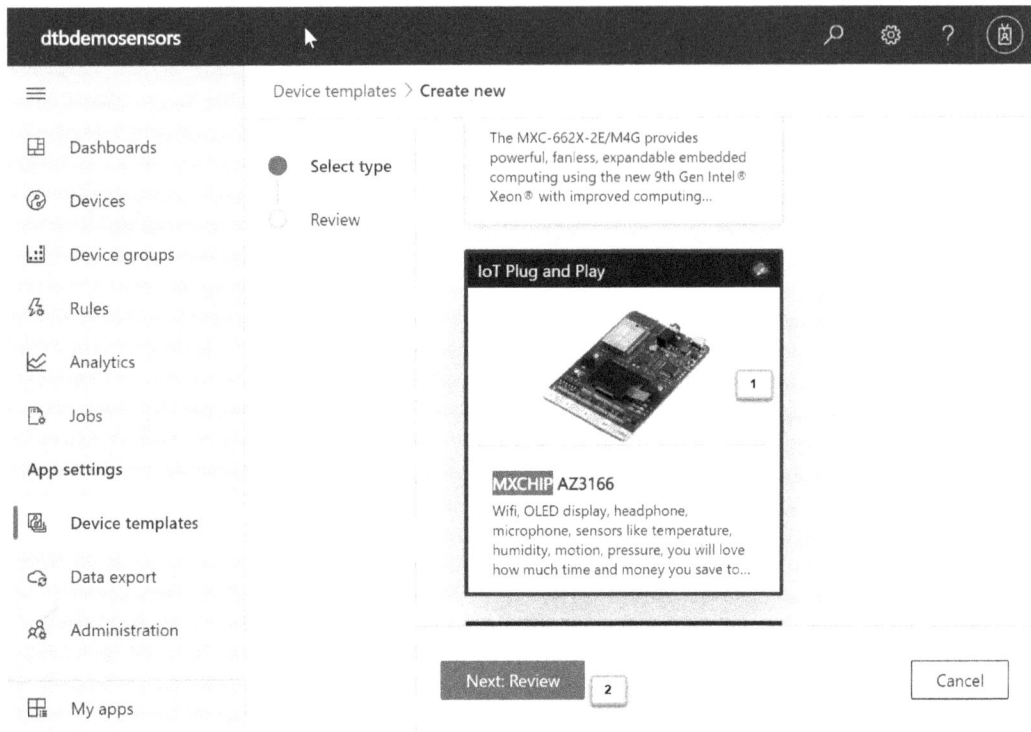

Figure 10.6 – Creating a new device template based on an actual device

Click the **Create** button to create the device template, as shown in *Figure 10.7*:

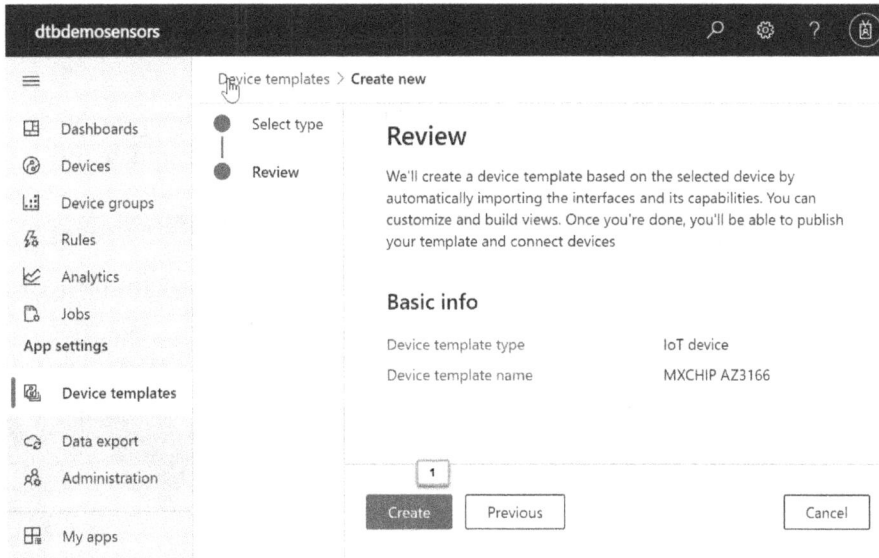

Figure 10.7 – Creating a new device template based on an actual device

This will create a new device template based on the specifications of the MXCHIP device. In the next step, we will create a device based on this device template. Execute the following steps, as shown in *Figure 10.8*:

1. Select **Devices** in the left menu.

2. Click on the **+ New** button:

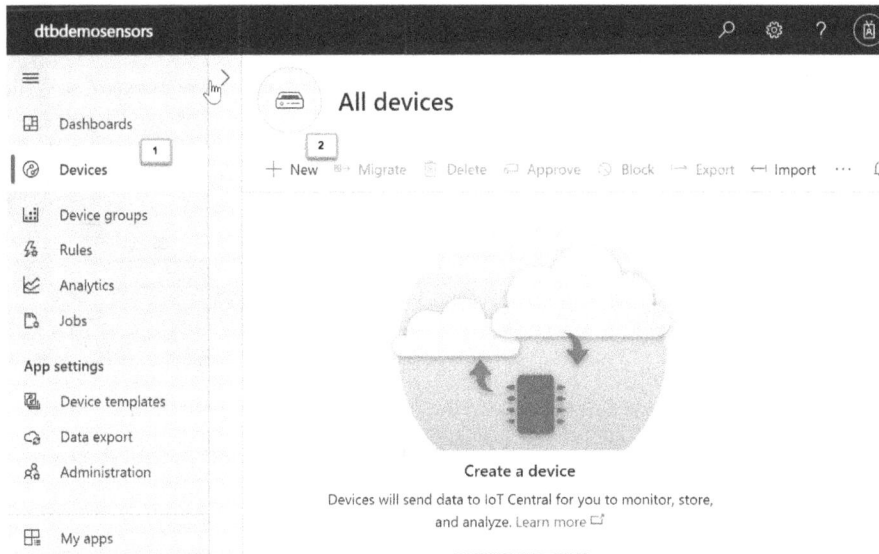

Figure 10.8 – Overview of all devices in the Azure IoT Central application

A new dialog box will appear, allowing us to create a new device. Execute the following steps, as shown in *Figure 10.9*:

1. Leave the device name and device ID as it is. This is a unique name generated for the device. We will need this name in the final stage when we update the digital twin.

2. Select **MXCHIP Getting Started Guide** as the device template.

3. Make sure to switch the **Simulate this device?** switch to on to have the device as a simulated device.

4. Click on the **Create** button:

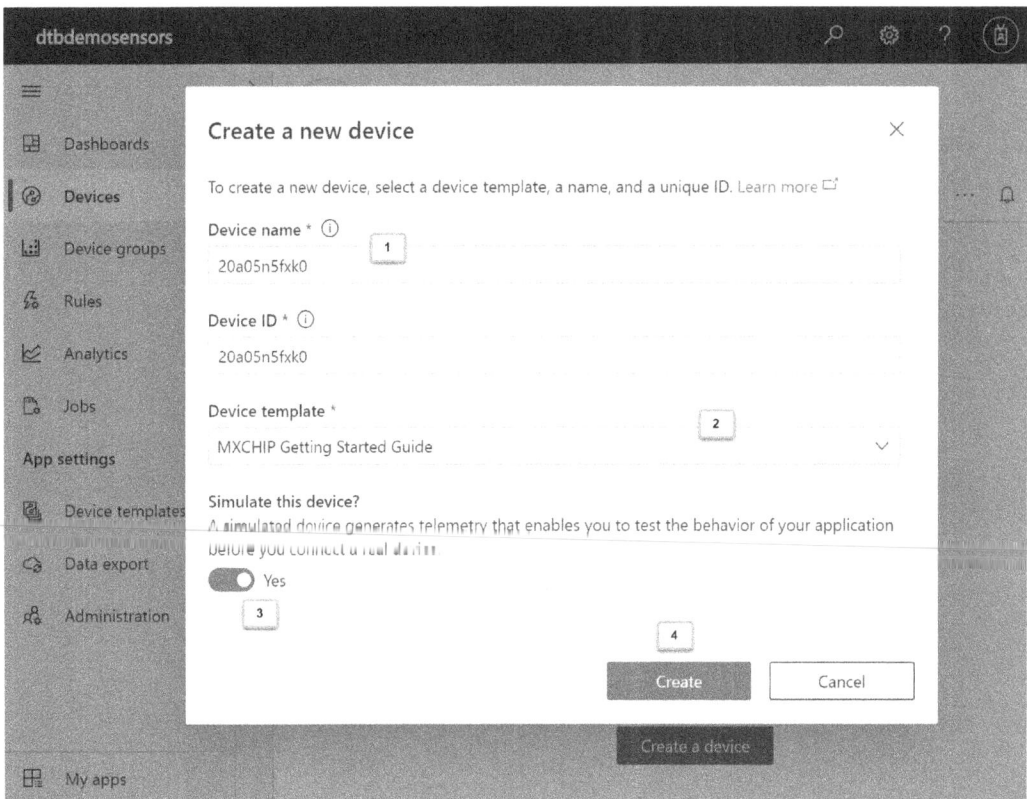

Figure 10.9 – Create a new device in the Azure IoT Central application

The new simulated device will be created. Click on **Devices** in the left menu. An overview of the devices containing our new device is shown in *Figure 10.10*:

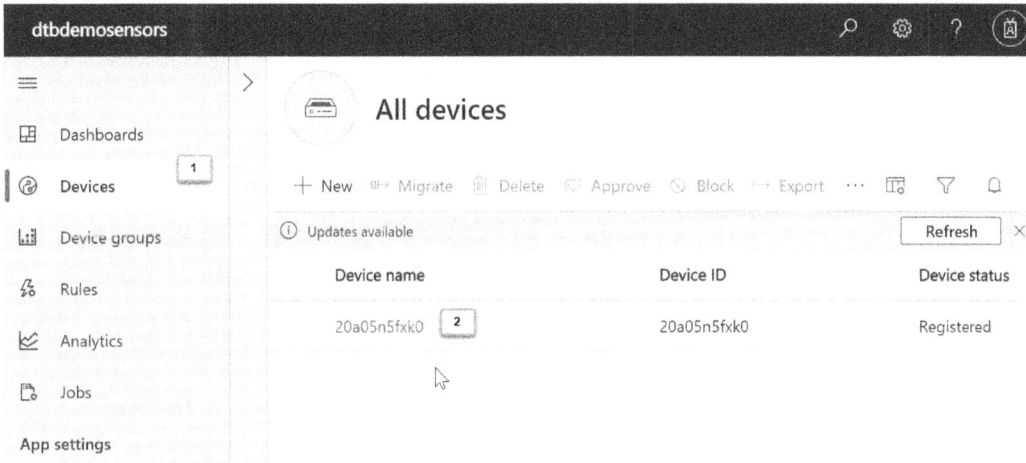

Figure 10.10 – The overview of devices containing our newly created device

Click on the name of the device, as shown in *Figure 10.10*. This will open a dashboard of the sensors of the device, as shown in *Figure 10.11*:

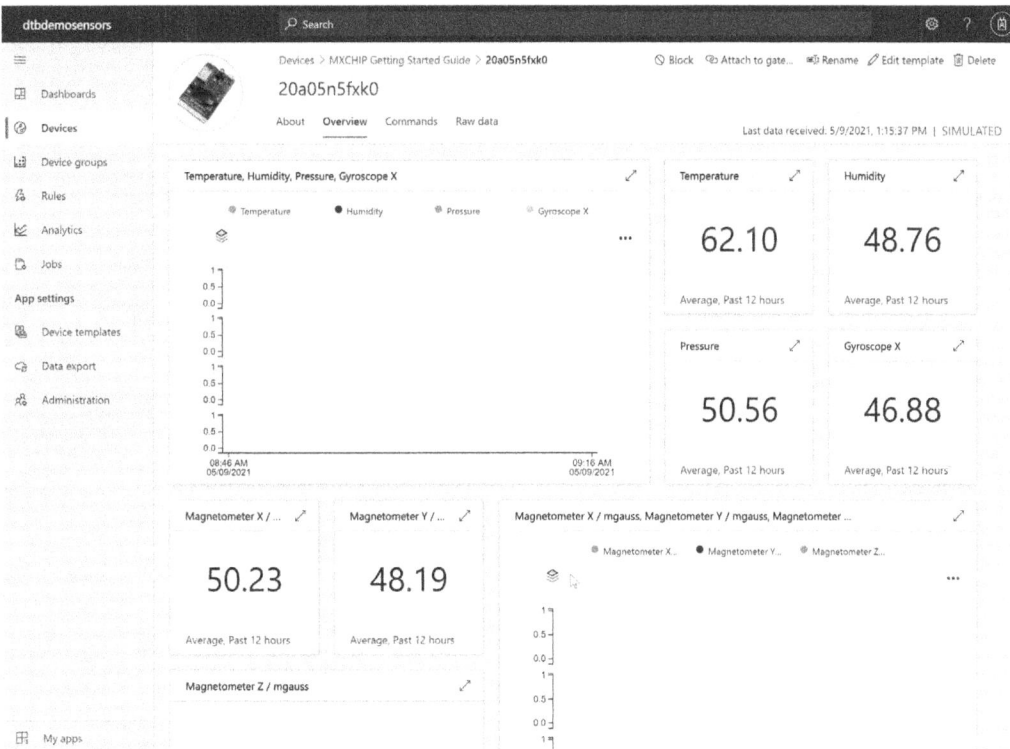

Figure 10.11 – Viewing the output of the demo sensor in the Azure IoT Central application

We have created an Azure IoT Central application resource. Inside the application, we have created a device template and a device based on the MXCHIP. In the next section, we will learn how to get sensor messages on Azure Service Bus.

Getting sensor messages on Azure Service Bus

In this section, we will set up an Azure Service Bus queue and configure the Azure IoT Central application to export telemetry to this queue. Go to the Azure portal using the following URL: `https://portal.azure.com`. We start by creating the Azure Service Bus instance. Execute the following steps, as shown in *Figure 10.12*:

1. Click on the hamburger and select **+ Create**. Enter `service bus` in the search field and press *Enter*.

2. Click on the **Create** button in the **Service Bus** box and select the **Service Bus** option:

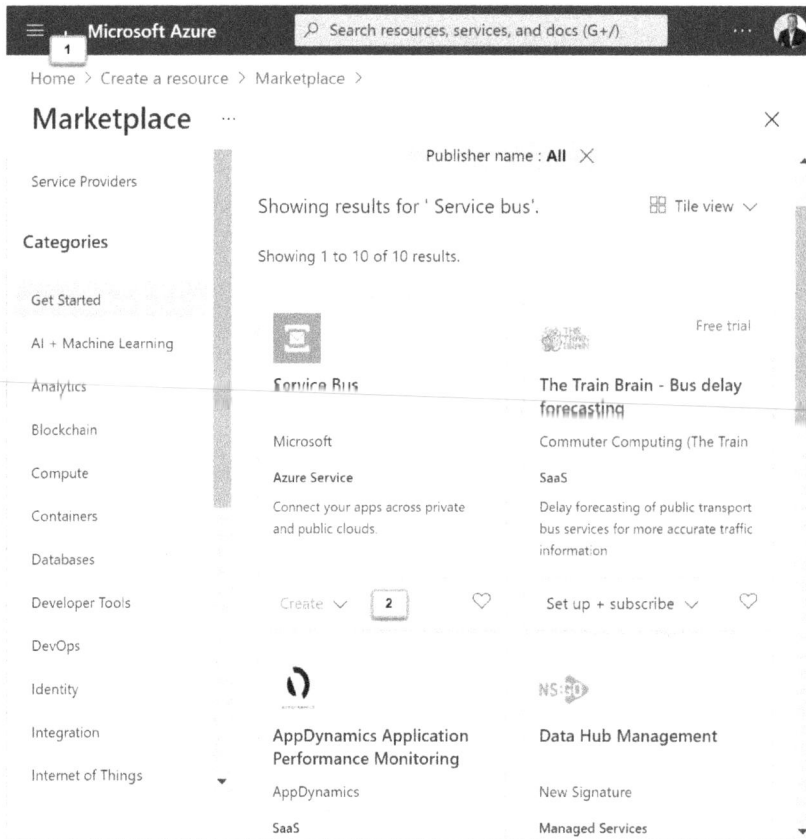

Figure 10.12 – The marketplace view used to find Azure Service Bus

Execute the following steps, as shown in *Figure 10.13*:

1. Select your subscription.
2. Select the **DigitalTwinsBook** resource group.
3. Enter the name `dtbservicebus` for our new Azure Service Bus instance. If the name is already in use, choose another unique name.
4. Select the location. Choose the location that is closest to you.
5. Select **Basic** for the pricing tier.
6. Click on the **Review + create** button:

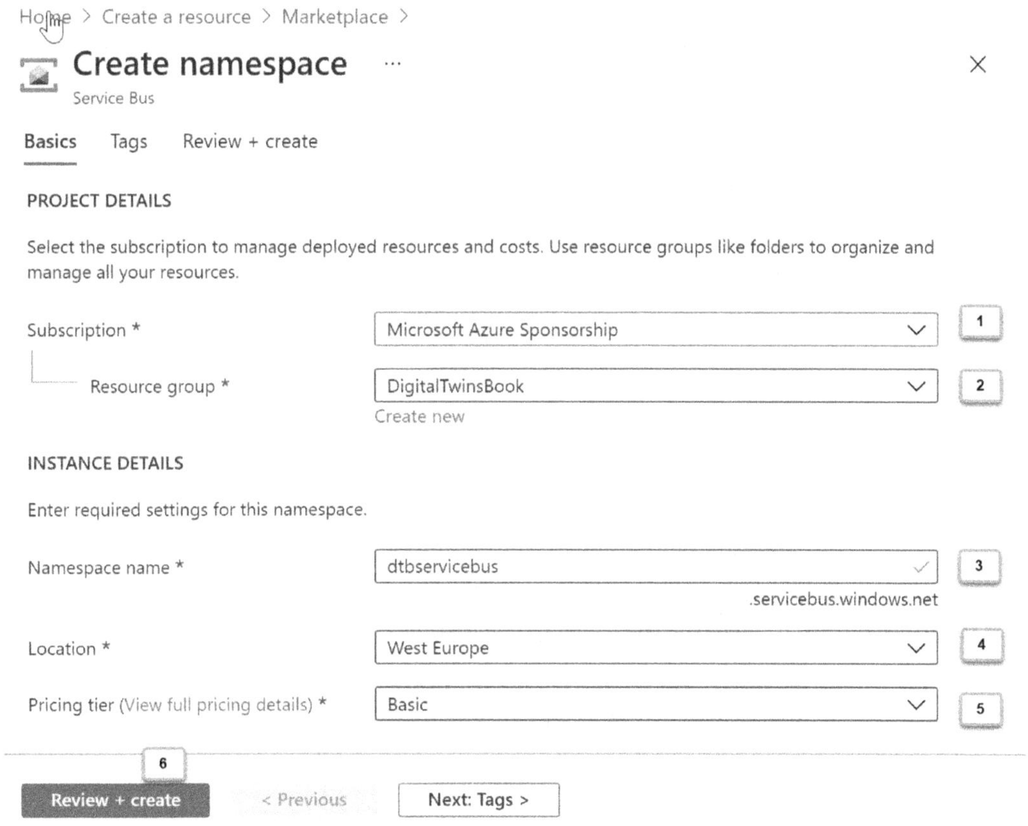

Figure 10.13 – Create an Azure Service Bus instance

Click on the **Create** button, as shown in *Figure 10.14,* to create the Azure Service Bus resource:

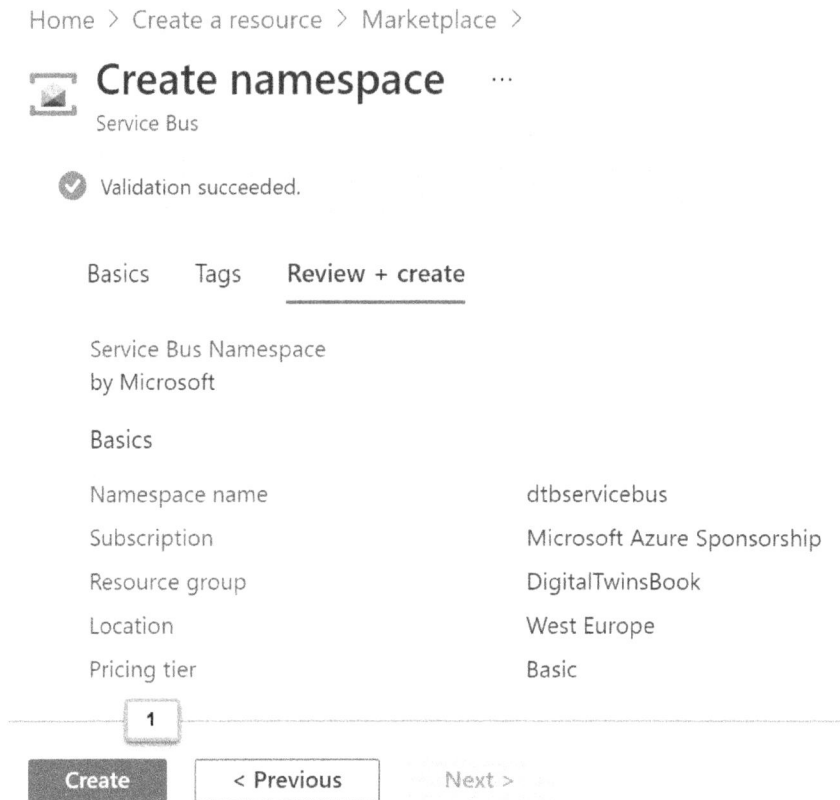

Home > Create a resource > Marketplace >

Create namespace ...

Service Bus

✓ Validation succeeded.

Basics　　Tags　　**Review + create**

Service Bus Namespace
by Microsoft

Basics

Namespace name	dtbservicebus
Subscription	Microsoft Azure Sponsorship
Resource group	DigitalTwinsBook
Location	West Europe
Pricing tier	Basic

[1]

Create　　< Previous　　Next >

Figure 10.14 – Create an Azure Service Bus resource

The Azure Service Bus resource will be deployed. When it is ready, you can click on the **Go to resource** option to view the overview of the resource.

The next step is to create a new queue to receive our messages from the Azure IoT Central application. Execute the following steps as shown in *Figure 10.15*:

1. Select **Queues** in the left menu of the resource.
2. Click on the **+ Queue** button.
3. Enter iotcentral in the **Name** field.
4. Click on the **Create** button:

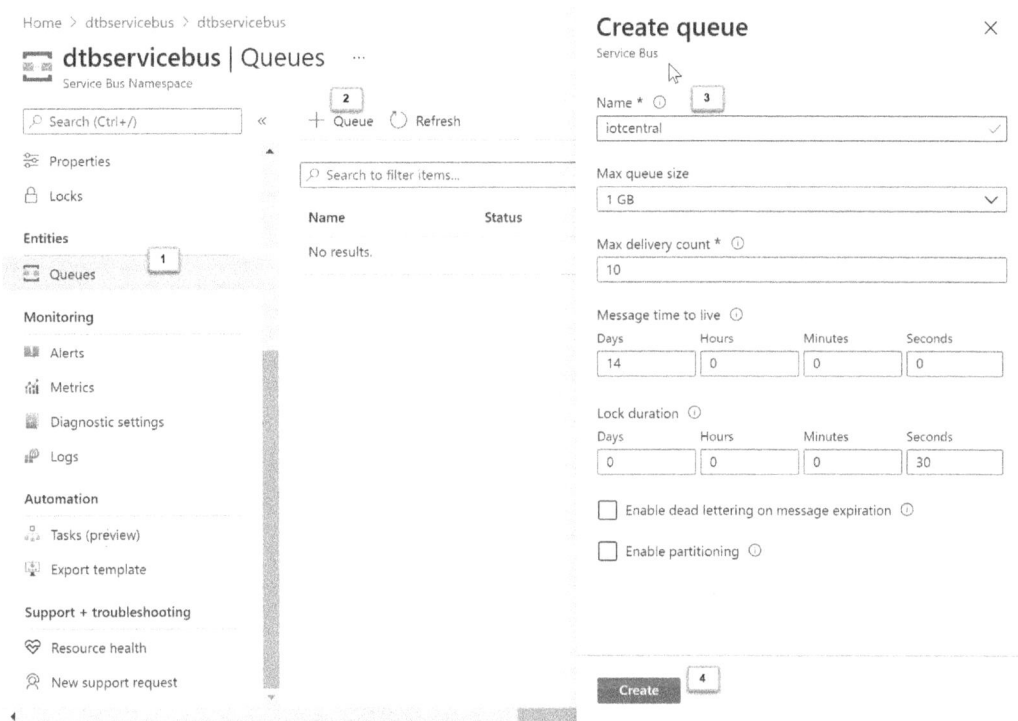

Figure 10.15 – Create a queue in Azure Service Bus

We need a connection string when we want to connect to the queue from the Azure IoT Central application. For that, we will create a shared access service policy to give access to the Azure Service Bus queue. Select the newly created queue by clicking on its name. Execute the following steps, as shown in *Figure 10.16*:

1. Select **Shared access policies** from the left menu.

2. Click on the **+ Add** button to add a new policy.

3. Enter the `iotcentrallistener` policy name.

4. Make sure that you have checked the **Listen** and **Send** checkboxes. The **Send** role is used by the Azure IoT Central application for sending messages to the queue. The **Listen** role is used for getting messages from the Azure Service Bus queue via Azure Functions.

5.   Click on the **Create** button:

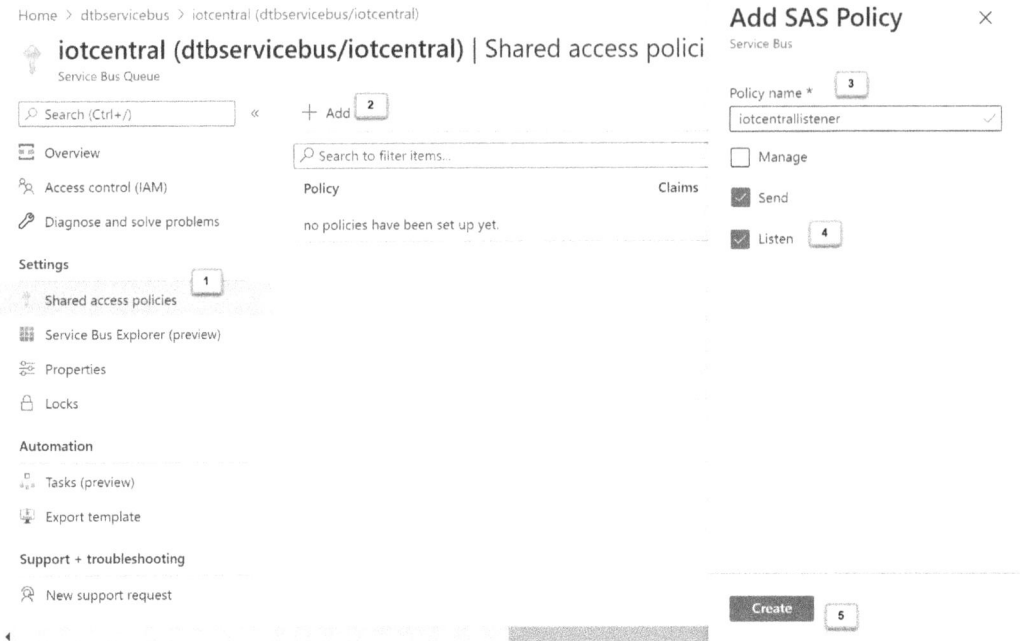

Figure 10.16 – Create a shared access service policy

Execute the following steps, as shown in *Figure 10.17*:

1.   Select the created policy to open it. This will show the primary and secondary keys and connection strings.

2.   Click on the copy symbol behind the primary or secondary connection string to copy it. Store this connection string in a notepad. We will need this connection string several steps later:

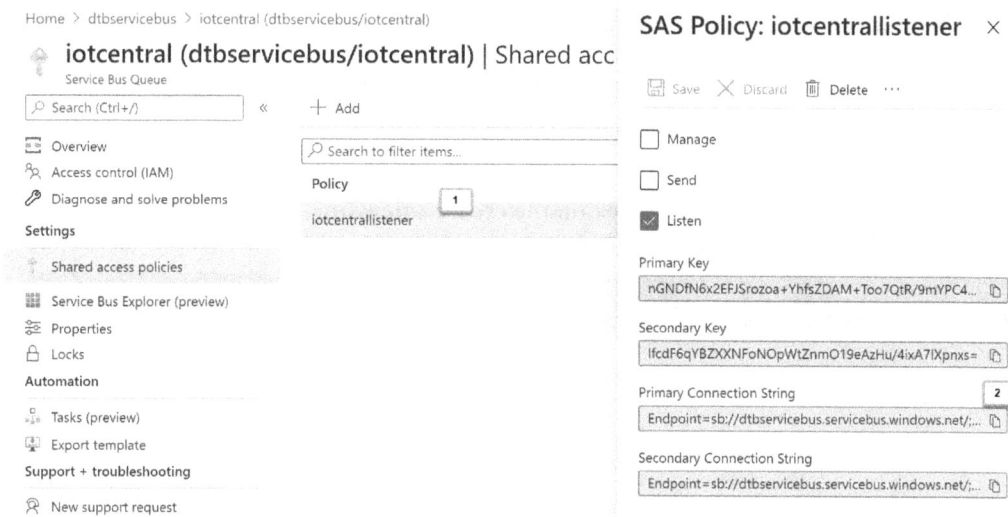

Figure 10.17 – Get the connection string from the shared access service policy

We have now created an Azure Service Bus queue, and we have the connection string to connect to it. The next step is to export telemetry data from the Azure IoT Central application to this queue. Go to the IoT Central application in the browser. Execute the following steps, as shown in *Figure 10.18*:

1. Select **Data export** under **App settings** in the left menu.

2. Click on the **+ New export** or **Add an export** button:

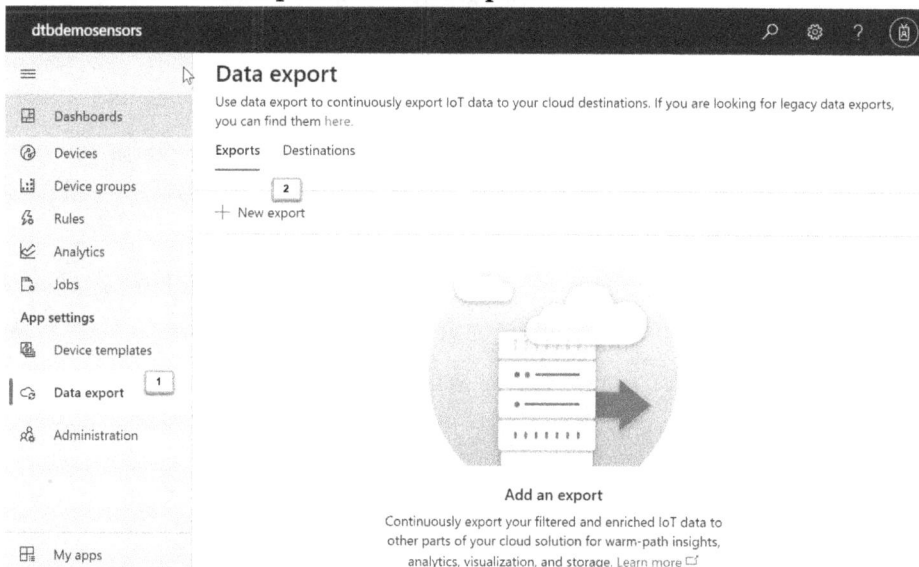

Figure 10.18 – Create a data export in the Azure IoT Central application

Execute the following steps, as shown in *Figure 10.19*:

1. Enter Azure Service Bus as the name for the export.
2. Select **Telemetry** as the type of data to export.
3. We have not yet created any destinations. An export configuration requires a destination to send data to. Click on the **create a new one** link. This will open a new dialog to create a destination:

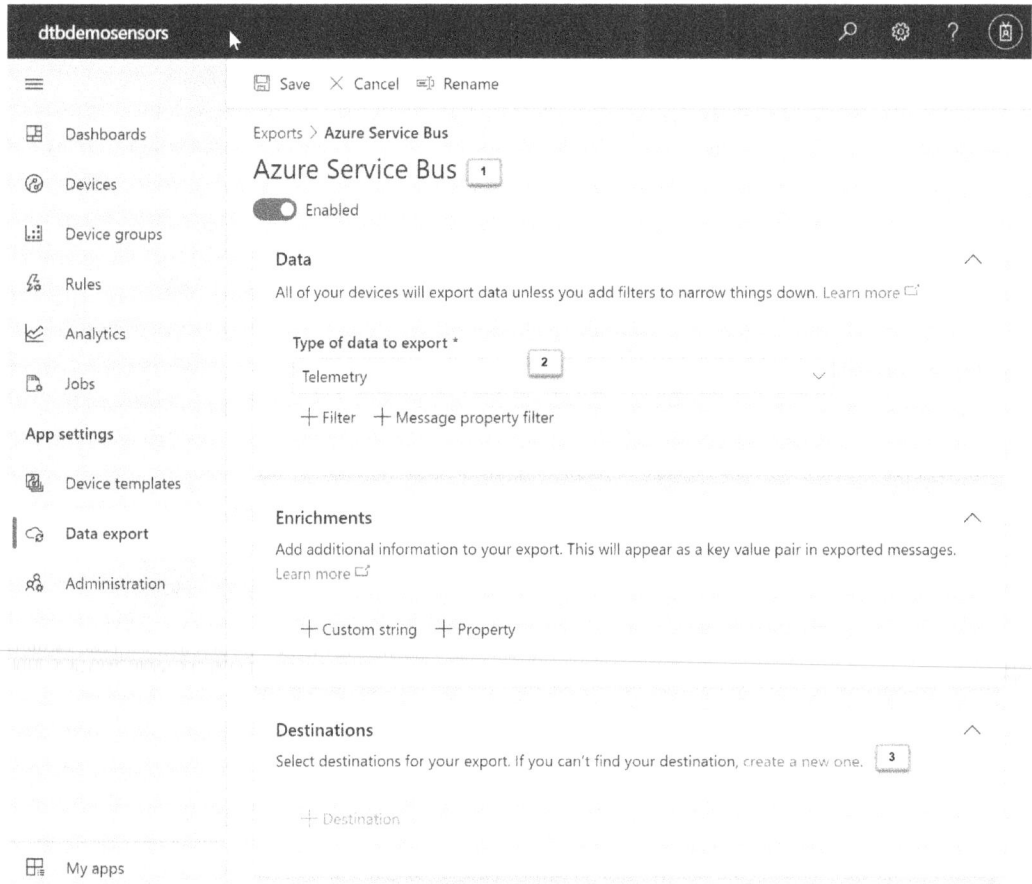

Figure 10.19 – Start creating a data export to an Azure Service Bus resource

Execute the following steps, as shown in *Figure 10.20*:

1. Enter Azure Service Bus as the name for the destination.
2. Select **Azure Service Bus Queue** as the destination type.

3. Copy the connection string you stored in the notepad earlier into the **Connection string** field.

4. Click on the **Create** button.

The dialog will disappear when the **Create** button is clicked. Make sure to click the **Save** button at the top of the dialog box shown in *Figure 10.19*.

Refer to the following screenshot:

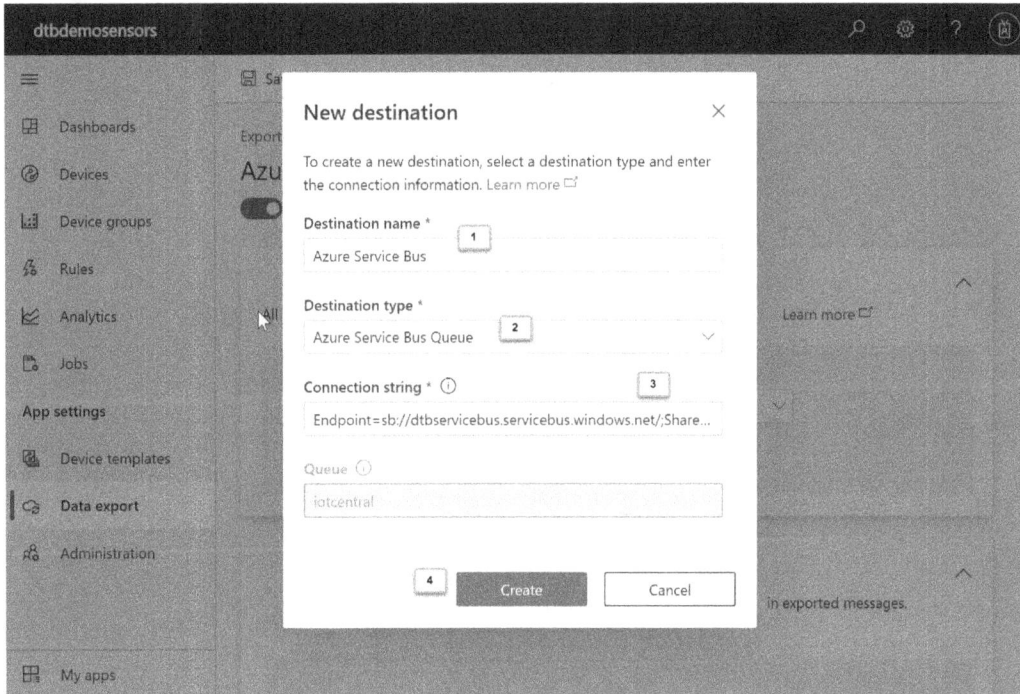

Figure 10.20 – Create a new destination for our data export

You will notice that the status of the export is set to **Starting**. It will take some time before the export moves its status to **Healthy**. From that moment, the data is flowing into the Azure Service Bus queue.

We can see the result of the export in the Service Bus Explorer. Go back to the Azure portal and open the Azure Service Bus resource. Make sure you have the queue selected. Execute the following steps, as shown in *Figure 10.21*:

1. Click on **Service Bus Explorer (Preview)** in the left menu. At the time of writing, this feature is still in preview.

2. Click on the **Peek** tab.

3. Click on the **Peek** button.

4. Select one of the messages from the list.

5. You will see information about the message:

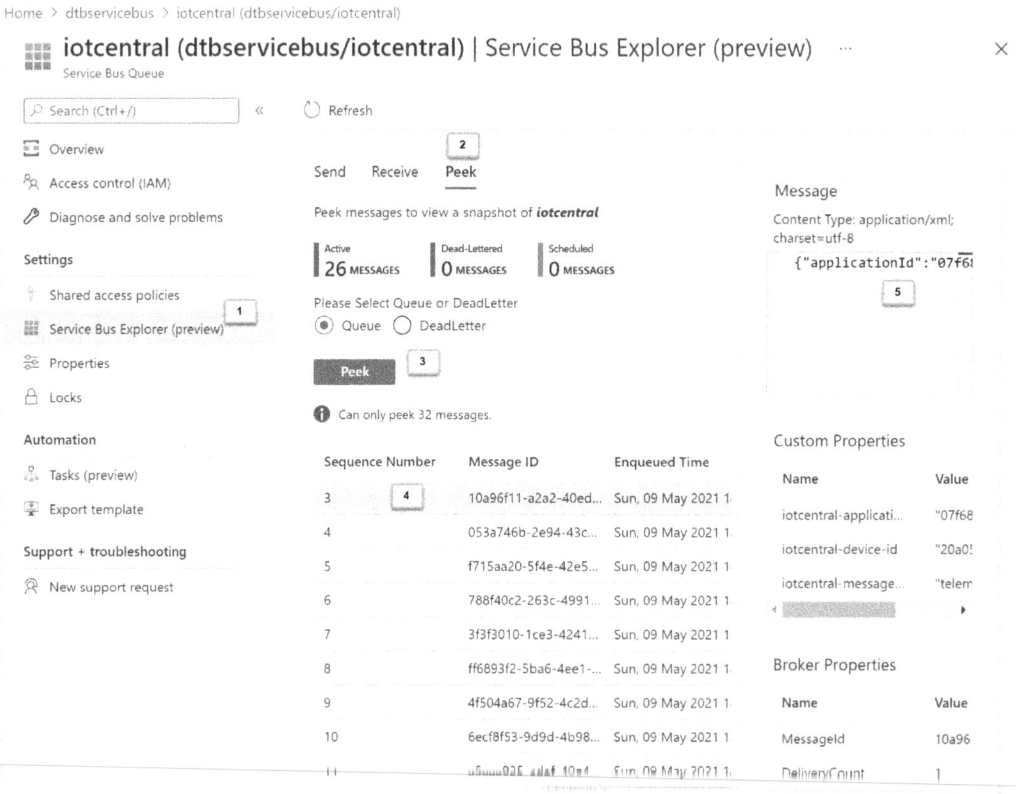

Figure 10.21 – Use the Service Bus Explorer to view the incoming messages

The message is built from the actual message and the additional *custom* and *broker* properties. The *broker* properties are mainly about the message status within the queue. The *custom* properties contain additional information provided by the export about the source. One of them, `iotcentral-device-id`, contains the name of the device. The message itself contains all the information about the measurement. The result is shown in the following snippet:

```
{
    "applicationId": "07f68e0d-8cdf-4140-97c5-004837c8c3a4",
    "messageSource": "telemetry",
    "deviceId": "20a05n5fxk0",
```

```
"schema": "default@v1",
"templateId": "urn:modelDefinition:tvmhajwhzkp:ftgkpldslxr",
"enqueuedTime": "2021-05-09T14:18:12.201Z",
"telemetry": {
    "magnetometerY": 20.587307793433844,
    "magnetometerZ": 77.30056665037554,
    "accelerometerX": 56.42830152911276,
    "accelerometerY": 51.49601755579991,
    "gyroscopeY": 61.6463566623173,
    "gyroscopeZ": 64.7649999722227,
    "temperature": 40.60607212053207,
    "humidity": 29.540769856538056,
    "accelerometerZ": 55.57901529177223,
    "gyroscopeX": 52.64513824622266,
    "pressure": 40.19060570643864,
    "magnetometerX": 51.12185084460027
},
"messageProperties": {},
"enrichments": {}
}
```

As you can see, it contains all the telemetry values of each of the sensors on the device. It also contains the date and time of the measurement.

In this section, we have set up an Azure Service Bus queue and connected the Azure IoT Central application by exporting the telemetry results to the queue.

Summary

We have learned how to set up the first part of a pipeline that can be used in a production scenario, using several Azure services to get telemetry data from an IoT device onto the Azure Service Bus queue and view the results. Each of the services has been explained and configured in a secure and scalable way. By reading this chapter, you have gained the skills necessary to set up a demo sensor using Azure IoT Central and connect it to Azure Service Bus for transporting the messages.

In the next chapter, we will be building the rest of the pipeline to transfer the sensor data from the Azure Service Bus queue to an Azure digital twin.

Questions

As we conclude, here is a list of questions for you to test your knowledge regarding this chapter's material. You will find the answers in the *Assessments* section of the *Appendix*:

1. What service application is being used to manage the IoT devices?

    a. Azure Functions

    b. Azure IoT Central application

    c. Azure Service Bus

2. What is the MXCHIP?

    a. An IoT Development Kit

    b. A single sensor device

    c. A demo sensor

3. How can we identify which sensor the data is coming from in Azure Service Bus?

    a. `messageid`

    b. `applicationid`

    c. `iotcentral-device-id`

Further reading

To learn more on the subject, see the following:

- What are application templates?: `https://docs.microsoft.com/en-us/azure/iot-central/core/concepts-app-templates`

- MXCHIP AZ3166: `https://en.mxchip.com/az3166`

11

Updating the Model

In this chapter, we will continue building our pipeline for getting data into your model. We have set up the first part of the pipeline, which gets messages with sensor data onto the Azure Service Bus queue. We will learn how to read the messages from the Azure Service Bus queue into an Azure digital twin using an Azure function. After this chapter, you will have a complete understanding of getting sensor data through a pipeline into an Azure digital twin.

In this chapter, we'll cover the following topics:

- Updating the digital twin
- Creating and publishing an Azure function
- Creating a digital twin for the sensor
- Viewing the results

Technical requirements

We will be using Visual Studio to create a new project for building an Azure function. We will again be using the Azure portal extensively for creating Azure services and connecting everything together. The Azure Digital Twins Explorer is used to make additional changes and to view the result.

Updating the digital twin

This section will describe what we need to get the digital twin updated in Azure Digital Twins, as shown in *Figure 11.1*. We will start by updating the current sensor model with a property for storing the sensor data coming from the demo sensor. Then we need to create a storage account that is used by an Azure function to get data from Azure Service Bus. The Azure function is created by using Visual Studio and is published to an Azure function placeholder in Azure. Finally, a digital twin is created from the model and will receive the sensor data from the demo sensor:

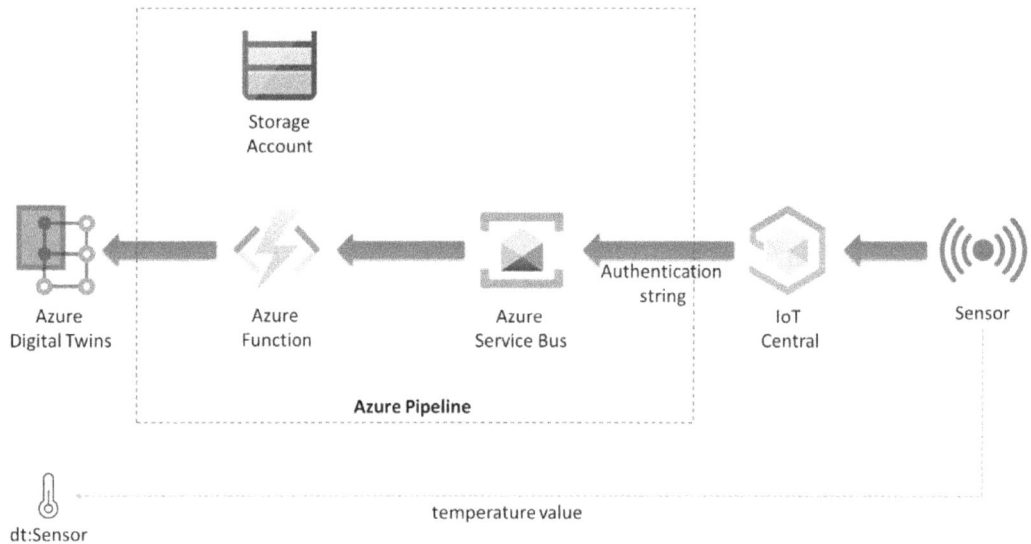

Figure 11.1 – Application architecture of a sensor connected to an Azure digital twin

Finally, a digital twin is created from the model and will receive the sensor data from the demo sensor.

Updating the sensor model

Our initial project contains a model for a sensor. But that model does not yet contain any defined properties. Since we need to update a digital twin based on this model with the temperature, we need to update the sensor model.

This requires us to open the initial SmartBuildingConsoleApp solution using Visual Studio. Go into the Solution Explorer and create a new subfolder named chapter9 in the Models folder.

Copy the `sensor.json` file from the `chapter7` folder into this `chapter9` folder. Open this copy of the `sensor.json` file and replace its contents with the following code:

```
{
    "@id": "dtmi:com:smartbuilding:Sensor;1",
    "@type": "Interface",
    "@context": "dtmi:dtdl:context;2",
    "displayName": "Sensor",
    "contents": [
        {
            "@type": "Property",
            "name": "temperature",
            "schema": "float",
            "writable": true
        }
    ]
}
```

As you can see, it contains a property of type `float` called `temperature`.

Now open the Azure Digital Twins Explorer, as you have done in previous chapters. Look up the sensor model in the **Model View** on the left. Remove the `Sensor` model and upload the new sensor model using the `sensor.json` file from the `Models/chapter9` path.

Creating a storage account

We will be using an Azure function to get data from the Azure Service Bus queue into the digital twin. An Azure function requires a storage account. We will need to go into the Azure portal via `https://portal.azure.com`. Execute the following steps, as shown in *Figure 11.2*:

1. Select the hamburger icon at the top of the screen.
2. Enter *storage account* in the search field and press *Enter*.

3.   Click on the **Create** button to create the Azure function:

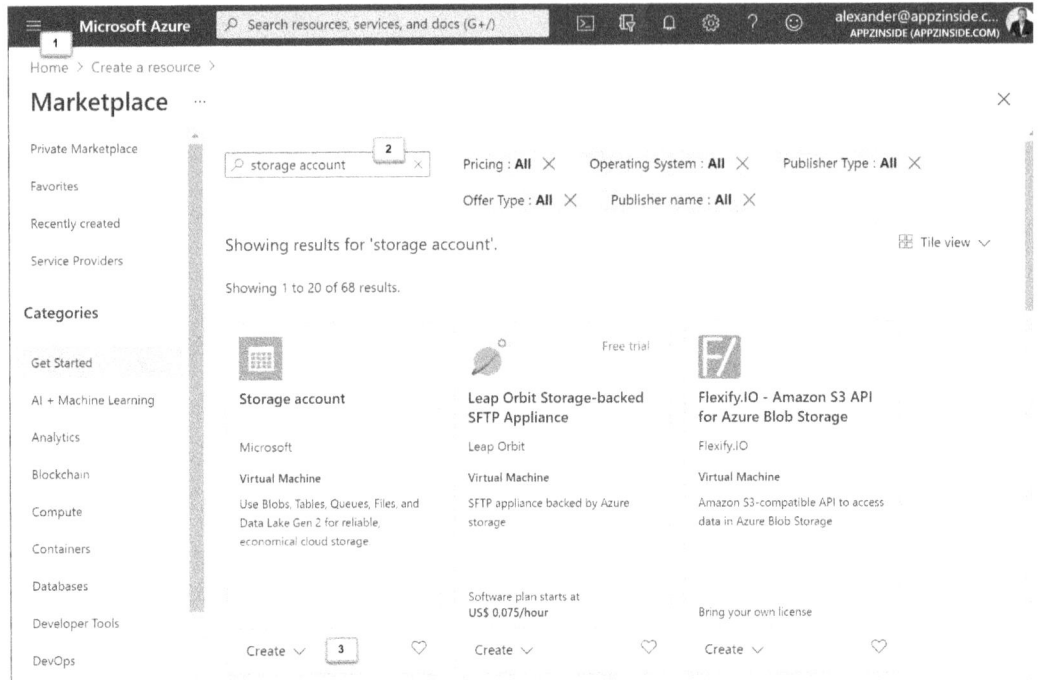

Figure 11.2 – Use the marketplace to create an Azure function

Execute the following steps, as shown in *Figure 11.3*:

1.   Select the right subscription.

2.   Select the **DigitalTwinsBook** resource group.

3.   Enter dtbiotcentraltelemetry as the name.

4.   Select a region. Choose the closest one to you.

5.   Click on the **Review + create** button:

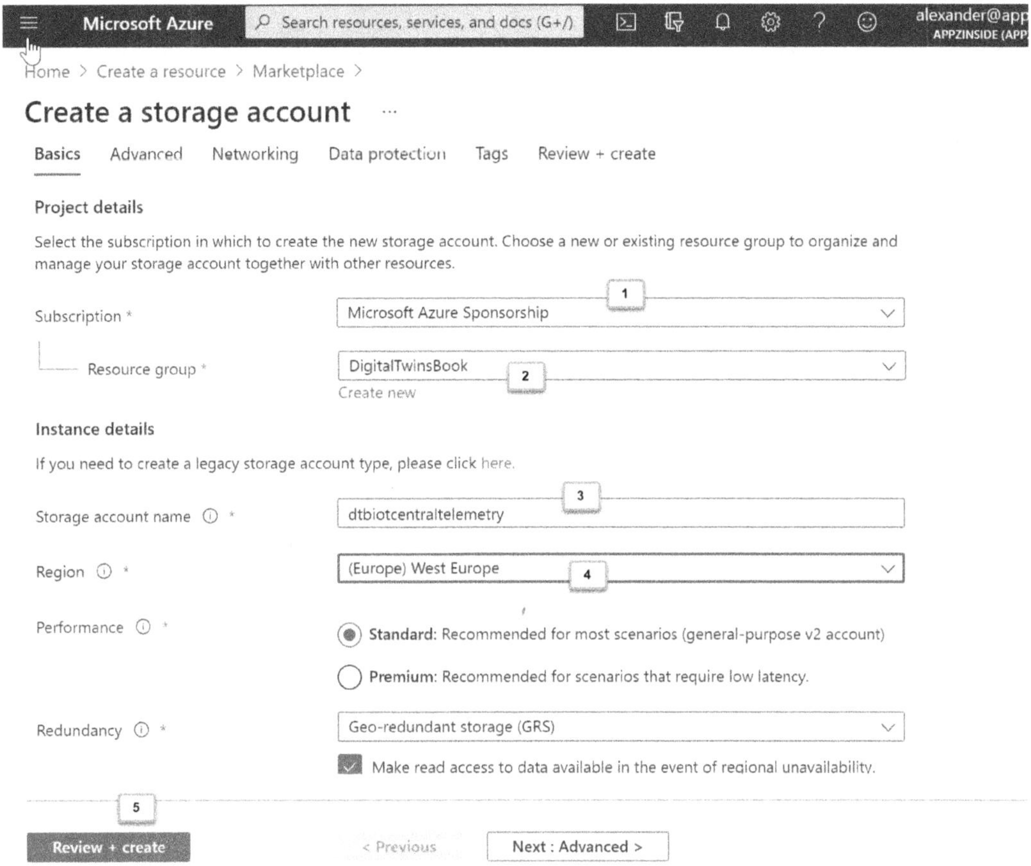

Figure 11.3 – Create a storage account

Click on the **Create** button, as shown in *Figure 11.4*, to create the storage account:

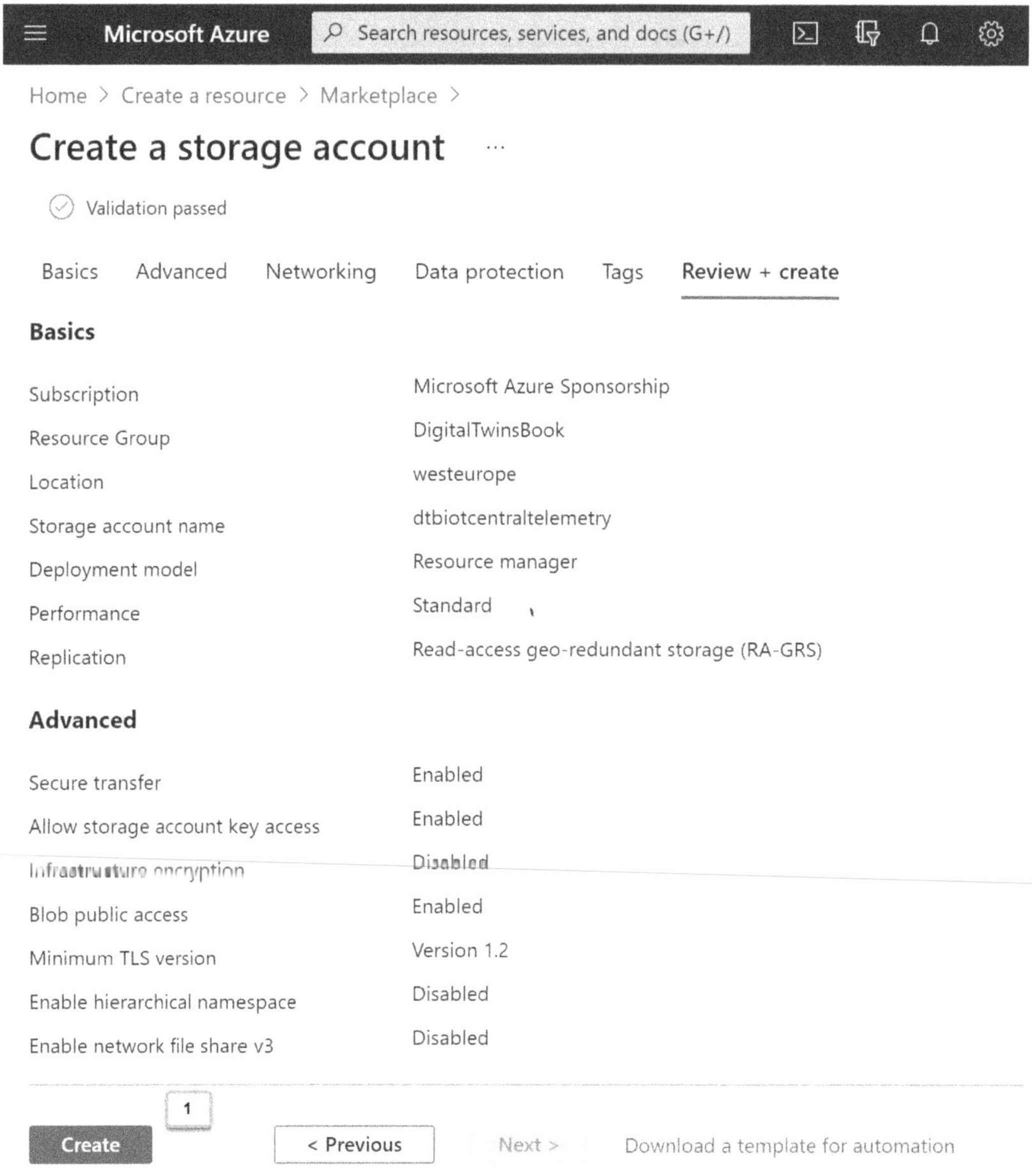

Figure 11.4 – Create a storage account

In this section, we have created a storage account. In the next part of this section, we will create an Azure function.

Creating an Azure function

An Azure function can be created by using Visual Studio. It is like creating a .NET web application that runs in the cloud based on a trigger. Start Visual Studio. Select **Create a new project**, as shown in *Figure 11.5*:

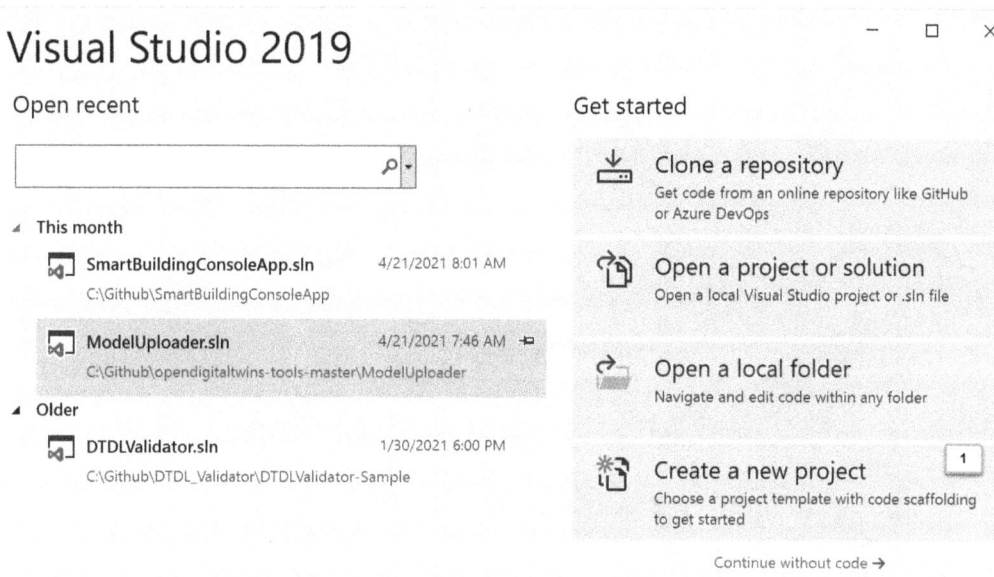

Figure 11.5 – Getting started with Visual Studio 2019

Execute the following steps, as shown in *Figure 11.6*:

1. Enter *Azure Function* in the search field.
2. Select the **Azure Functions** template.
3. Click on the **Next** button:

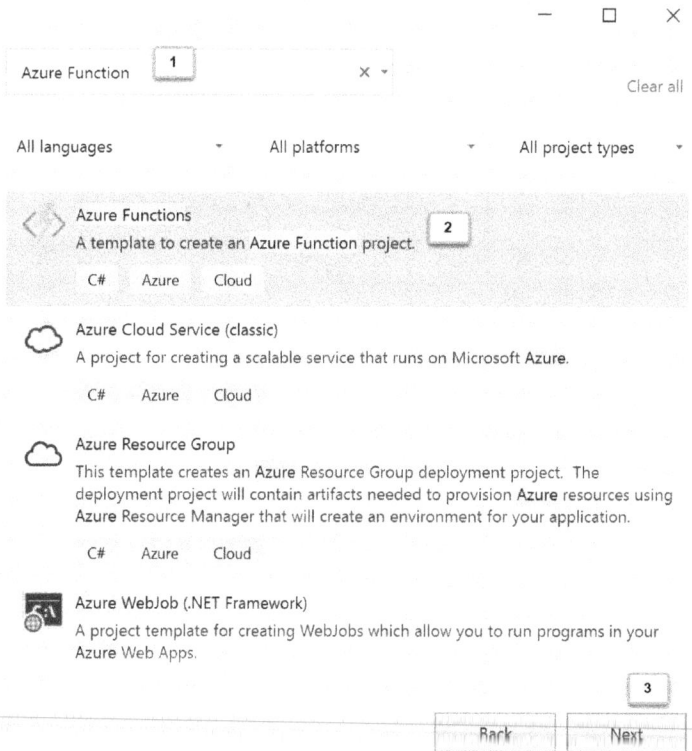

Figure 11.6 – Select Azure Functions as the template for our project

Execute the following steps, as shown in *Figure 11.7*:

1. Enter the project name `SmartBuildingSensorUpdater`.
2. Set the location to `c:\Github`.
3. Click on the **Create** button:

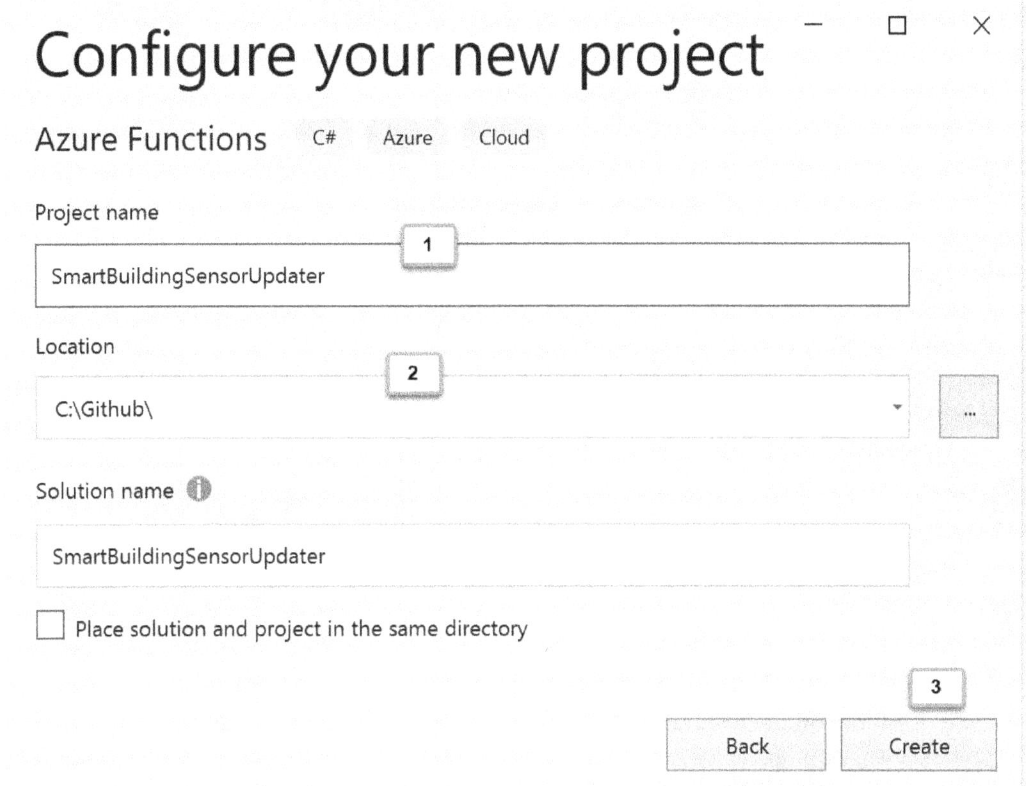

Figure 11.7 – Enter the project and location name for our project

Execute the following steps, as shown in *Figure 11.8*:

1. Select **Service Bus Queue trigger**.

2. Select a storage account. Choose **browse** from the dropdown to find and select our previously created storage account:

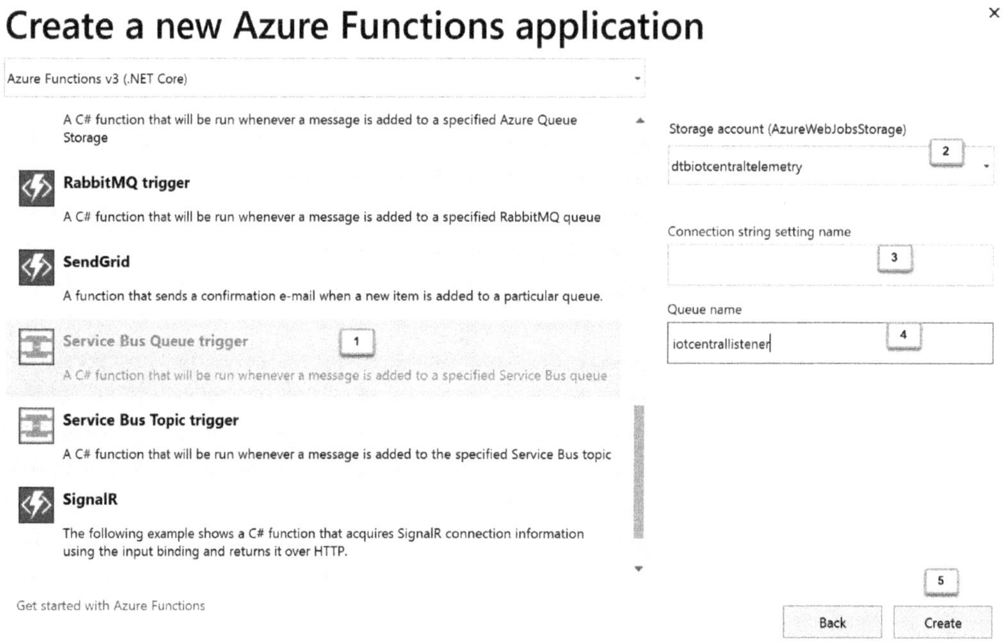

Create a new Azure Functions application   ✕

Azure Functions v3 (.NET Core)   ▾

A C# function that will be run whenever a message is added to a specified Azure Queue Storage

RabbitMQ trigger

A C# function that will be run whenever a message is added to a specified RabbitMQ queue

SendGrid

A function that sends a confirmation e-mail when a new item is added to a particular queue.

Service Bus Queue trigger   [1]

A C# function that will be run whenever a message is added to a specified Service Bus queue

Service Bus Topic trigger

A C# function that will be run whenever a message is added to the specified Service Bus topic

SignalR

The following example shows a C# function that acquires SignalR connection information using the input binding and returns it over HTTP.

Storage account (AzureWebJobsStorage)

[2]

dtbiotcentraltelemetry   ▾

Connection string setting name

[3]

Queue name

iotcentrallistener   [4]

Get started with Azure Functions

[5]

Back   Create

Figure 11.8 – Select the type of trigger we want to use for our Azure function

Before we finish the other steps, we need to select the storage account. The dialog shown in *Figure 11.9* will pop up. Make sure that you are logged in with the same credentials as you have been using before. Execute the following steps, as shown in *Figure 11.9*:

3. Select the subscription.

4. Select the `dtbiotcentraltelemetry` storage account.

5. Click on the **OK** button to select the storage account:

Azure Storage
Select existing or create new

AppzInside
alexander@appzinside.com

✕

⚠ Reenter your credentials

Subscription

[1]

Microsoft Azure Sponsorship

Storage accounts   ✛ ↻

Name	Resource group	Location
iotcentraltelemetrytoadt	cloud-shell-storage-westeurope	East US
cityinsightsmodel	CityInsights	West Europe
csb10030000955aeb44	cloud-shell-storage-westeurope	West Europe
cyclomediamodels	Cyclomedia	West Europe
dtbiotcentraltelemetry [2]	DigitalTwinsBook	West Europe
mixedrealityresourcegrou	MixedRealityResourceGroup	West Europe
mixedrealityusergroupres	MixedRealityUserGroupResources	West Europe
remoterenderingmodel	RemoteRenderingGroup	West Europe

[3]

OK   Cancel

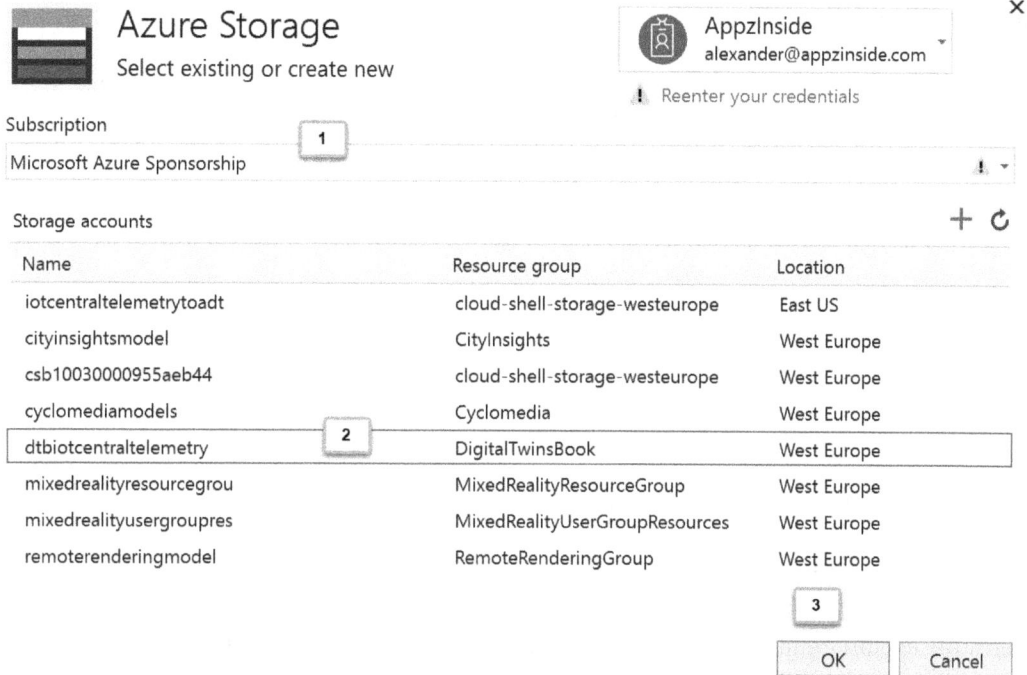

Figure 11.9 – Select the storage account for the Azure function

We need to continue with the steps as shown in *Figure 11.8*:

1. Keep the connection string empty. We don't want to have that embedded in code. We will add this later in the local configuration file and in the application configuration when the Azure function is deployed to Azure.

2. Enter the queue name `iotcentrallistener`.

3. Click on the **Create** button.

The Azure Functions project is generated, and the project is opened in Visual Studio. We start with renaming the function class. By default, this class is named `Function1`. Rename the `Function1.cs` class to `IoTCentralTrigger.cs` and make sure that the class name is also replaced by `IoTCentralTrigger`. The result is shown in *Figure 11.10*:

Figure 11.10 – Rename the Function1 class to IoTCentralTrigger

We need to use `DigitalTwinsManager` from the previous console application project. Open a file explorer and copy the `DigitalTwins` folder and its contents from `c:\github\SmartBuildingConsoleApp\SmartBuildingConsoleApp\` to `c:\github\SmartBuildingSensorUpdater\SmartBuildingSensorUpdater\`.

We need to add the `Azure.DigitalTwins.Core` and `Azure.Identity` Nuget packages to this project. This can be done in the same way as we did with the console application.

The `DigitalTwinsManager` class contains a `Connect` method, which is used to connect to our Azure Digital Twins instance using the logged-on user account. But an Azure function runs as a service inside Azure. It does not contain a logged-on user. We need an additional way of authenticating against the Azure Digital Twins service using an application ID.

Add the following method to the `DigitalTwinsManager` class:

```
public void ManagedConnect(string appId)
{
    HttpClient httpClient = new HttpClient();

    var cred = new ManagedIdentityCredential(appId);
```

```
    client = new DigitalTwinsClient(new Uri(adtInstanceUrl),
cred, new DigitalTwinsClientOptions { Transport = new
HttpClientTransport(httpClient) });
}
```

This method will use a managed identity based on an application ID to build up the credentials needed to authenticate against the Azure Digital Twins service. Add the following method, an additional constructor, to the `DigitalTwinsManager` class:

```
public DigitalTwinsManager(string appId)
{

    ManagedConnect(appId);

}
```

The next step is calling `DigitalTwinsManager` to update a property on a digital twin. We need to make sure that all the right namespaces are included. Replace the existing namespaces at the top of the `IoTCentralTrigger.cs` file with the following namespaces:

```
using System.Text;
using Microsoft.Azure.ServiceBus;
using Microsoft.Azure.WebJobs;
using Microsoft.Extensions.Logging;
using Newtonsoft.Json;
using Newtonsoft.Json.Linq;
using SmartBuildingConsoleApp.DigitalTwins;
```

Replace the contents of the `IoTCentralTrigger` class with the following code:

```
const string adtAppId = "https://digitaltwins.azure.net";
const string queueName = "iotcentral";


[FunctionName("IoTCentralTrigger")]
public static void Run([ServiceBusTrigger(queueName,
Connection = "ServiceBusConnection")] Message message, ILogger
log)
{
```

```
    string sensorId = message.UserProperties["iotcentral-
device-id"].ToString();
    string value = Encoding.ASCII.GetString(message.Body, 0,
message.Body.Length);
    var bodyProperty = (JObject)JsonConvert.
DeserializeObject(value);
    JToken temperatureToken = bodyProperty["telemetry"]
["temperature"];
    float temperature = temperatureToken.Value<float>();
    log.LogInformation(string.Format("Sensor Id:{0}",
sensorId));
    log.LogInformation(string.Format("Sensor Temperature:{0}",
temperature));

    DigitalTwinsManager manager = new
DigitalTwinsManager(adtAppId);
    manager.UpdateDigitalTwinProperty(sensorId, "temperature",
temperature);
}
```

The code retrieves the `iotcentral-device-id` property from the *Custom* properties of the message. This contains the ID of the sensor. We also retrieve the message body containing the telemetry and get the temperature from that. We log the information of the found sensor ID and temperature.

We set up a connection to the Azure Digital Twins instance using the `DigitalTwinsManager` class using the application ID set to `iotcentral`. And finally, we update the `temperature` property of the digital twin, which has the same ID as the sensor.

We have created an Azure Functions project and updated it with the code to access the Azure Digital Twins instance. The next part will explain how to set the connection string.

Setting the connection string

We need to set the connection string from the notepad in the `local.settings.json` file. This connection string is used when we run the Azure function in the debugger locally. Our connection string will look somewhat like the following snippet:

```
Endpoint=sb://dtbservicebus.servicebus.windows.net/;Shared
AccessKeyName=iotcentrallistener;SharedAccessKey=nGNDfN6xgtugf
JSrozoa+YhfsZDAM+Too7QtR/9mYPC40=;EntityPath=iotcentral
```

We need to remove the `;EntityPath=iotcentral` part of this connection string before we can use it in our application. Not doing so will result in an error:

```
Microsoft.Azure.ServiceBus: NamespaceConnectionString should
not contain EntityPath.
```

Add the connection string to the `local.settings.json` file as follows:

```
"ServiceBusConnection": "Endpoint=sb://dtbservicebus.
servicebus.windows.net/;SharedAccessKeyName=iotcentrallistener;
SharedAccessKey=nGNDfN6xgtugfJSrozoa+YhfsZDAM+Too7QtR/9mYPC40="
```

We need to add the same connection string as an application configuration in the Azure function after it is deployed to Azure. The `local.settings.json` file is only used locally.

Creating an Azure function placeholder

When an Azure function is being deployed to Azure, it needs an existing Azure function in Azure, or one needs to be created during the deployment process. In this example, we will create the Azure function up front as a placeholder for our Azure function project. This step is only done once. Go to the Azure portal at `https://portal.azure.com` and log in.

Execute the following steps, as shown in *Figure 11.11*:

1. Click on the hamburger icon in the top-left corner and select + **Create a resource**.
2. Type in the search field *Function App*.
3. Click on the **Create** button in the **Function App** box:

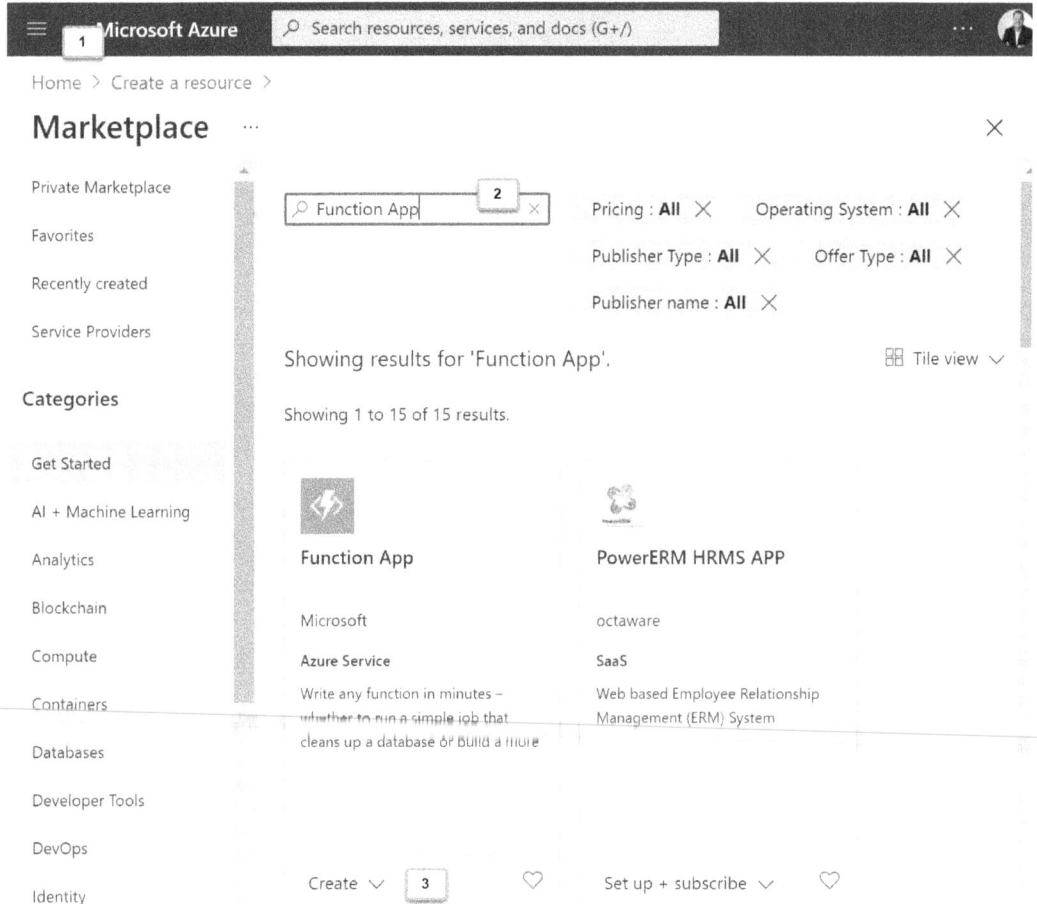

Figure 11.11 – Search for the Azure function in the marketplace

A dialog opens to create an Azure function app. Execute the following steps, as shown in *Figure 11.12*:

1. Select the subscription.
2. Select the **DigitalTwinsBook** resource group.

3. Enter the name *IoTCentralTrigger*. Select another unique name if this name is already taken.

4. Select **Code** and make sure that the runtime stack is set to **.NET**. The version is automatically selected and does not have to be changed.

5. Select the region. Choose the one that's closest to you.

6. Click on the **Review + create** button:

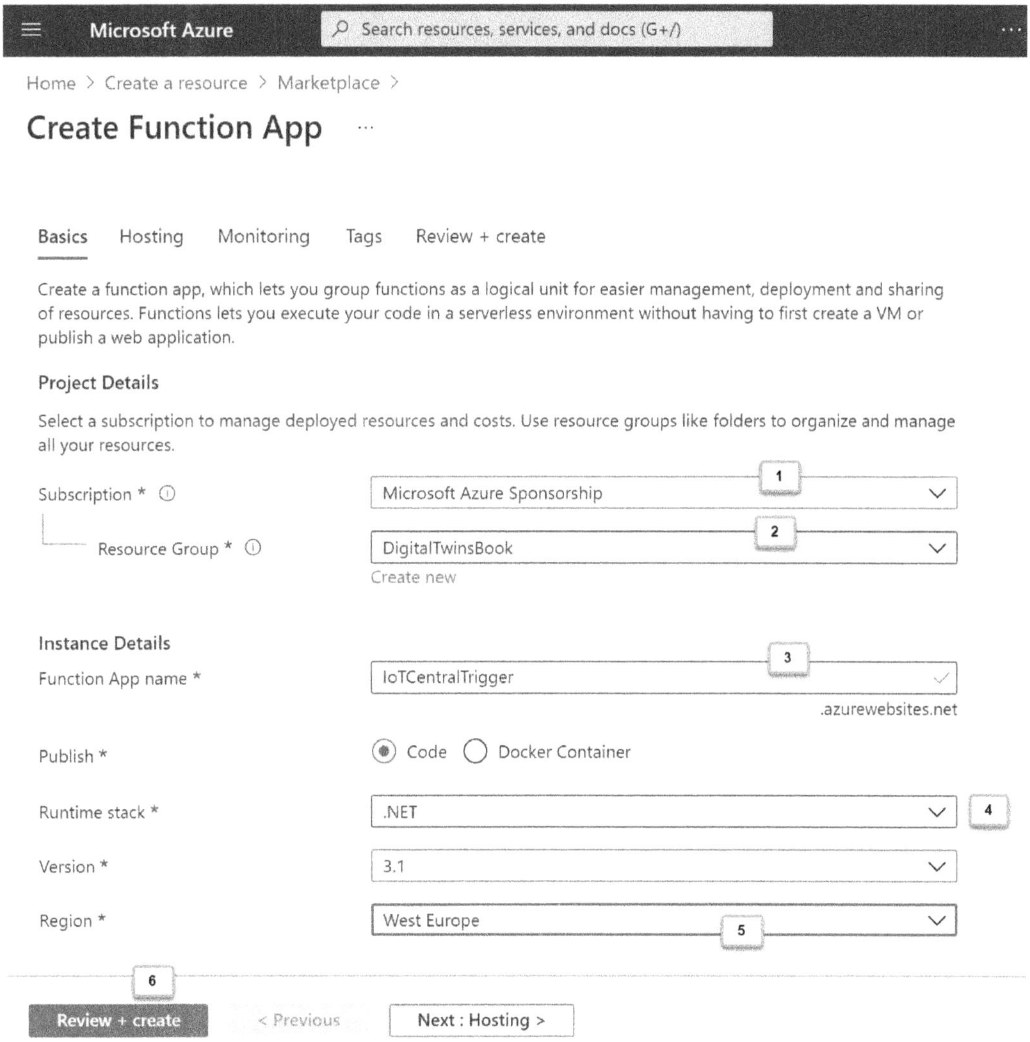

Figure 11.12 – Create an Azure function app in Azure

Click on the **Create** button, as shown in *Figure 11.13*, to create the Azure function in Azure:

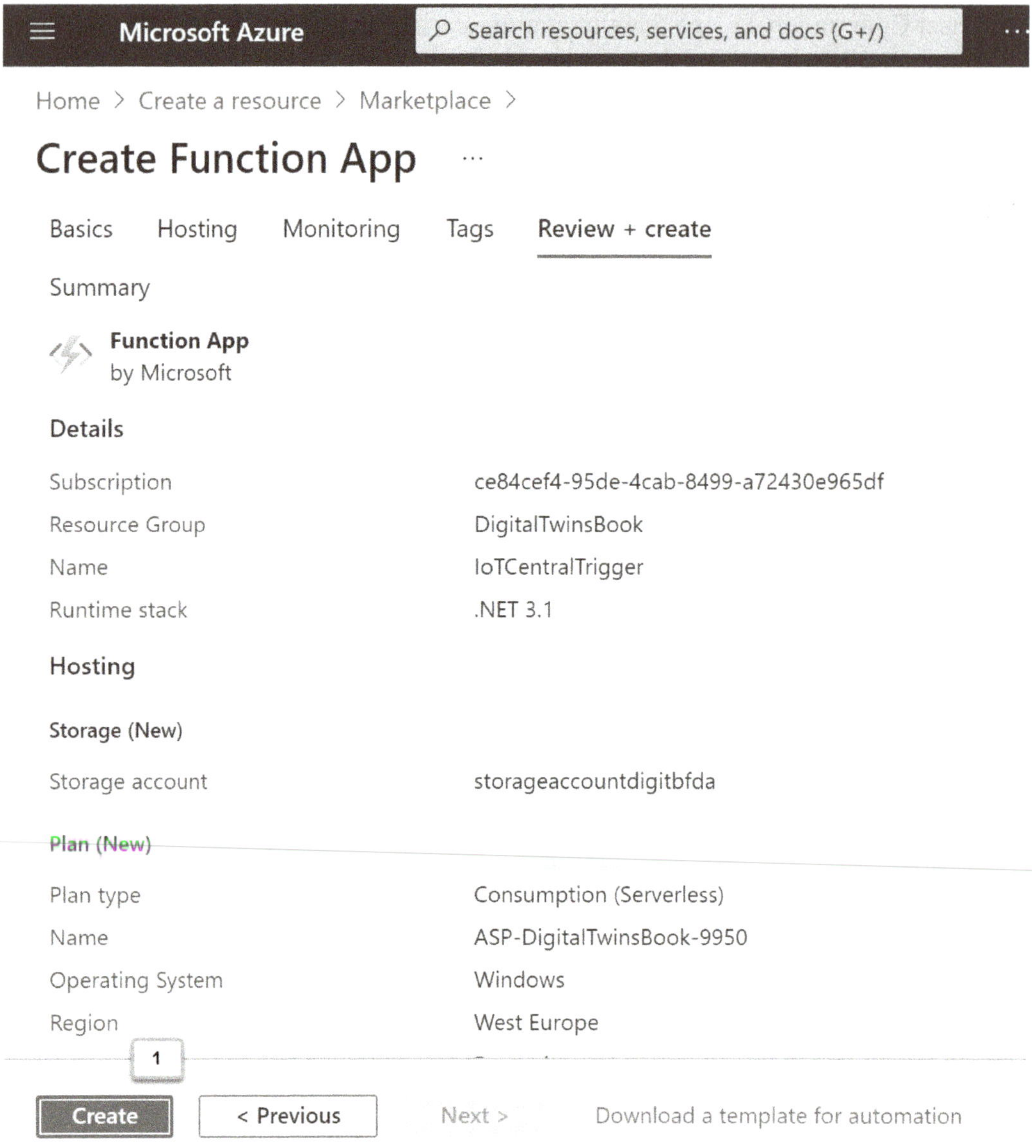

☰   **Microsoft Azure**   🔍 Search resources, services, and docs (G+/)   ⋯

Home > Create a resource > Marketplace >

Create Function App   ⋯

Basics   Hosting   Monitoring   Tags   **Review + create**

Summary

Function App
by Microsoft

Details

Subscription   ce84cef4-95de-4cab-8499-a72430e965df

Resource Group   DigitalTwinsBook

Name   IoTCentralTrigger

Runtime stack   .NET 3.1

Hosting

Storage (New)

Storage account   storageaccountdigitbfda

Plan (New)

Plan type   Consumption (Serverless)

Name   ASP-DigitalTwinsBook-9950

Operating System   Windows

Region   West Europe

1

Create   < Previous   Next >   Download a template for automation

Figure 11.13 – Create the Azure function

The deployment of the Azure function can take some time. When the Azure function is deployed, you can click on the **Go to resource** button to open the resource.

Granting the Azure function permissions

We need to grant the Azure function permission to access the Azure Digital Twins service. This step is executed using two commands in Windows PowerShell. Open PowerShell. Use the following command to log in with your credentials:

```
az login
```

We need to assign an identity to the Azure function. To do that, we need to call the following function:

```
az functionapp identity assign -g "<resource group name>" -n
"<azure function name>"
```

The name of the resource group and the name of the function can be found on the overview page of the function app in the Azure portal. The command will look like this:

```
az functionapp identity assign -g "DigitalTwinsBook" -n
"IoTCentralTrigger"
```

Assigning an identity will generate a principal ID, which is shown in *Figure 11.14*:

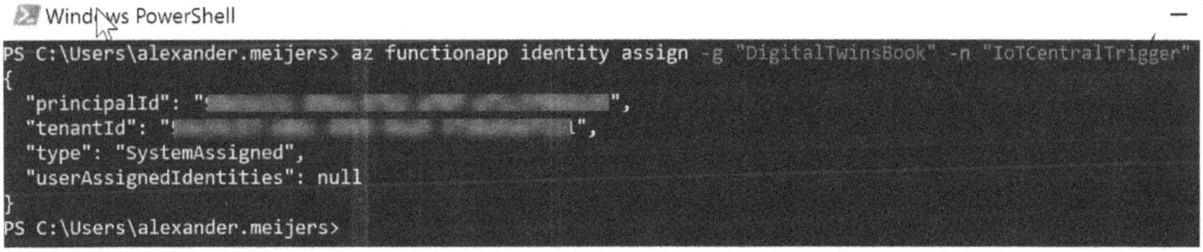

Figure 11.14 – Assign an identity to the Azure function

Execute the following command with `principalId`:

```
az dt role-assignment create --dt-name "DTBDigitalTwins"
--assignee "<principalId>" --role "Azure Digital Twins Data
Owner"
```

The output will look like this:

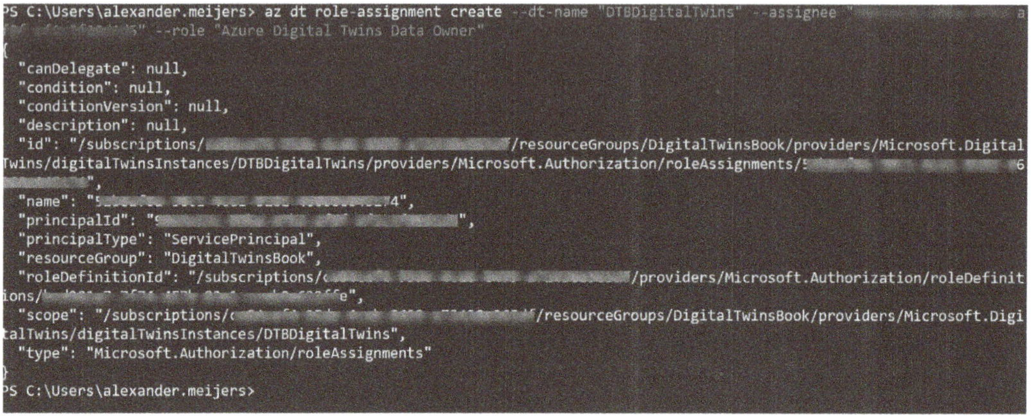

Figure 11.15 – Assign the Digital Twins Data Owner role to the identity

This identity and assignment need to be done only once. In the next section, we will deploy the Azure function from Visual Studio.

Publishing the Azure function

In this step, the Azure function project is published to the Azure function location in Azure. Right-click on the SmartBuildingSensorUpdater project. Select the **Publish…** menu option, as shown in *Figure 11.16*:

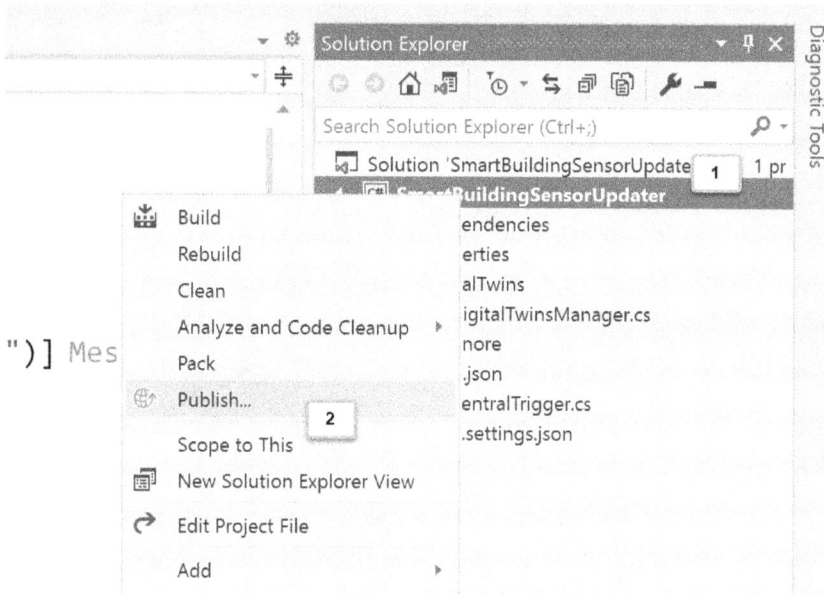

Figure 11.16 – Publishing the Azure function

Execute the following steps, as shown in *Figure 11.17*:

1.   Select **Azure** as the publishing target.

2.   Click on the **Next** button.

Publish   ×

Where are you publishing today?

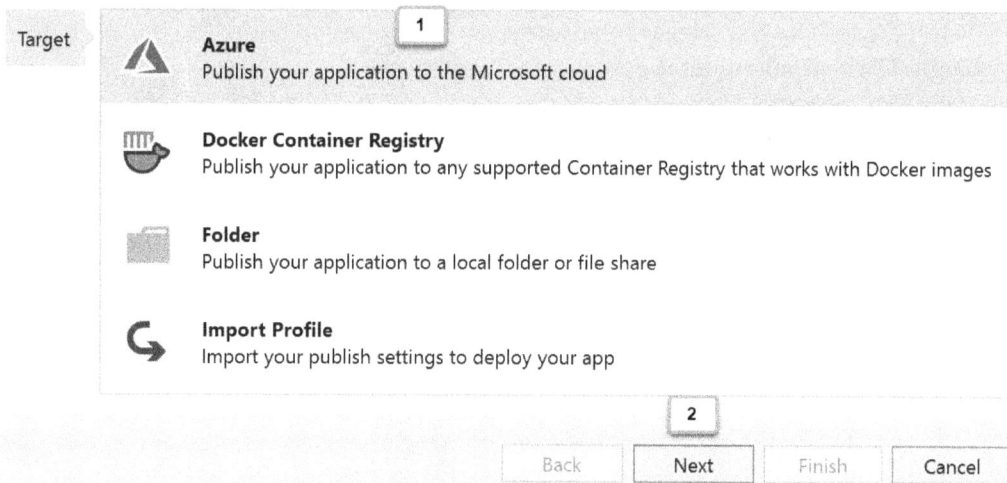

Figure 11.17 – Select the publishing target

Execute the following steps, as shown in *Figure 11.18*:

1. Select **Azure function App (Windows)** as the specific target.

2. Click on the **Next** button.

Publish   ✕

Which Azure service would you like to use to host your application?

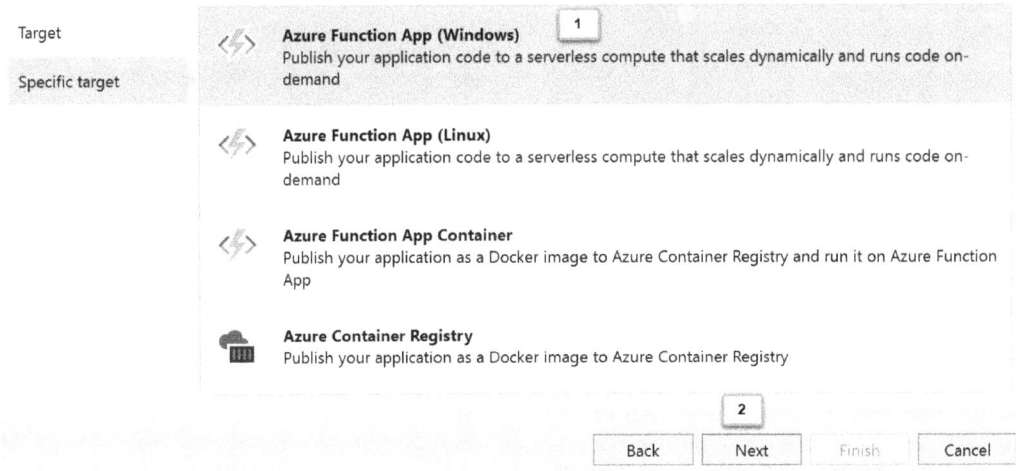

Target		Azure Function App (Windows) [1]
		Publish your application code to a serverless compute that scales dynamically and runs code on-demand
Specific target		

Azure Function App (Linux)
Publish your application code to a serverless compute that scales dynamically and runs code on-demand

Azure Function App Container
Publish your application as a Docker image to Azure Container Registry and run it on Azure Function App

Azure Container Registry
Publish your application as a Docker image to Azure Container Registry

[2]

Back   Next   Finish   Cancel

Figure 11.18 – Select the target type for publishing the Azure function

Execute the following steps, as shown in *Figure 11.19*:

1. Select the subscription.

2. Select the **IoTCentralTrigger (Consumption)** function app under the **DigitalTwinsBook** resource group in the hierarchy.

3.   Click on the **Finish** button.

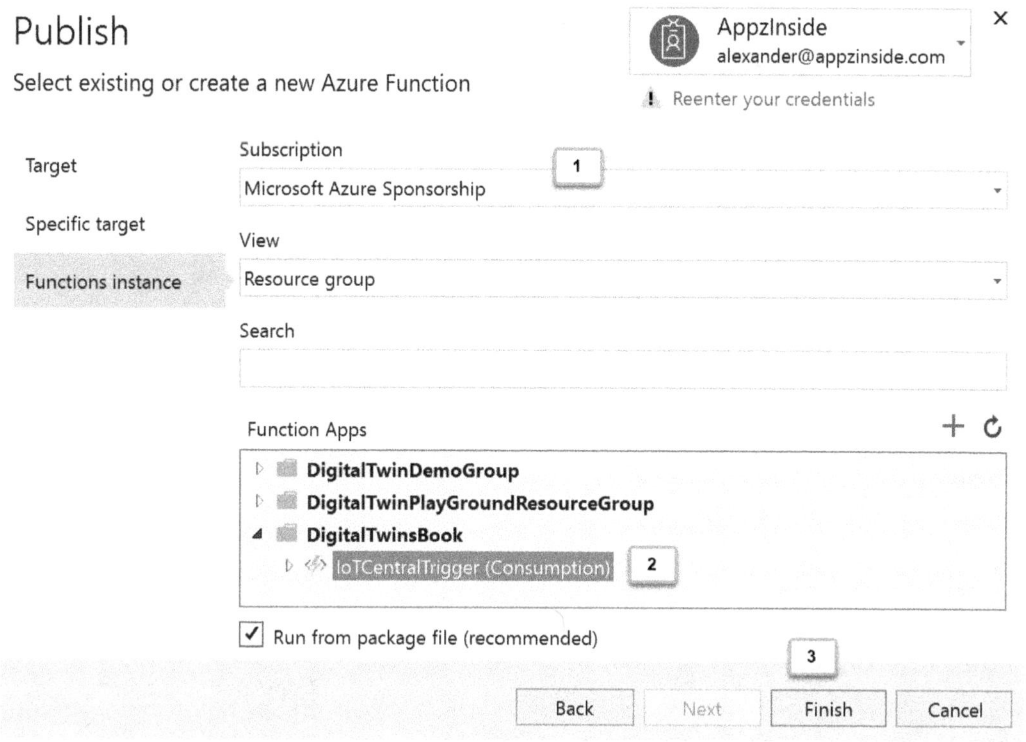

Figure 11.19 – Specify the function instance for publishing the Azure function

Click on the **Publish** button to start the deployment. The deployment will take some time before it is finished. There will be several warnings in the **Service Dependencies** section, which you can ignore. The messages are mainly caused by not being logged in properly with some of the credentials:

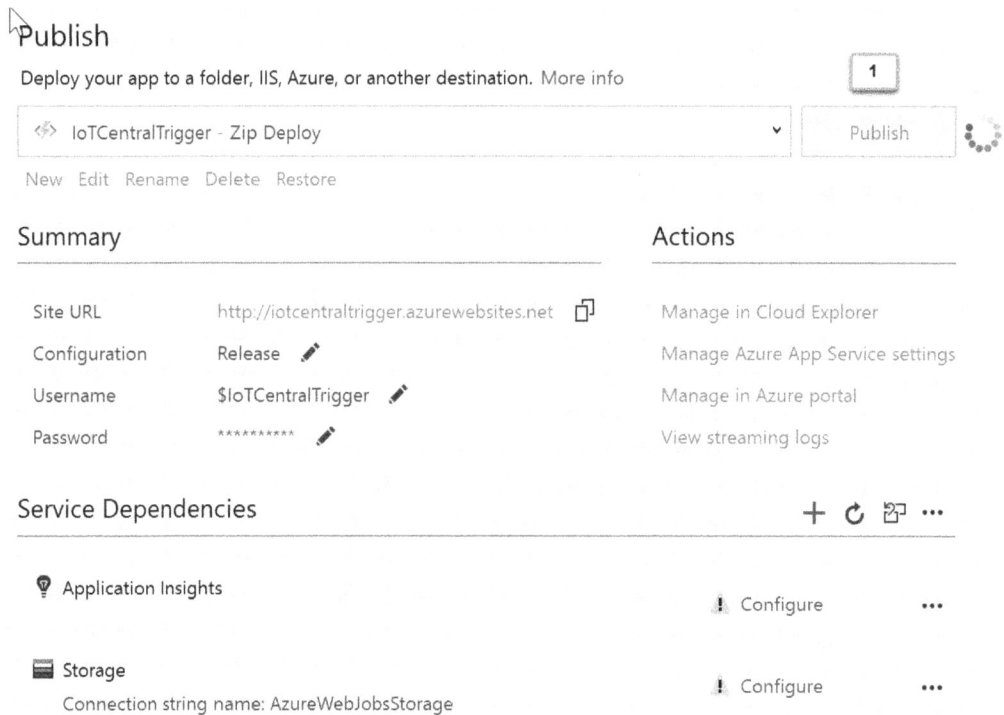

Publish

Deploy your app to a folder, IIS, Azure, or another destination. More info   [1]

</>   IoTCentralTrigger - Zip Deploy   ∨   Publish

New Edit Rename Delete Restore

Summary   Actions

Site URL   http://iotcentraltrigger.azurewebsites.net 🗐   Manage in Cloud Explorer

Configuration   Release 🖉   Manage Azure App Service settings

Username $IoTCentralTrigger 🖉   Manage in Azure portal

Password   ********** 🖉   View streaming logs

Service Dependencies   + ⟳ 🗗 ⋯

💡 Application Insights   ⚠ Configure   ⋯

🖿 Storage   ⚠ Configure   ⋯
     Connection string name: AzureWebJobsStorage

Figure 11.20 – Publish the Azure function

We have published the Azure function in Azure. In the next part, we need to set the connection string in the application settings of the Azure function.

Setting the connection string

The connecting string defined in the `local.settings.json` file is only used for running the Azure function locally in Visual Studio. We need to add the same connection string as an application setting in the published Azure function. This can be done by adding an additional settings file for production. But we want to show how to set an application setting via the Azure portal.

Go to the Azure portal via `https://portal.azure.com` and log in. Open the **IoTCentralTrigger** Azure function resource. Execute the following steps, as shown in *Figure 11.21*:

1. Select **Configuration** from the left menu.

2. Click on the **+ New application setting** button.

3. Fill in the name *ServiceBusConnection*.

4. Fill in the connection string that we had stored in a notepad. Make sure that the last part, `;EntityPath=iotcentral`, has been removed.

5. Click on the **OK** button.

6. Click on the **Save** button:

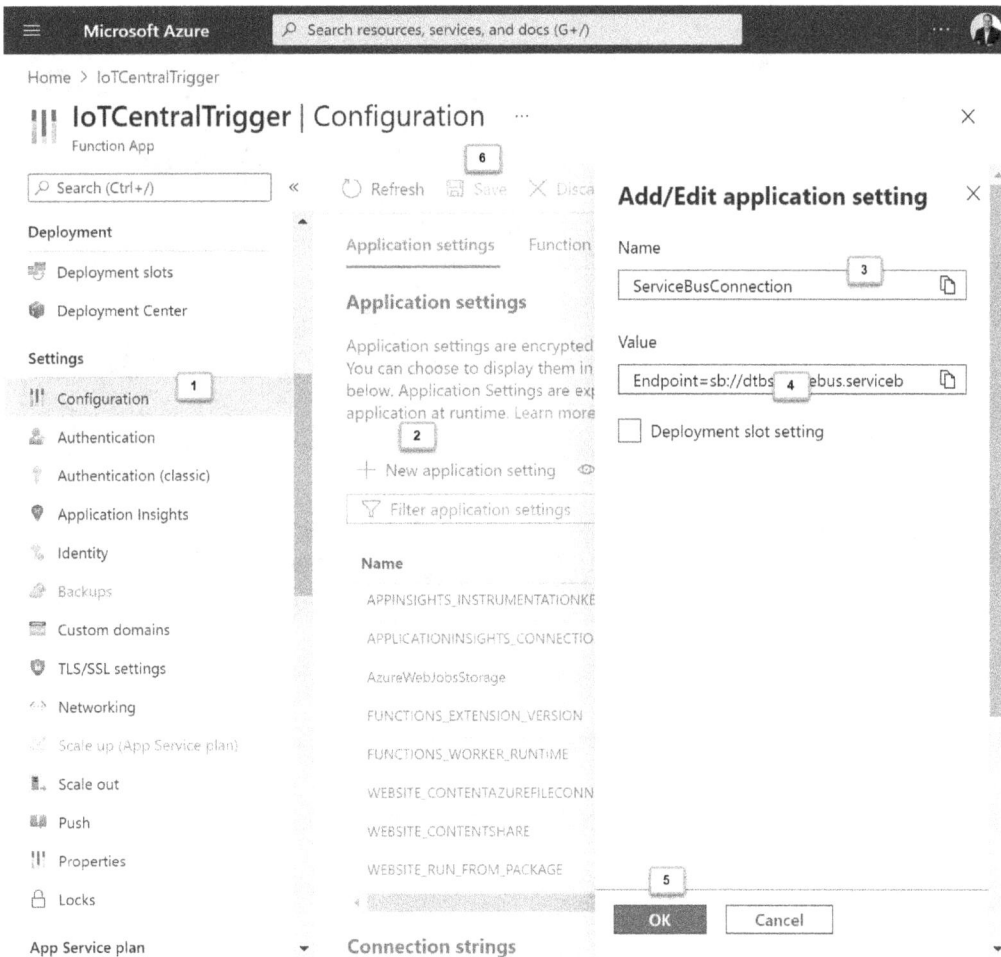

Figure 11.21 – Add an application setting for the Azure function

We have added *ServiceBusConnection* to the application settings. In the next part, we create a digital twin based on the sensor model.

Creating a digital twin for the sensor

We need to create a digital twin for the sensor model to receive the temperature updates from the sensor. Open the Azure Digital Twins Explorer. Execute the following steps, as shown in *Figure 11.22*:

1. Press the + symbol next to the **Sensor** model in the **Model View**. A **New Twin Name** dialog will appear.

2. Fill in the ID of the sensor. This should be the same ID as we get in our messages.

3. Click on the **Save** button:

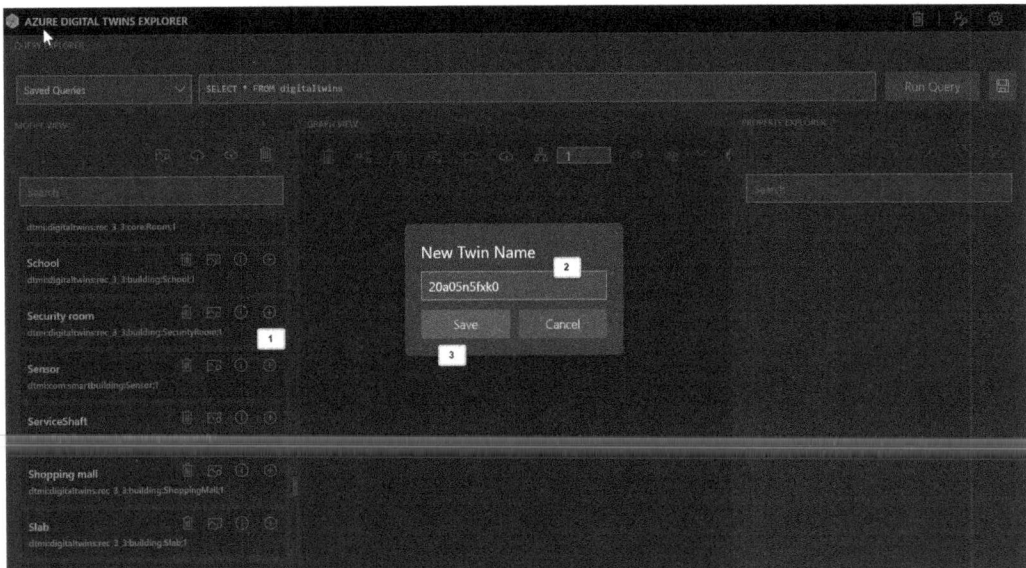

Figure 11.22 – Create a new digital twin for the sensor model

We have created a digital twin with the exact same name as the actual sensor. In the last part of this section, we will view the result of the pipeline.

Viewing the result

There are several ways of viewing the result of this pipeline, which transports the temperature value from a sensor to our digital twin. One of them is looking at the log stream of the Azure function trigger. Open the Azure portal and go to the **IoTCentralTrigger** Azure function. Execute the following step, as shown in *Figure 11.23*.

1. Select **Log stream** in the left menu.

    It will take some time before the log stream shows any information. But at some point, it starts showing the messages that are handled by the Azure function:

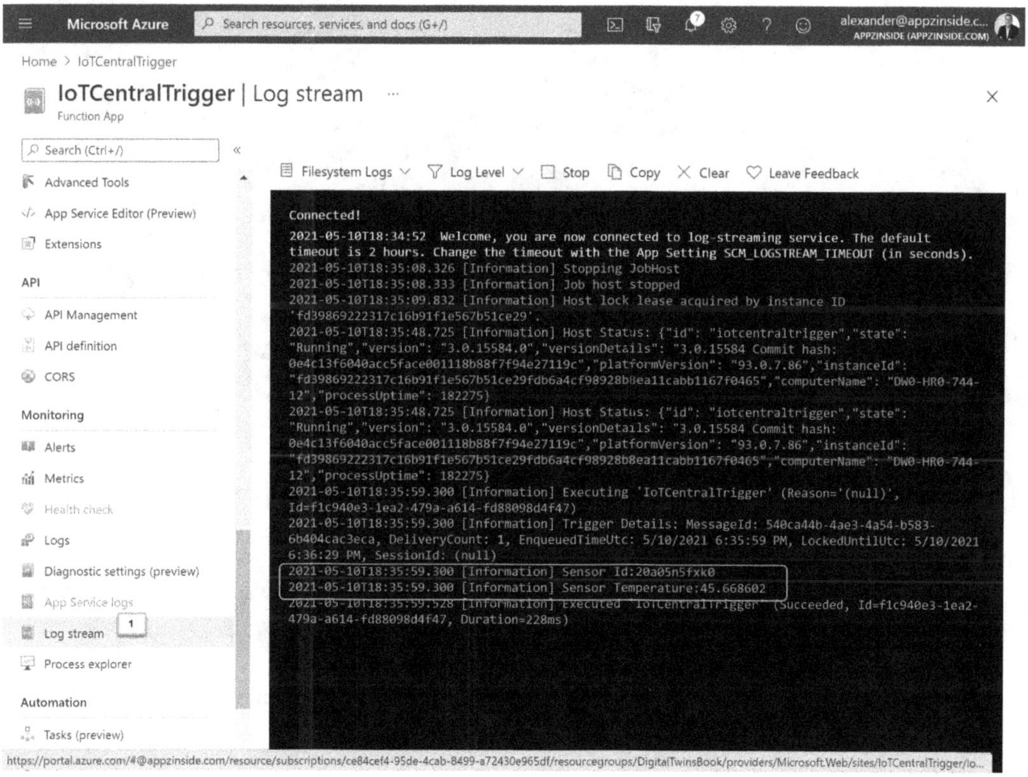

Figure 11.23 – View the logs of the Azure function trigger

Another way is opening the Azure Digital Twins Explorer and viewing the digital twin. Execute the following steps, as shown in *Figure 11.24*:

1. Click on the **Run Query** button to refresh the **Graph View**.

2. Select the digital twin with the sensor ID.

3. View the current set temperature property:

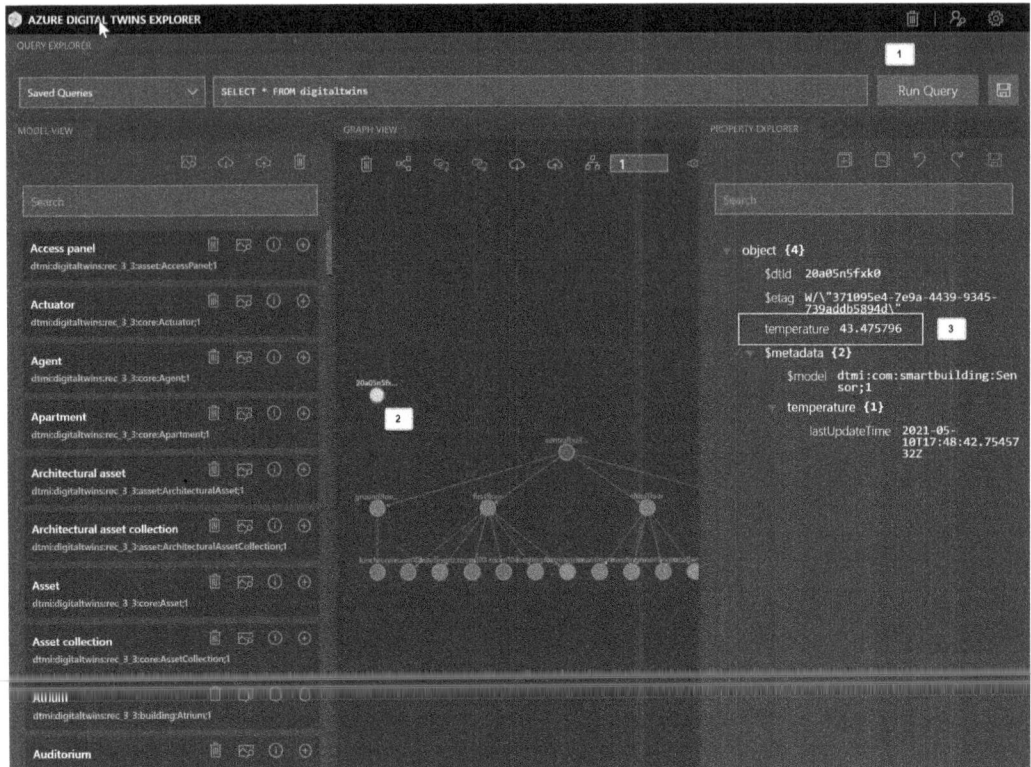

Figure 11.24 – View the result of the digital twin getting an updated temperature value

You will need to reload the digital twins in the **Graph View** and reselect the digital twin to see the difference in temperature after an update.

We have created a pipeline in Azure using several Azure components to move telemetry data from a demo sensor into a digital twin in Azure Digital Twins. The digital twin is related to the demo sensor by using the same name as the demo sensor.

Summary

We have learned how to set up a pipeline that can be used in a production scenario using several Azure services to get telemetry data from an IoT device into a digital twin using IoT Central. Each of the services has been explained and configured in a secure and scalable way. You will now be able to build your own pipeline to connect demo or real-life sensors to several digital twins inside an Azure Digital Twins instance.

In the next chapter, we will have a more detailed look at the available integrations with Azure services, which will allow you to send digital twin data to connected endpoints.

Questions

As we conclude, here is a list of questions for you to test your knowledge regarding this chapter's material. You will find the answers in the *Assessments* section of the *Appendix*:

1. Which menu option can we use to monitor the debug messages of the running Azure function?

    a. Alerts

    b. Logstream

    c. Metrics

2. Which file contains the configuration settings that are used locally?

    a. `local.settings.json`

    b. `settings.json`

    c. `local.json`

3. Which part of the connection string needs to be removed for it to work in the Azure function?

    a. `SharedAccessKeyName`

    b. `SharedAccessKey`

    c. `EntityPath`

Further reading

To learn more on the subject, check out the following links:

- Write client app authentication code: `https://docs.microsoft.com/en-us/azure/digital-twins/how-to-authenticate-client`

- Azure Functions documentation: `https://docs.microsoft.com/en-us/azure/azure-functions/`

12
Event Routing

Event routing allows you to send data to consumers outside of the service. Think about sending data between two different Azure digital twins or sending data to additional downstream services to storage or other processes to handle the data. In this chapter, you will learn how notifications are triggered in the Azure Digital Twins instance and how messages are routed through an endpoint to finally reach their destination in another Azure service.

In this chapter, we will cover the following topics:

- Data ingress and egress
- Event notifications
- Understanding event routes
- Creating an event grid topic
- Creating an endpoint
- Creating an event route
- Subscribing to event messages
- Monitoring event route messages

Technical requirements

We will continue in the .NET console application while using the .NET SDK to build up our Azure Digital Twins instances using models. Azure Digital Twins Explorer will be used to view the result of the .NET calls made by the console application.

Data ingress and egress

When working with integrations around Azure Digital Twins, the terms data ingress and data egress will often appear. But what do they mean and what is the difference? Let's take a look:

- **Data ingress**: This is about ingesting data into an Azure digital twin by receiving data and events from *upstream services*.

- **Data egress**: This is about routing data from an Azure digital twin to connected endpoints such as **Azure Event Hub**, **Azure Event Grid**, and **Azure Service Bus**. *Downstream services* will be connected to one of those endpoints to get the data for different purposes.

We will explain event routes and endpoints later in this chapter. Data ingress and data egress have a lot to do with upstream and downstream services. Let's explain upstream and downstream services.

Upstream services such as **Azure IoT Hub** and **Azure Logic Apps** allow us to receive data and update an Azure digital twin. To ingest data from any of these sources, it is common to use an **Azure Function**. A good example is the pipeline we built in the previous chapters, where data was received through an **Azure Service Bus** from a sensor defined within **IoT Central**. In that situation, we used an Azure Function to move the data from an Azure Service Bus to an Azure digital twin. We used IoT Central instead of IoT Hub since IoT Central allows us to create demo sensors, is easier to set up, and is highly scalable. Upstream services are often used to deliver telemetry and to generate notifications.

Downstream services such as **Azure Time Series Insights** and **Azure Maps** allow us to move data from an Azure digital twin to other services to store data, generate analytic information, display layers of data on a map, and more.

In this section, we have mentioned several Azure Services that are probably new to you. Let's explain them before continuing:

Azure Service	Description	Type
Azure Event Hub	This is a large streaming platform that can receive and process millions of events per second. The platform allows you to transform and store the data by using various storage and analytics providers and adapters. Azure Event Hub allows us to perform application logging, archive data, stream device telemetry, and much more. This platform can be used as an endpoint for data egress with Azure Digital Twins. More information can be found here: `https://docs.microsoft.com/en-us/azure/event-hubs/event-hubs-about.`	Endpoint
Azure Event Grid	This Azure service allows us to build event-based solutions. In principle, the service allows us to route event(s) from incoming connections to outcoming ones using filters. The service has built-in support for events coming for a large list of Azure services and supports event sources including Blob storage, Event Hub, Service Bus, machine learning, and more. The output goes to event handlers and webhooks such as Azure Functions, Service Bus, Logic Apps, and more. More information can be found here: `https://docs.microsoft.com/en-us/azure/event-grid/overview.`	Endpoint
Azure Service Bus	This service was used in our pipeline in the previous chapters. It is a fully managed message broker that contains one or more message queues. More information can be found here: `https://docs.microsoft.com/en-us/azure/service-bus-messaging/service-bus-messaging-overview.`	Endpoint

Azure Service	Description	Type
Azure IoT Hub	This service is a central message hub for bi-directional communication between an IoT solution and its attached sensors. The service is highly scalable and secure, allowing us to work with millions of devices at the same time. It also supports ways of monitoring the attached devices and reporting failures. IoT Hub is often used when building Azure digital twin solutions for the enterprise. More information can be found here: `https://docs.microsoft.com/en-us/azure/iot-hub/about-iot-hub.`	Upstream
Azure Logic Apps	This service allows us to create and run workflows within the cloud. Logic Apps can integrate with a variety of apps, data, and services. Workflows are created in a user-friendly interface. The interface allows you to use a large set of connectors to connect to over 100+ apps and services. Workflows can run through the initiation of another application or run in an automated or scheduled process. More information can be found here: `https://docs.microsoft.com/en-us/azure/logic-apps/logic-apps-overview.`	Upstream
Azure Maps	This service provides geographical context to web and mobile applications to create solutions that integrate location information. The service can be used to create internal and external maps. More information can be found here: `https://docs.microsoft.com/en-us/azure/azure-maps/about-azure-maps.`	Downstream
Azure Time Series Insights	This service provides a scalable IoT analytics platform. The platform allows you to manage, store, and visualize IoT data that is highly contextualized. It enables us to perform several types of analysis on data, including spotting trends and anomalies. More information can be found here: `https://docs.microsoft.com/en-us/azure/time-series-insights/overview-what-is-tsi.`	Downstream

More services can be seen as upstream and downstream. However, the preceding list contains the most important and well-known services that are used in combination with Azure Digital Twins. Some of these services will be discussed and shown in more detail later in this chapter.

So far, we've learned about the differences between data ingress and data egress. We've also learned about some of the upstream and downstream services that are used with Azure Digital Twins. Before we start learning about event routing in more depth, we need to learn about event notifications. These will be explained in the next section.

Event notifications

Event notifications are an important part of event routing with Azure Digital Twins. Azure digital twins produce notifications based on different events. These events are routed to different locations within the same Azure digital twin or outside the Azure digital twin to an endpoint. Event notifications describe the notifications for these events. The following events will generate notifications:

Event Type	Triggered On	Notification
Digital Twin change notification	• Property value change • Metadata change • Digital Twin change • Component change • Model change	Change
Digital Twin life cycle notification	• Digital Twin created • Digital Twin deleted	Life Cycle
Digital Twin relationship change notification	• Relationship created • Relationship updated • Relationship deleted	Relationship Change
Digital Twin telemetry message	• Connected device sends telemetry	Telemetry Message

Each notification is built up from a body and a header. The header contains several key-value pairs that differ based on the protocol being used. Notifications from Azure Digital Twins conform to the *CloudEvents* standard.

> **Understanding the CloudEvents Standard**
>
> *The CloudEvents* standard is a specification for describing event metadata in a standardized way. Their purpose is to simplify the standard model across services and platforms to provide seamless integration between the services of different platforms. *CloudEvents* is part of the CNCP Serverless workgroup. More information can be found on the following GitHub page: `https://github.com/cncf/wg-serverless`.

Each notification contains a sequence number specifying the order of the notifications that were produced by its source. The header will also contain additional extension attributes. These extension attributes are determined by the connected service. An example would be that when an event is sent to Azure Event Grid, additional attributes are added specifically for that service.

The body of the notification message contains detailed information about the notification itself. In most cases, the body is a JSON Payload. However, it could also be possible that it has a different format, depending on the usage of the message. In this book, we will only focus on the JSON payload. The structure of a JSON payload differs between the different types of notifications.

Let's explain notifications with the following example. We will react to the notification when a single property of an Azure digital twin is changed. Let's assume that we are updating the temperature of our digital twin called *Sensor*.

The digital twin, representing a sensor, can be updated in different ways. Our digital twin sensor, which is connected to our demo sensor in IoT Central, is automatically updated through the pipeline we built in the previous chapters. It could also be that you create a digital twin sensor that is not connected to an actual sensor yet. In that case, it is possible to change the value by using Azure Digital Twins Explorer, as shown in the following screenshot. Perform the following steps:

1. Press the **Run Query** button to update the Graph View.
2. Select the digital twin sensor that you want to change.
3. Change the temperature value.
4. Press the **Save** button to save the change.
5. A dialog window will appear containing the patch information regarding the temperature value change:

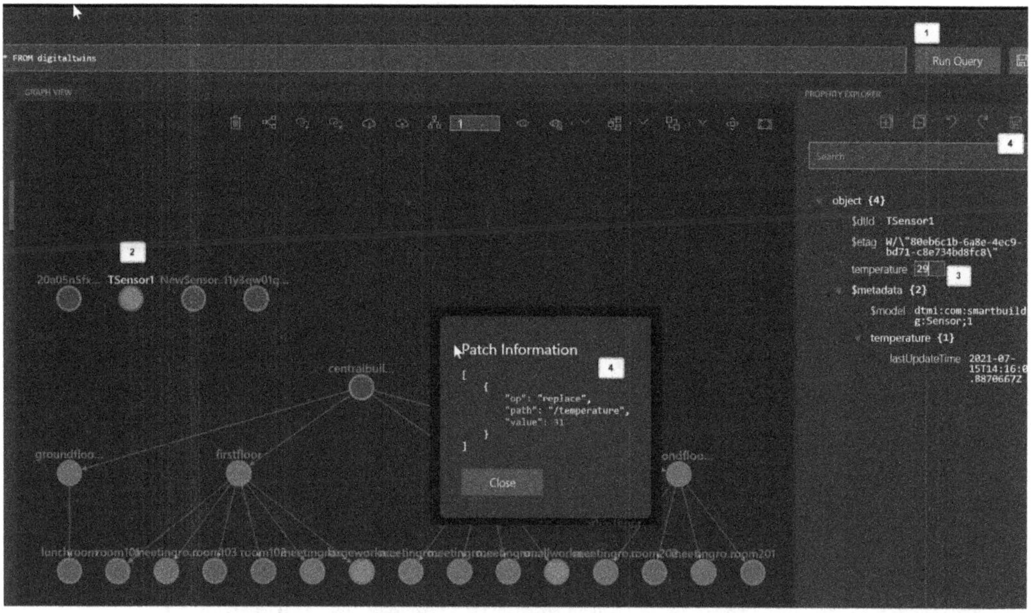

Figure 12.1 – Updating the value of a digital twin sensor

It is also possible to update the temperature value of the sensor using .NET code. This can be accomplished using the following method from the `DigitalTwinsManager` class:

```
public void UpdateDigitalTwinProperty(string twinId, string
property, object value)
```

When calling the method, it will generate a JSON payload containing the following:

```
[
    {
        "op": "replace",
        "path": "/temperature",
        "value": 21.3421
    }
]
```

This will change the temperature to `21.3421` on the sensor specified by its Digital Twin ID of `twinId`. The result of the body of the event notification will be as follows:

```
{
    "modelId": "dtmi:example:com:Sensor;1",
    "patch": [
```

```
    {
        "value": 21.3421,
        "path": "/Temperature",
        "op": "replace"
    }
    ]
}
```

With that, you have learned about which event notifications are available with Azure Digital Twins. We have walked through an example that shows the event notification body when a property of an Azure digital twin is updated. In the next section, we will look at event routes.

Understanding event routes

Event routes allow us to send data between two different Azure digital twins within the same Azure digital twin or to another service outside the Azure digital twin.

Routing an event within the Azure digital twin is done to update a specific Digital Twin based on changing another Digital Twin. Some example scenarios are shown here:

- Updating a digital twin that represents an *Alert* containing the maximum values of sensors. This *Alert* can be accessed to get a high-level overview of the maxima of the sensors in the system.

- Updating a parent digital twin object with an average. Let's assume you have a *Floor* containing several *Room* digital twins with temperature sensors. An average temperature is then stored at the *Floor* level.

Routing an event outside the Azure digital twin is done to send data to another service using one of the event endpoints. Some example scenarios are shown here:

- Storing temperature values in an Azure storage account. This could be done to save historical data in Blob Storag.

- Sending data from digital twins to Azure Maps to visualize overlays of data on a map.

- Routing notification messages using Azure Service Bus to other services. This gives us the ability to extend our current pipeline to different services, each using the data to perform additional business processes.

We can route events by defining endpoints and routes in the Azure Digital Twins service. The Azure Digital Twins service has an event stream and dispatch that redirects events coming from notifications to endpoints using event routes, as shown in the following diagram:

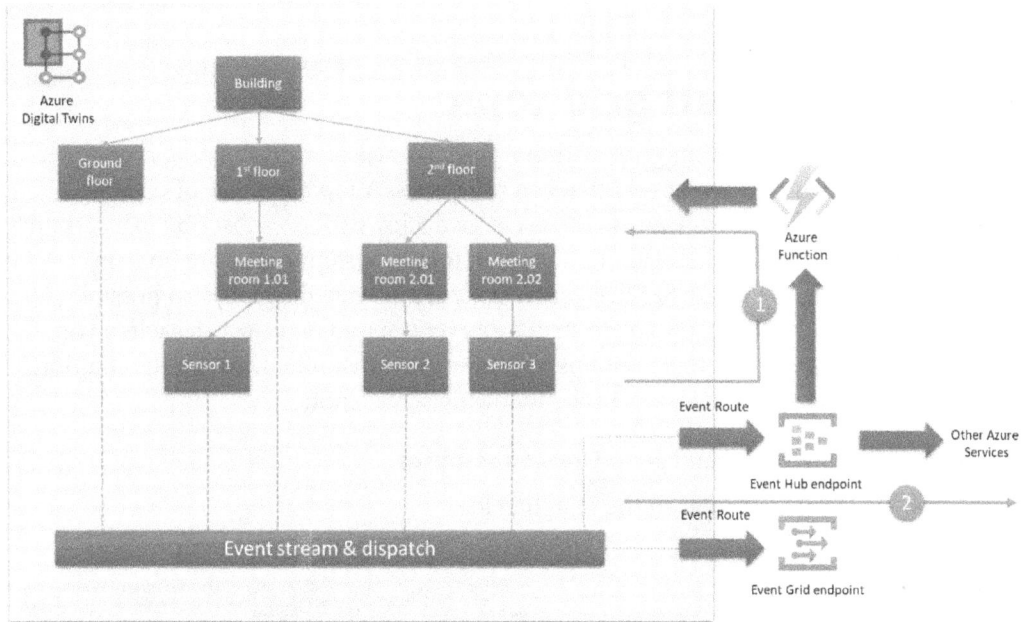

Figure 12.2 – An overview of the flow of events from notifications

Endpoints define where the event needs to be routed. Only **Event Grid**, **Event Hub**, and **Service Bus** are supported. An event route will route the event to one of the defined endpoints. Each event route can have a filter. The filter will be applied to each event and determine whether the event needs to be sent forward.

The route, indicated by number *1*, shows how an event is routed to another digital twin within the same Azure Digital Twins instance. The event is routed back through an **Event Hub** and an **Azure Function**.

The route, indicated by number *2*, shows how events are routed outside the Azure Digital Twins instance to other services through predefined endpoints. One of these endpoints is called the **Event Grid**. This Event Grid allows us to route events to one or more other Azure services.

In this section, you have learned how events are routed from an Azure Digital Twins instance to itself or other services.

In the next few sections of this chapter, we will be routing events to an Event Grid. The Event Grid has no way of showing the messages it receives. Therefore, we will set up a message queue in our existing Azure Service Bus that will receive the messages from the Event Grid. Azure Service Bus allows us to view these messages using the Service Bus Explorer.

Creating an event grid topic

There are different forms of event grids within Azure. We will start by creating an event grid topic service. This is an event grid that can use topics. A topic is a way of creating collections with similar types of events. The event grid allows you to subscribe to a certain topic to receive those events. There are two types of topics. System topics are built-in topics specifically used by Azure Services, such as Azure Service Bus. It is also possible to create custom topics. This allows us to structure events based on categories when you have a large enterprise system that contains a lot of events streaming through the solution. Execute the following steps, as shown in the following screenshot:

1. Open the Azure portal and go to **Create a resource**.
2. Search for *Event Grid topic* in the search field.
3. Press the **Create** button:

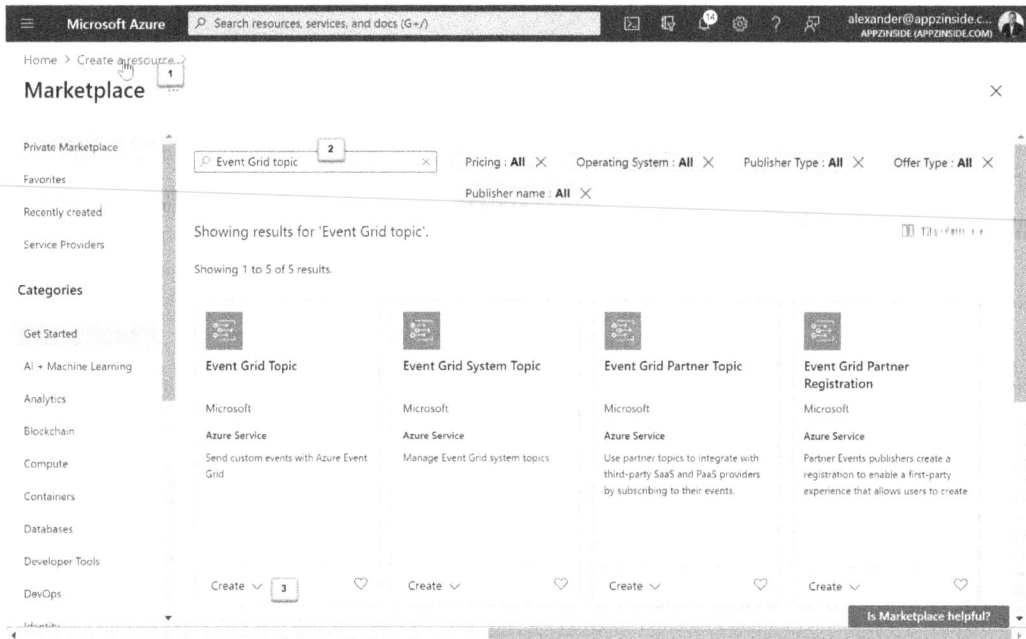

Figure 12.3 – Creating an Event Grid topic service

Execute the following steps, as shown in the following screenshot:

1. Select your subscription.
2. Select our resource group, which is called **DigitalTwinsBook**.
3. Name the Event Grid topic `DTBEventGrid`.
4. Select the closest region. In my case, it is **West Europe**.
5. Press the **Review + create** button:

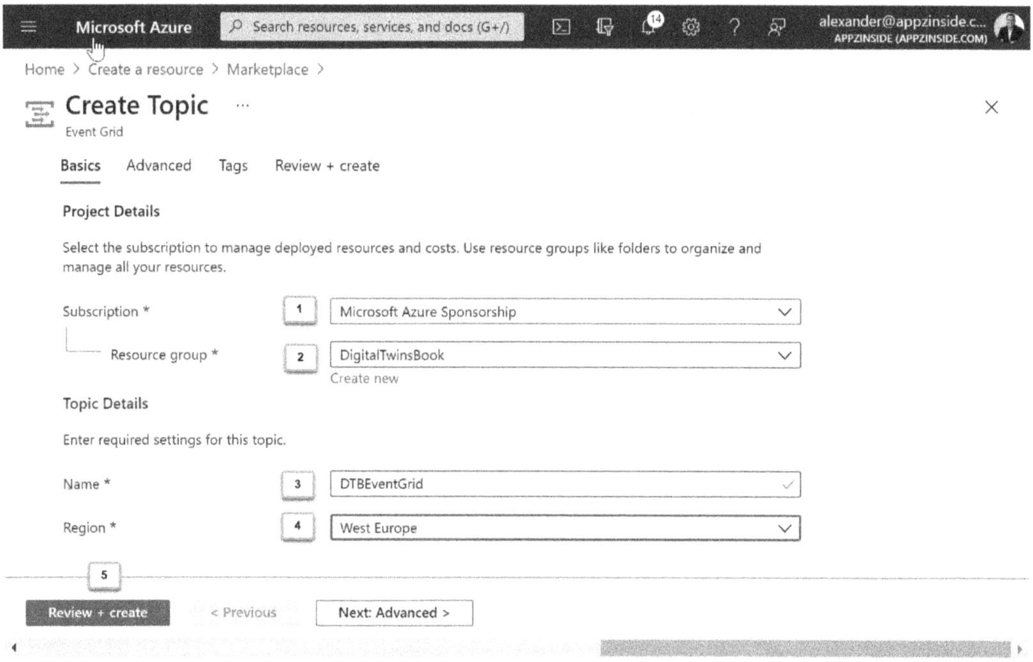

Figure 12.4 – Entering information in the required fields to create an Event Grid Topic

Check all the fields you have entered in the previous dialog, as shown in the following screenshot. Press the **Create** button to create the Event Grid Topic.

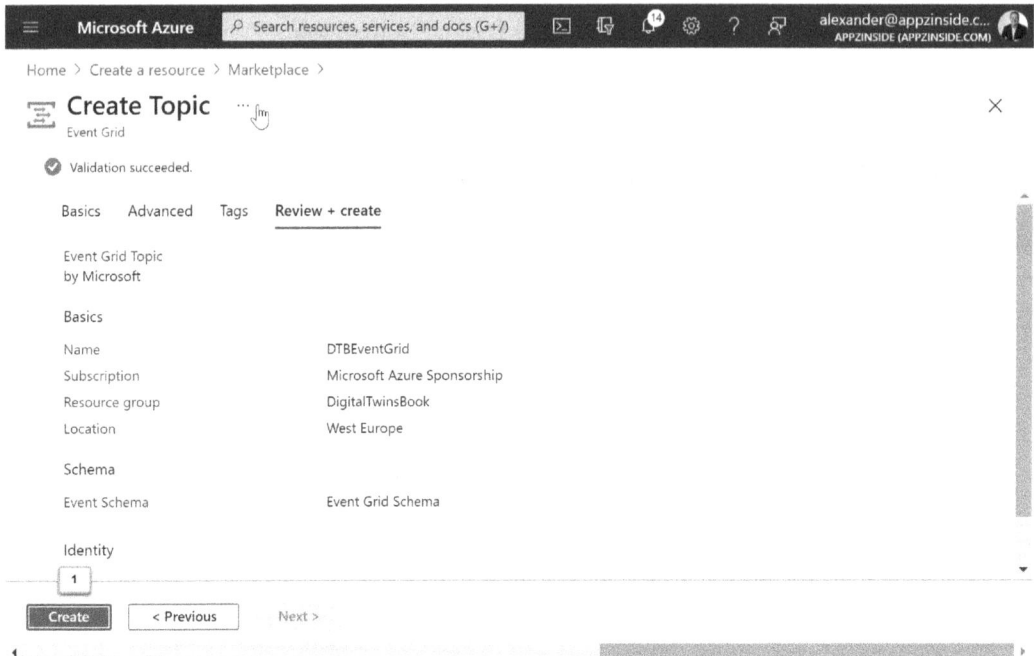

Figure 12.5 – Creating the Event Grid Topic

In this section, we created an Event Grid Topic called DTBEventGrid. In the next section, we will create an endpoint for this Event Grid Topic.

Creating an endpoint

To route messages, you must have an endpoint defined. This endpoint defines where the messages are routed to. In our example, we want to route the messages to our Event Grid Topic, which is called DTBEventGrid.

Endpoints are created within the Azure Digital Twins instance. Execute the following steps to create an endpoint, as shown in the following screenshot:

1. Open the existing Azure Digital Twins instance; that is, DTBDigitalTwins.

2. Select the **Endpoints** option from the left menu.

3. Press the + **Create an endpoint** button at the top.

4. Enter `TemperatureWarnings` as the name for our endpoint.

5. Select **Event Grid** for **Endpoint type**.

6. Select your subscription where you have defined `DTBEventGrid`.

7. Select `DTBEventGrid` for **Event Grid topic**.

8. Press the **Save** button to create the endpoint:

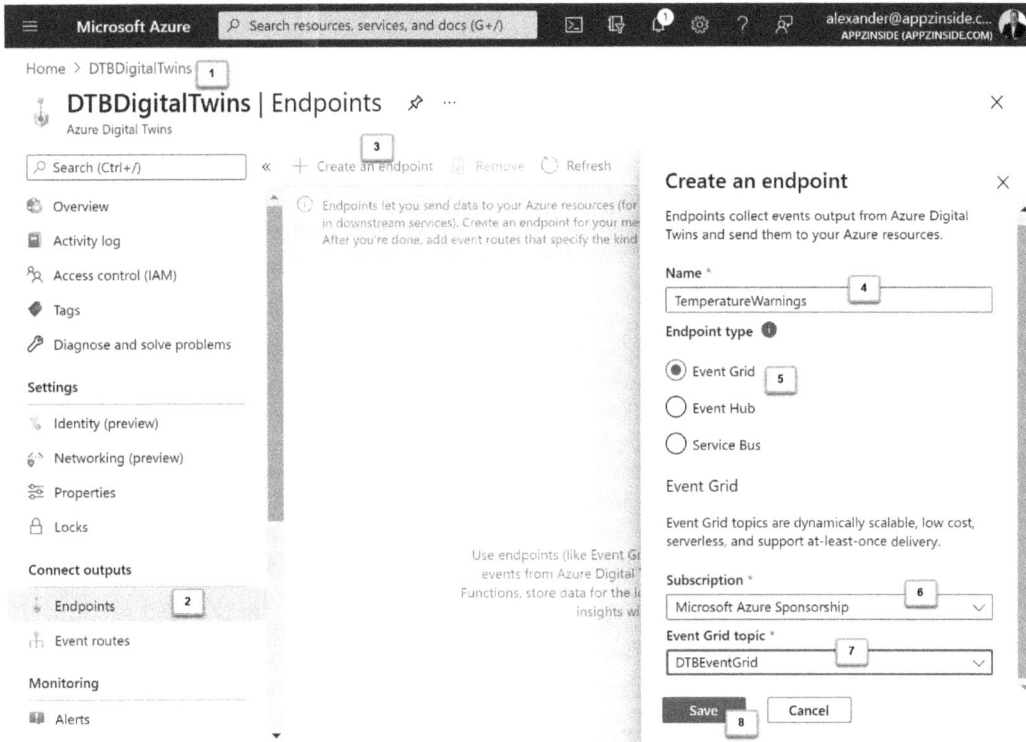

Figure 12.6 – Creating an endpoint in our Azure Digital Twins instance

Now, let's return to our overview of the endpoints, as shown in the following screenshot. This overview shows all the endpoints that have been defined within the Azure Digital Twins instance. The status of the endpoint is set to **Provisioning** when it is created. The endpoint will be available for use when its status moves to *active*:

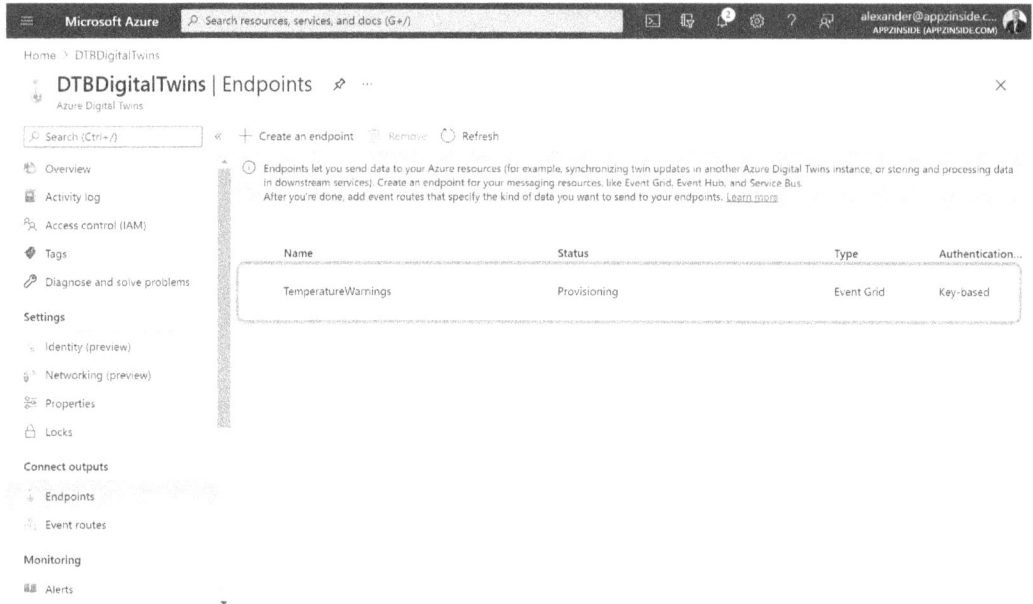

Figure 12.7 – Overview of the endpoints defined in the Azure Digital Twins instance

With that, we have created our first endpoint for an Event Grid Topic. In the next section, we will be creating an event route to get messages into the Event Grid Topic.

Creating an event route

Event routes are also created in the Azure Digital Twins instance. Follow these steps, as shown in the following screenshot:

1. Open the existing Azure Digital Twins instance; that is, DTBDigitalTwins.
2. Select the **Event routes** option from the left menu.
3. Press the + **Create an event route** button at the top.
4. Enter SensorTemperatureHigh as the name for our endpoint.
5. Select the **TemperatureWarnings** endpoint.

6.   Enable the advanced editor and enter the following filter:

```
type = 'Microsoft.DigitalTwins.Twin.Update' AND
$body.modelId = 'dtmi:com:smartbuilding:Sensor;1' AND
$body.temperature > 25.0
```

Before we continue and save the event route, it would be good to explain what the filter does. The first part of the filter will only allow events that are generated through an update using the `Microsoft.DigitalTwins.Twin.Update` type. The second part of the filter defines the model type of the digital twin that we want to receive messages from. Since we only want to receive messages from sensors, we compare `modelId` with `dtmi:com:smartbuilding:Sensor;1`. The last part of the filter defines that we only want to have messages from digital twins that have a property called `temperature` with a value above `25.0`.

This filter will only send messages to the endpoint from *Sensor* digital twins with a temperature value higher than `25.0`.

More information about creating filters can be found here: `https://docs.microsoft.com/en-us/azure/digital-twins/how-to-manage-routes?tabs=portal%2Cportal2%2Cportal3#filter-events`.

7.   Press the **Save** button to create the event route:

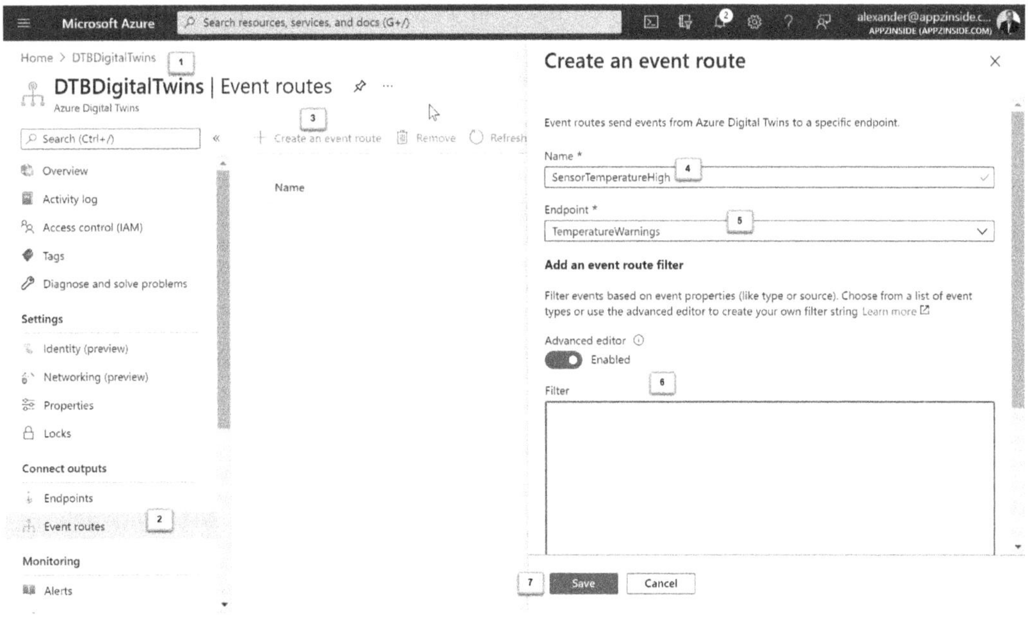

Figure 12.8 – Creating an event route

With that, we have created an event route that uses a filter to filter out only messages from sensors with a temperature higher than 25.0. The messages are routed to the endpoint defined by the Event Grid Topic. In the next section, we will explain how to monitor and view these messages.

Subscribing to event messages

The Event Grid Topic service is great for transporting messages to other services. Unfortunately, it is not possible to view the messages through the Azure portal. But there is a trick to view the messages and, at the same time, show how we can transport the messages to another service. The queue of an Azure Service Bus can be viewed using the Service Bus Explorer.

We will start by creating a new queue in our existing Azure Service Bus. Execute the following steps, as shown in the following screenshot:

1. Open our existing Azure Service Bus; that is, dtbservicebus.
2. Select the **Queues** option from the left menu.
3. Press the + **Queue** button.
4. Enter monitoringeventgridmessages for **Name**.
5. Tick the **Enable dead lettering on message expiration** checkbox.

6. Press the **Create** button to create the queue:

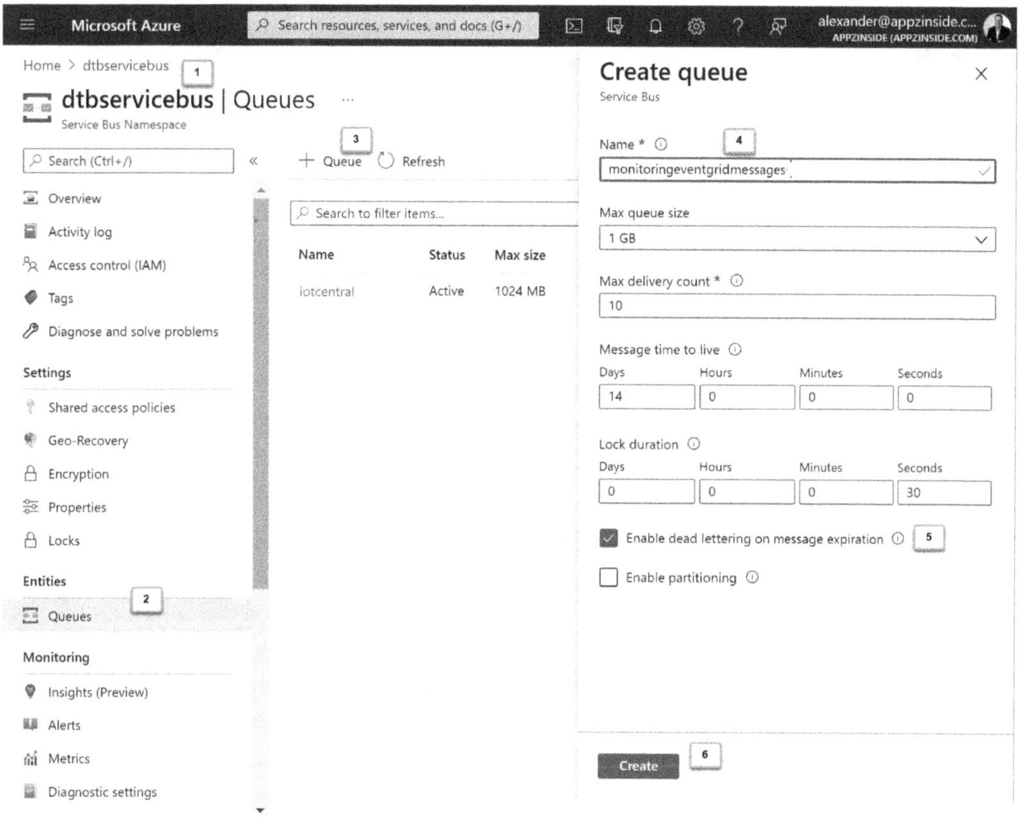

Figure 12.9 – Creating a queue in Azure Service Bus

The newly created queue is visible in the overview of queues, as shown in the following screenshot:

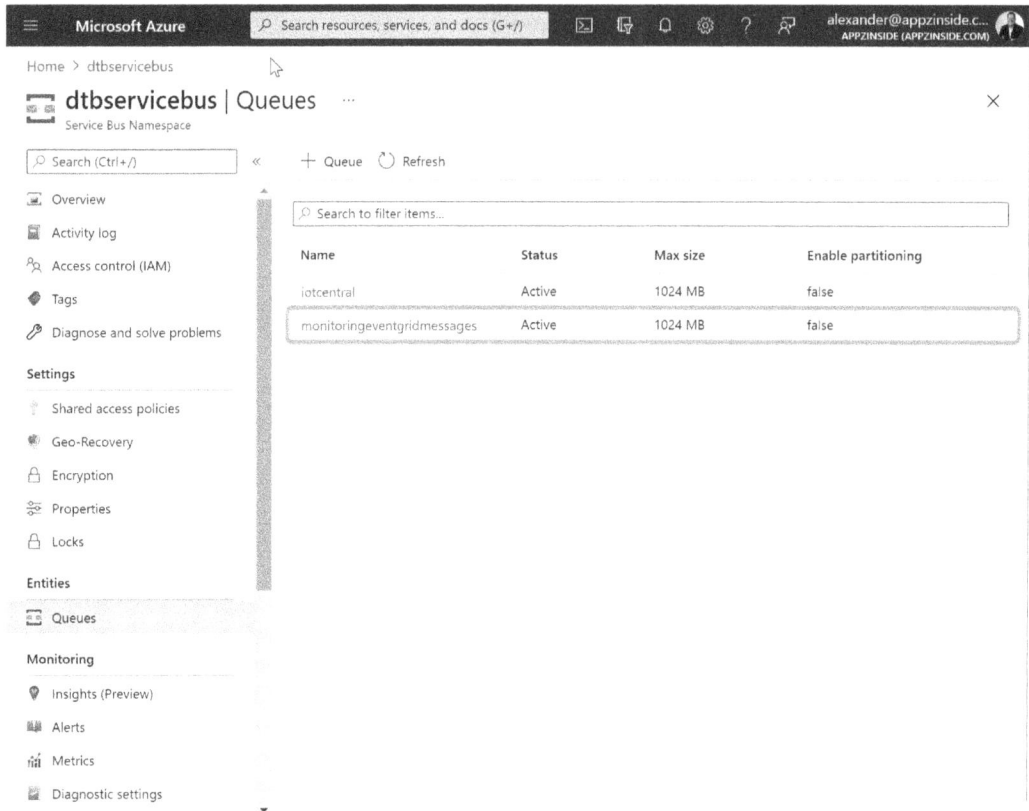

Figure 12.10 – Overview of the queues in Azure Service Bus

Now, we need to subscribe the Azure Service Bus Queue to messages coming from the Event Grid Topic. The portal allows us to create an event subscription. But to fully understand how the event subscription is created, we will be creating the event subscription using PowerShell. This time, we will be using Cloud Shell, which is available in the Azure portal. Press the **Cloud Shell** button in the top bar of the Azure portal, as shown in the following screenshot. This will open a Cloud Shell window at the bottom. You will be logged in with the same credentials that you used for the Azure portal.

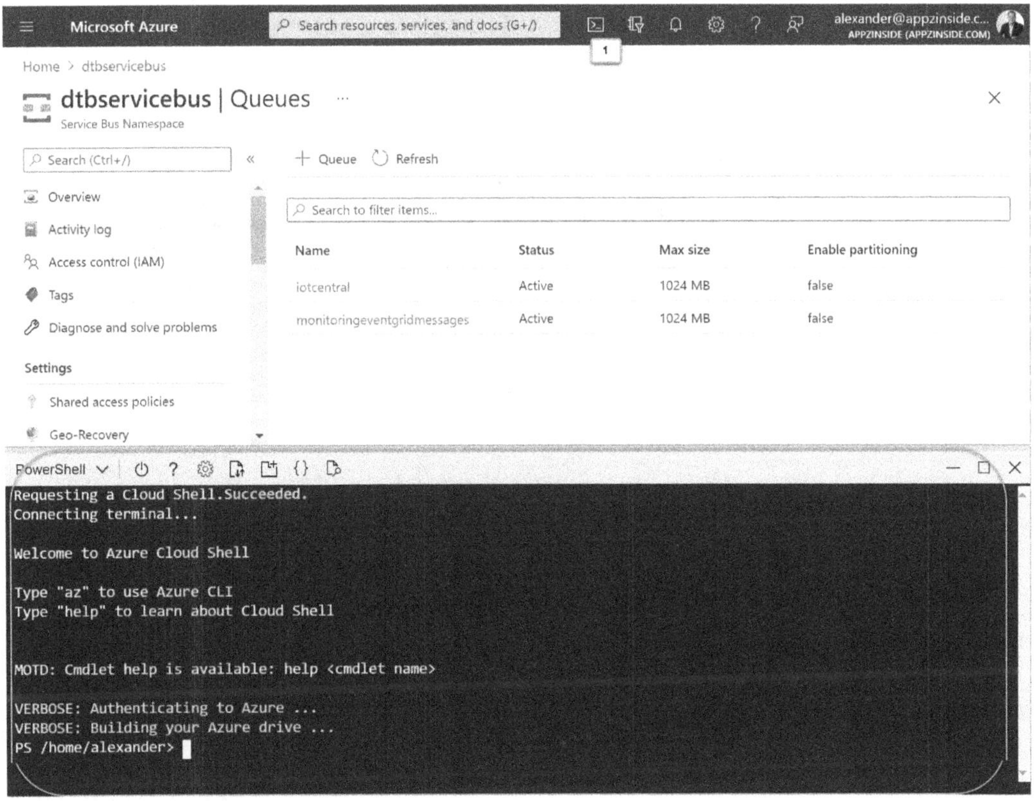

Figure 12.11 – Opening Cloud Shell to create an event subscription

An event subscription for an Event Grid is created based on the Event Grid ID and the Service Bus Queue ID. These IDs look somewhat different than what you would expect. We start by getting the Event Grid ID using the following PowerShell command:

```
az eventgrid topic show -n DTBEventGrid -g DigitalTwinsBook
--query "id" --output tsv
```

The result of executing the preceding PowerShell command is shown in the following screenshot:

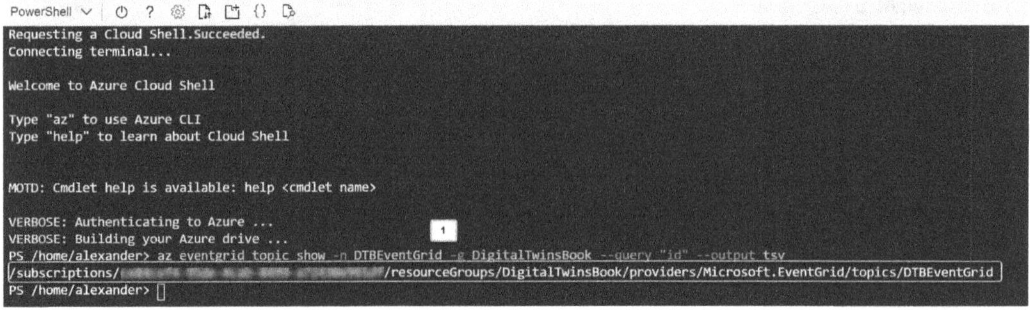

Figure 12.12 – The result of the PowerShell command to get the Event Grid ID

The Event Grid ID contains the full path to the Event Grid Topic. The following is an example showing what this looks like:

```
/subscriptions/5bd208a1-6861-4c10-8f1b-7a5b63f52a20/
resourceGroups/DigitalTwinsBook/providers/Microsoft.EventGrid/
topics/DTBEventGrid
```

Now, let's get the Service Bus ID. Execute the following PowerShell command in Cloud Shell:

```
az servicebus queue show --namespace-name dtbservicebus --name
monitoringeventgridmessages --resource-group DigitalTwinsBook
--query id --output tsv
```

The result of executing the preceding PowerShell command is shown in the following screenshot:

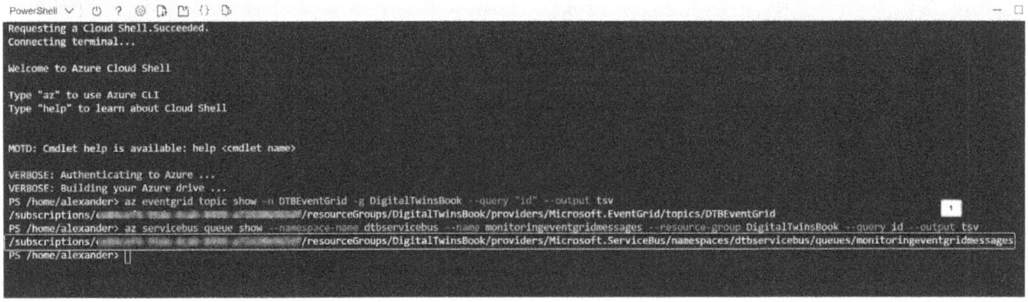

Figure 12.13 – The result of the PowerShell command to get the Service Bus Queue ID

The Service Bus ID contains the full path to the Service Bus Queue. The following shows an example of what this looks like:

```
/subscriptions/5bd208a1-6861-4c10-8f1b-7a5b63f52a20/
resourceGroups/DigitalTwinsBook/providers/Microsoft.ServiceBus/
namespaces/dtbservicebus/queues/monitoringeventgridmessages
```

The final step is creating the event subscription. This requires specifying a --name, --source-resource-id, --endpoint-type, and --endpoint. –source-resource-id is the full path we got back for the Event Grid ID. –endpoint-type is set to servicebusqueue. --endpoint is the full path we got back from the Service Bus Queue ID.

Execute the following PowerShell command in Cloud Shell:

```
az eventgrid event-subscription create --source-resource-
id /subscriptions/5bd208a1-6861-4c10-8f1b-7a5b63f52a20/
resourceGroups/DigitalTwinsBook/providers/Microsoft.EventGrid/
topics/DTBEventGrid --name monitoringeventgridmessages
--endpoint-type servicebusqueue --endpoint /
subscriptions/5bd208a1-6861-4c10-8f1b-7a5b63f52a20/
resourceGroups/DigitalTwinsBook/providers/Microsoft.ServiceBus/
namespaces/dtbservicebus/queues/monitoringeventgridmessages
```

The result of executing the preceding PowerShell command is shown in the following screenshot:

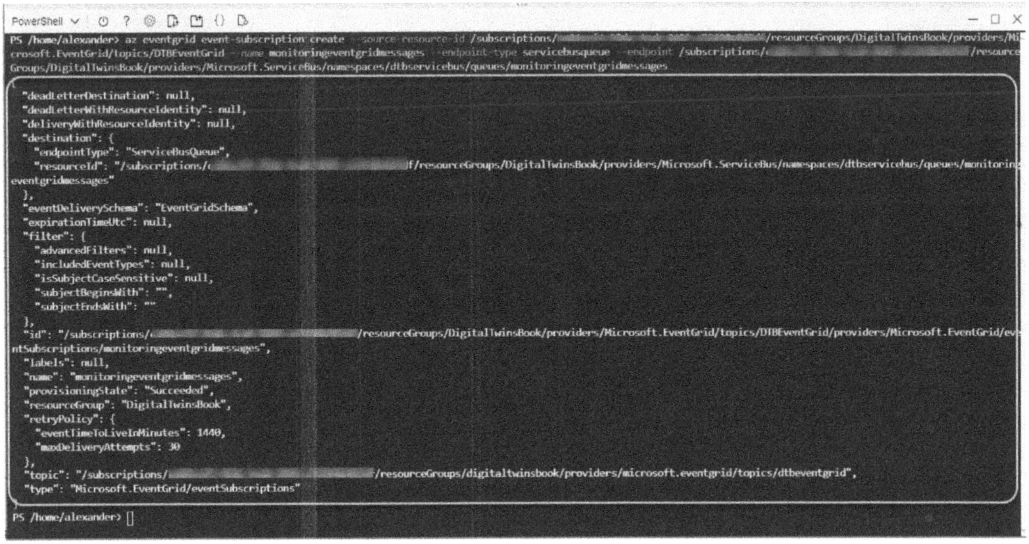

Figure 12.14 – The result of creating the event subscription

With that, we have created an event subscription between the Event Grid Topic and an Azure Service Bus Queue. In the next section, we will use this subscription to monitor and view the messages.

Monitoring event route messages

As we stated earlier, we will be using the subscription we've created to monitor the event messages in the Azure Service Bus Queue. Execute the following steps to open the Service Bus Queue, as shown in the following screenshot:

1. Open the `dtbservicebus` Azure service.
2. Select the **Queues** option from the left menu.
3. Click on the queue named **monitoringeventgridmessages**:

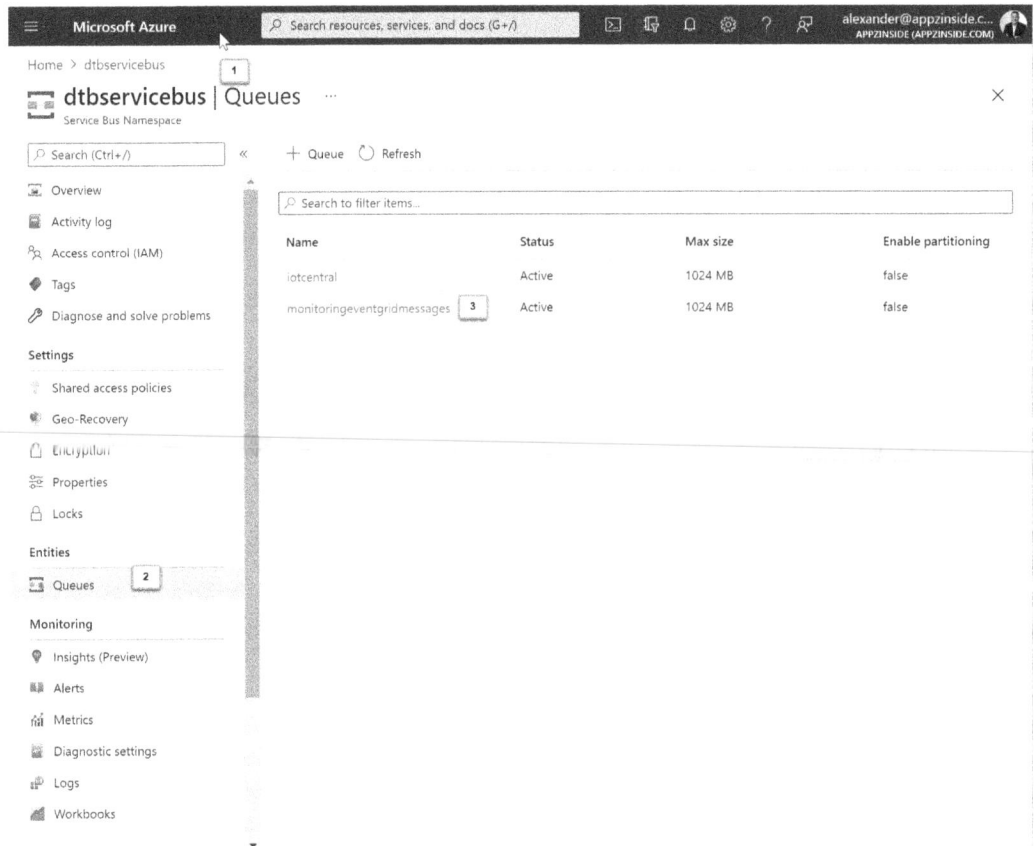

Figure 12.15 – Opening the Service Bus Queue

This will open the `monitoringeventgridmessages` Service Bus Queue. Execute the following steps, as shown in the following screenshot:

1. Select the **Service Bus Explorer (preview)** option from the left menu. At the time of writing, this function is still in preview.

2. Select the **Peek** tab. The number of messages will appear in the view. It can take some time to get messages in there if the subscription has just been created.

3. Press the **Peek** button.

4. Select one of the messages from the list.

5. An overview will open containing the message information.

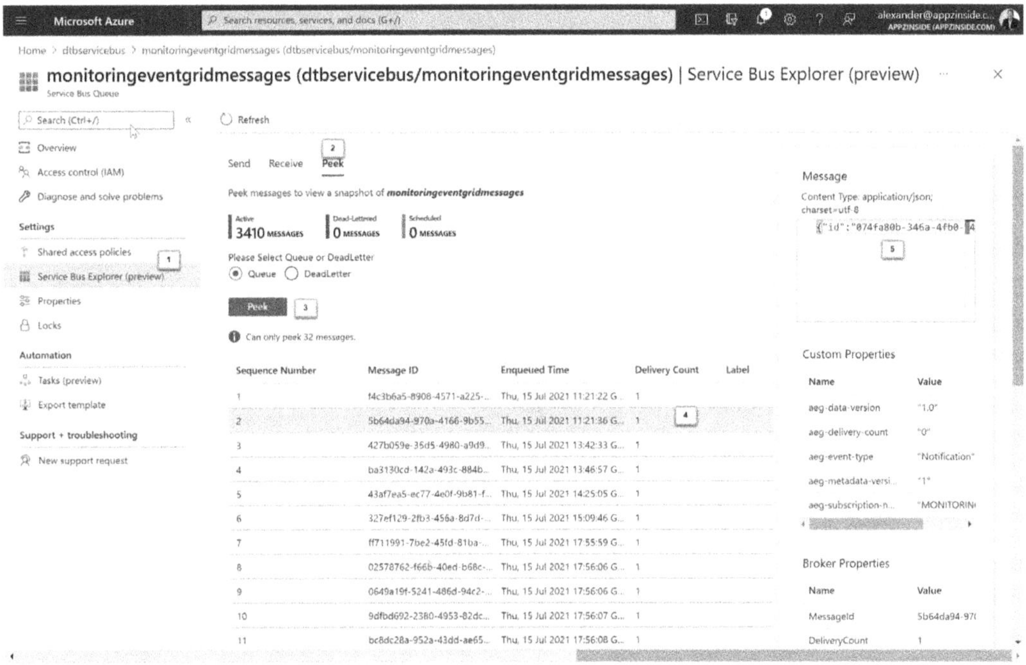

Figure 12.16 – Viewing messages in the Service Bus Queue

The **Message** section contains the actual message that was sent. This message came from an update from a digital twin sensor and was routed to the Event Grid Topic. The subscription between the Service Bus Queue and the Event Grid Topic will transport the message to the Service Bus Queue:

```
{
    "id": "f24e0da3-1e2c-4707-872a-83969fac5c24",
    "subject": "20a05n5fxk0",
    "data": {
```

```
    "data": {
      "modelId": "dtmi:com:smartbuilding:Sensor;1",
      "patch": [
        {
          "value": 36,
          "path": "/temperature",
          "op": "replace"
        }
      ]
    },
    "contenttype": "application/json",
    "traceparent": "00-b31ae470c52b888bbe5547262e7dc80e-
736b2a14e4fd744c-01"
    },
  "eventType": "Microsoft.DigitalTwins.Twin.Update",
  "dataVersion": "1.0",
  "metadataVersion": "1",
  "eventTime": "2021-07-15T11:21:35.5417519Z",
  "topic": "/subscriptions/ 5bd208a1-6861-4c10-8f1b-
7a5b63f52a20/resourceGroups/DigitalTwinsBook/providers/
Microsoft.EventGrid/topics/DTBEventGrid"
}
```

The message tells us a lot. The subject field contains the name of the digital twin and, thus, the ID of the actual sensor we used. This message came from the sensor with an ID of 20a05n5fxk0. The data part of the message contains the actual patch information, which specified what and how something was changed. In this case, the temperature value has been replaced with a value of 36.0. Finally, the message contains an eventType of update, as well as the source of the message. The source is the Event Grid Topic and is specified by the topic field.

As we can imagine, this message can look somewhat different, depending on which source the message was sent from.

Summary

In this chapter, we learned how notifications are triggered. We also learned how a message can be routed to an Event Grid Topic endpoint. A subscription was created to receive the messages from the Event Grid Topic and put them in an Azure Service Bus Queue to monitor and view the contents of the messages. This technique will help integrate Azure Digital Twins with a variety of different Azure services. This allows you to store data, process data through a pipeline, or even use the data as an overlay on a map.

In the next chapter, we will learn about other possible integrations, such as Azure Maps and Azure Time Series Insights with Azure Digital Twins. We will be using Event Grid Topics to connect these services to the Azure Digital Twins instance.

Questions

As we conclude, here is a list of questions for you to test your knowledge regarding this chapter's material. You will find the answers in the *Assessments* section of the *Appendix*:

1. Data ingress is receiving data and events from what?

    a. Upstream services

    b. Downstream services

    c. Azure services

2. How is it possible to update another digital twin within the same Azure Digital Twins instance based on a notification?

    a. No additional Azure services are required.

    b. Event Grid and Event Hub.

    c. EventHub and Azure Function.

3. How can we monitor event messages?

    a. It is not possible to monitor event messages.

    b. Using the metrics and logs on the Event Grid Topic.

    c. Subscribing with an Azure Service Bus Queue.

Further reading

To learn more about the topics covered in this chapter, please refer to the following links:

- Data ingress and egress for Azure Digital Twins: `https://docs.microsoft.com/en-us/azure/digital-twins/concepts-data-ingress-egress`

- Event notifications: `https://docs.microsoft.com/en-us/azure/digital-twins/concepts-event-notifications`

- Route events within and outside of Azure Digital Twins: `https://docs.microsoft.com/en-us/azure/digital-twins/concepts-route-events#about-event-routes`

- Managing endpoints and routes in Azure Digital Twins (APIs and the CLI): `https://docs.microsoft.com/en-us/azure/digital-twins/how-to-manage-routes-apis-cli`

13
Setting up Azure Maps

In this chapter, we will set up the Azure Maps account service to create an indoor map. As part of that, we will need to upload and convert a drawing package that describes the indoor map in detail. We will need to preconfigure several elements of an Azure map to be able to connect a digital twin to automatically update an area on the indoor map.

You will learn how to configure and set up an Azure Maps account and find out what you need to do to update a certain area on a map.

This chapter contains the following sections:

- Understanding Azure Maps
- Creating an Azure Maps account
- Creating a Creator resource
- Building a map
- Uploading a map
- Converting a map
- Creating and validating a dataset

- Creating and validating a tileset
- Creating and validating feature stateset

Technical requirements

We will be using the Azure portal to create several Azure services to support an indoor map. To preconfigure the elements, we will be using **application programming interface (API)** calls via the Postman application.

Understanding Azure Maps

We focused on getting digital twin events through Event Grid in the previous chapter. Event Grid allows us to send these events to other Azure services. One of those services is Azure Maps. Azure Maps allows us to create location-aware applications based on geospatial information. We can build rich visualizations for any type of device using this Azure service. This service can be enriched with rich datasets, allowing you to build all kinds of different solutions for web, mobile, and even other platforms, solutions that present layers of data on top of a map providing services—for example, improving travel time and avoiding traffic congestion, finding the shortest and quickest available routes, gaining insights into elevations and specific terrains, customer geographical information, weather, and much more.

Azure Maps can be used in conjunction with *TomTom*, *AccuWeather*, and *Moovit* to build solutions that use real-time locations. Azure Maps also supports **role-based access control (RBAC)**. This allows us to define different roles, each with specific rights, allowing us to use the content. In conjunction with that, it enhances privacy and protection, as required to comply with current **General Data Protection Regulation (GDPR)** legislation.

We will be receiving events from Event Grid and will use Azure Maps to update certain areas based on the value of a `temperature` property of the sensor digital twin. An overview of the architecture can be seen here:

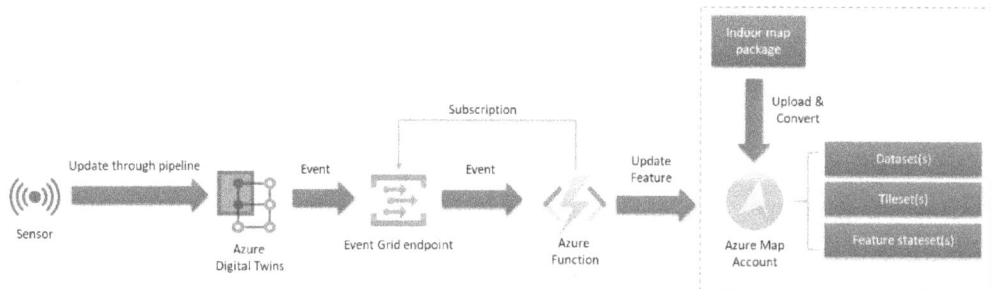

Figure 13.1 – Overview of the architecture and the mechanism of updating the Azure map

An Azure Maps account can be updated by using an Azure Function. The Azure Function will have a subscription on one of the Event Grid endpoints. But before we build this architecture, we will need to set up the Azure Maps account and corresponding elements such as **dataset**, **tileset**, and **feature stateset**. Each of these elements will be explained in more depth during the sections in this chapter.

We will start in the next section by creating an Azure Maps account.

Creating an Azure Maps account

To create an Azure Map account, open the Azure portal via `https://portal.azure.com`. Follow these next steps to create a new resource:

1. Open **Marketplace** to create a new resource.

2. Enter `Azure Maps` in the search box.

3. Click on the **Create** link to create an Azure Maps resource.

    The following screenshot highlights this process:

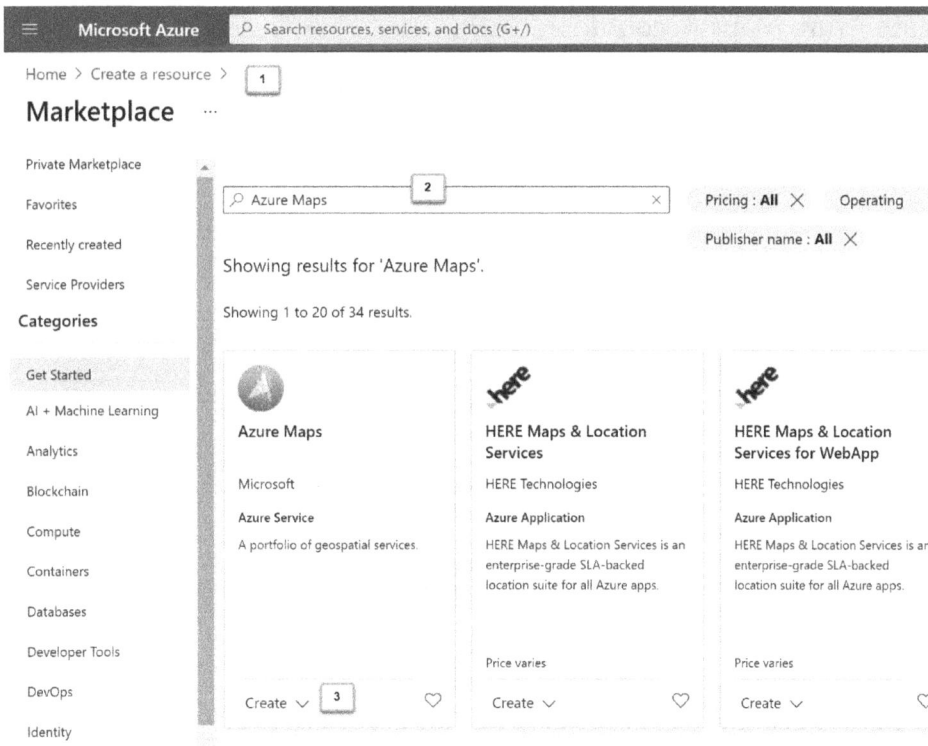

Figure 13.2 – Creating an Azure Maps account

This will open a new window for creating an Azure Maps account. Now, execute the following steps:

1. Select your subscription.

2. Select our `DigitalTwinsBook` resource group.

3. Enter the name `DTBMaps`.

4. Select the `Gen2 (Maps & Location Insights)` pricing tier. This pricing tier is required to use the Azure Maps account in conjunction with TomTom. More information about pricing tiers can be found here: `https://docs.microsoft.com/en-us/azure/azure-maps/choose-pricing-tier`.

5. Azure Maps shares locations with the third-party TomTom. Therefore, we will need to select the checkbox to confirm that we have understood the license and privacy statement.

6. Press the **Create** button.

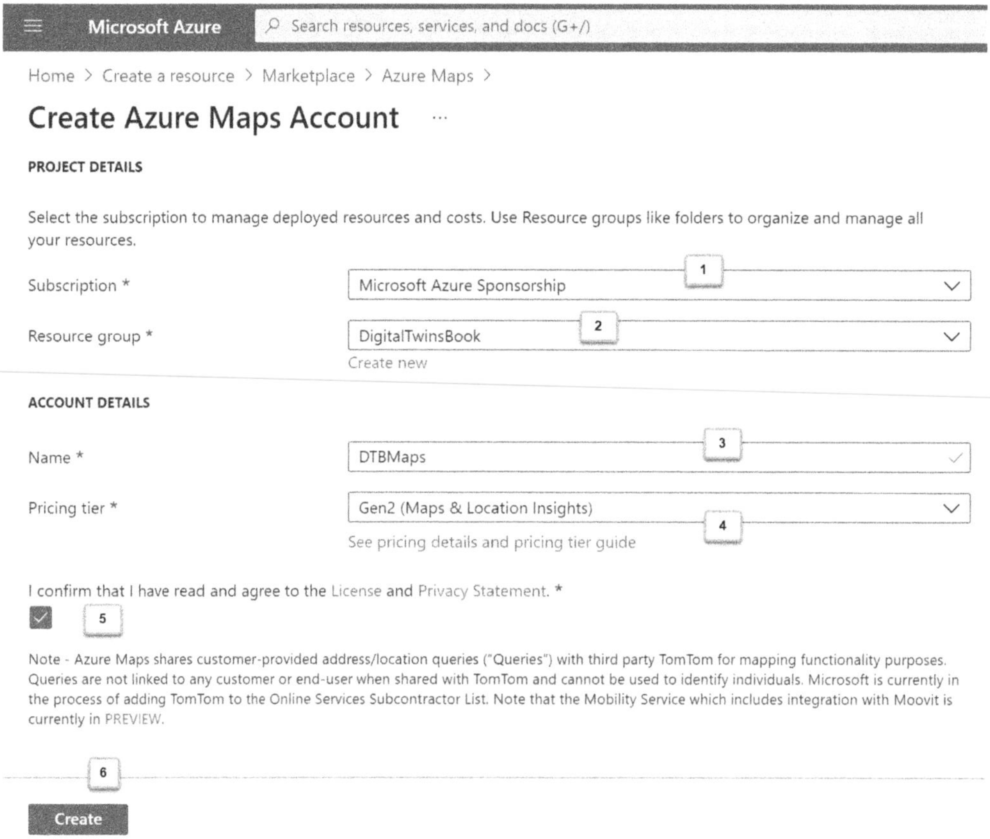

Figure 13.3 – Entering information for creating an Azure Maps account

It will take some time to create this Azure resource. Open the `DTBMaps` resource as soon as it is created and ready. We will need to get an authentication key. The authentication key will later in this chapter be used as the `subscription-key` value for making calls with several APIs to the Azure Maps resource. Execute the following steps:

1. Open the `DTBMaps` resource.

2. Select the **Authentication** option in the left menu.

3. Copy the **Primary Key** value or the **Secondary Key** value. Make sure to store this key temporarily in a notepad. We will be using this key later during several API calls.

   The following screenshot highlights this process:

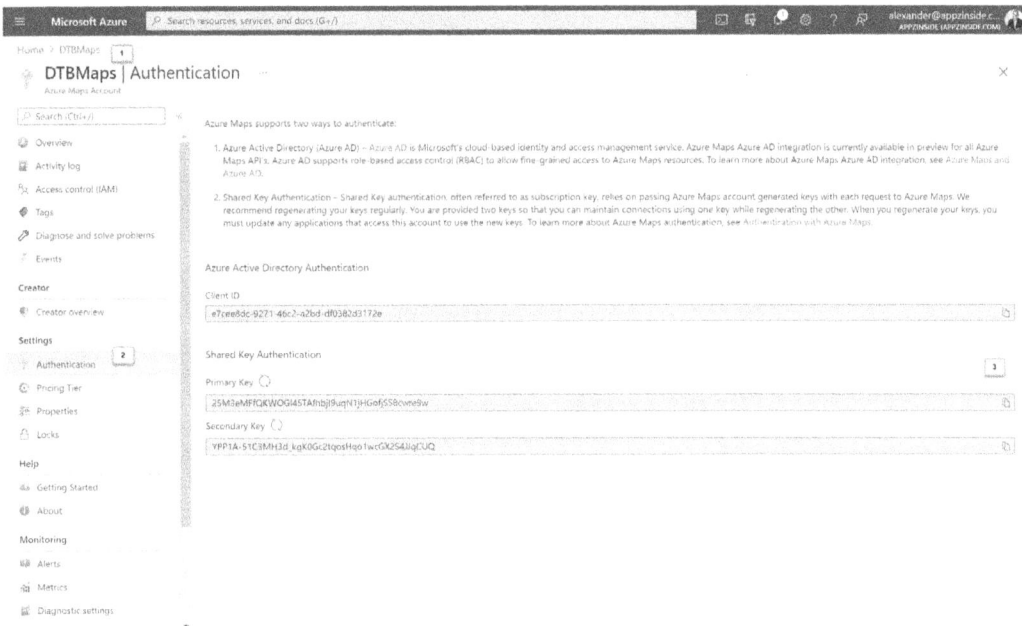

Figure 13.4 – Getting the authentication key from the Azure Maps account

We have created an Azure Maps account and retrieved the authentication key. In the next section, we will be creating a Creator resource.

Creating a Creator resource

A Creator resource allows us to create private indoor map data. Private indoor map data can be used to create an interactive and dynamic indoor map by connecting digital twins to different parts of the map. This allows us to update a certain area of the map with a color based on a temperature sensor. But before we can do that, we need to create a Creator resource in the Azure Maps account. Follow the next steps to do this:

1. Open the DTBMaps resource.
2. Select **Creator overview** in the left menu.
3. Press the **Create** button.

   The following screenshot highlights this process:

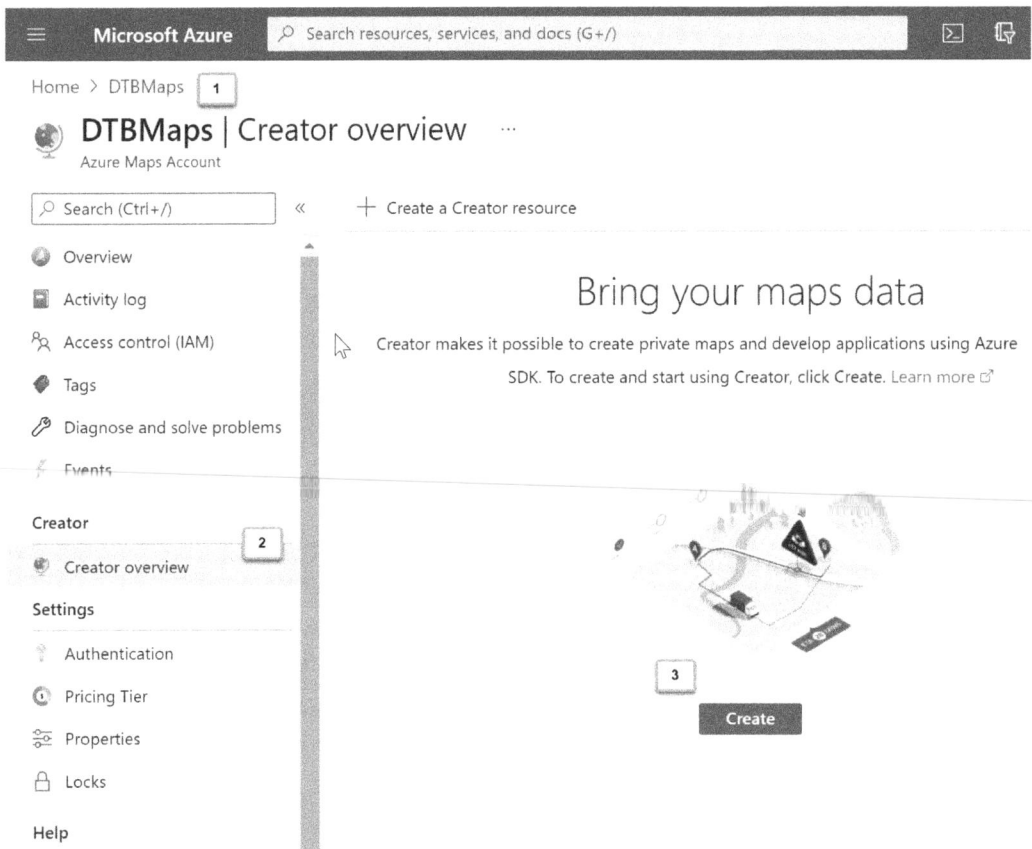

Figure 13.5 – Opening Creator overview in the Azure Maps account resource

This will open the next step. Execute the following steps to configure the Creator resource:

1. Enter `IndoorMap` as the name for the Creator resource.

2. Enter the value `1` in the **Storage Units** field. Each storage unit represents `100` MiB. **MiB** stands for **mebibyte** and is a multiple of the unit byte for storing digital information. We can specify up to `100` storage units. The total available storage units are shared across all Creator services and are used to store the indoor map.

3. Select the location. It is imperative to know which location you have selected, as this can influence the **Uniform Resource Locators (URLs)** of the API calls to the Azure Maps account resource. I have selected **(Europe) West Europe**, which will cause the URLs to start with `https://eu.atlas.microsoft.com/` instead of `https://us.atlas.microsoft.com/`. More information can be found about creating a service geographic scope at `https://docs.microsoft.com/en-us/azure/azure-maps/creator-geographic-scope`. This becomes a bit clearer in the next section of this chapter.

4. Click on the **Review + Create** button.

   The following screenshot highlights this process:

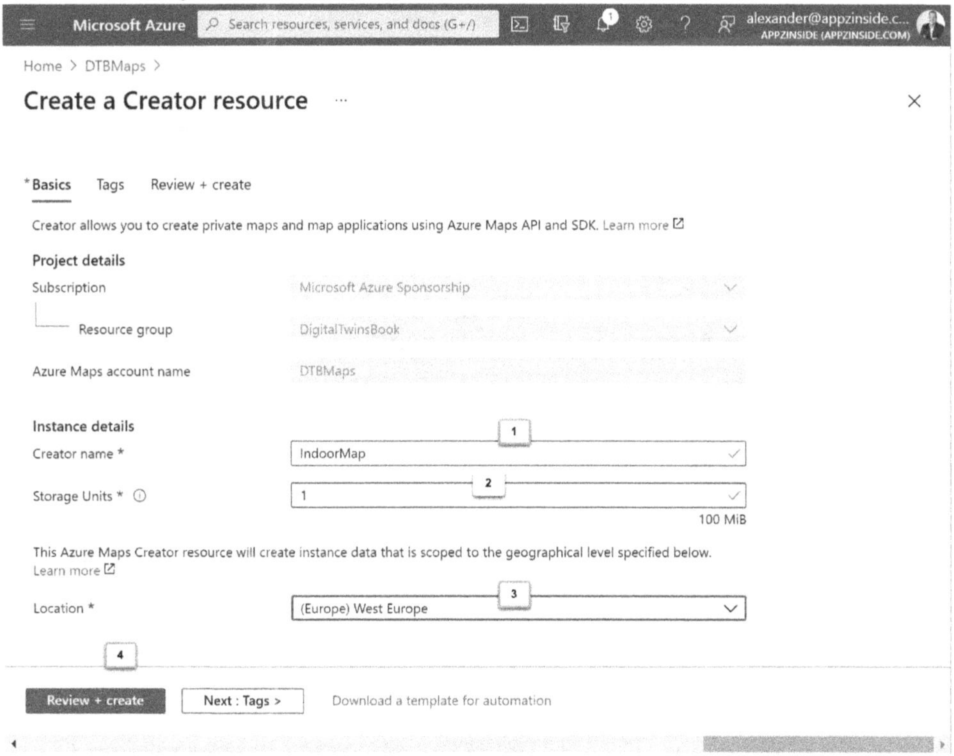

Figure 13.6 – Filling in several values to create a Creator resource

This will open the final step, which shows all the configuration values selected in the previous step. Proceed as follows:

1. Press the **Create** button to create a Creator resource, as illustrated in the following screenshot:

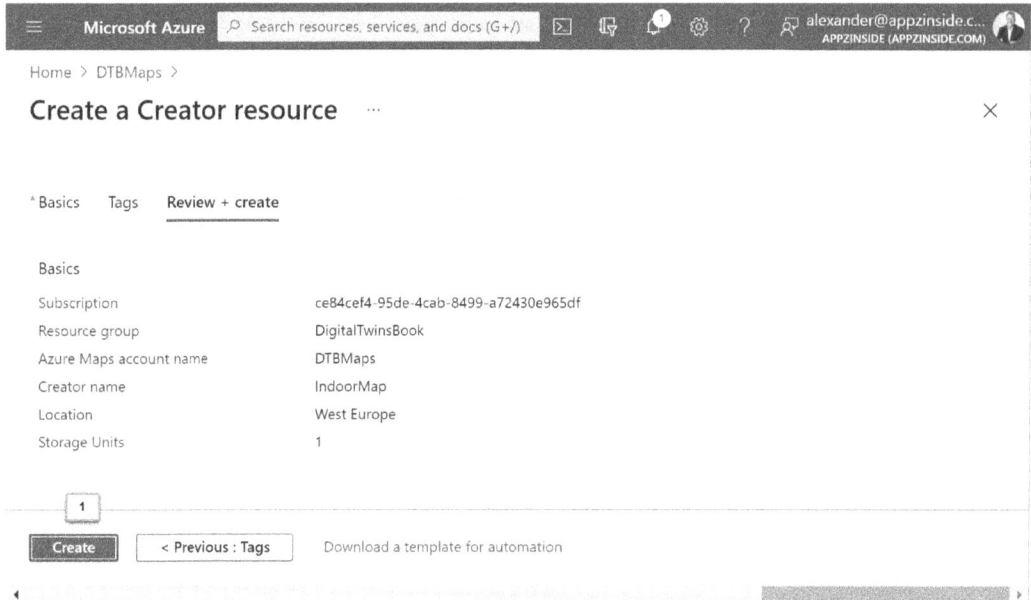

Figure 13.7 – Creating a Creator resource

The Creator resource is being created. This will take some time. After creation, an overview is shown. We can always make configuration changes to the Creator resource by opening it, as shown in *Figura 13.5*.

We have created a Creator resource. In the next section, we will start using the Creator resource to upload a map.

Building a map

Indoor maps are based on the DWG format. DWG is a binary file format that contains **two-dimensional (2D)** and **three-dimensional (3D) computer-aided design (CAD)** designs with additional metadata. DWG is a native format that is accepted by many CAD applications such as AutoCAD. CAD drawings are vector-based.

The Creator service expects a drawing package. We will be using one of the sample packages of Microsoft. Download the following package: `https://github.com/Azure-Samples/am-creator-indoor-data-examples/blob/master/Sample%20-%20Contoso%20Drawing%20Package.zip`.

Unzip the package to a temporary folder and view the contents to have an idea of its structure.

A drawing package consists of one or more DWG files and a manifest file. In the case of an indoor map, each of the DWG files represents a floor of the building. The manifest file contains information that describes the building, its floors, and other metadata—for example, the physical location of the building, the geolocation, each of the building levels, and each of the layers used in the DWG files. It also contains unit properties. Each area on a floor is a unit. Each unit can have one or more properties that describe the unit.

We need to have a CAD application to create our own DWG files. Most CAD applications, such as AutoCAD, are expensive. It is possible to view the contents of a DWG file with DWG TrueView. This application is free to use and allows you also to convert DWG files to other formats.

Download the DWG TrueView application via the following website: `https://www.autodesk.com/viewers`. The application is large and will take some time to download and install. After the download, unpack the solution to the `c:\autodesk` folder. This is a temporarily used folder. Install the application to the `c:\program files\Autodesk` folder. The following screenshot shows the application being installed:

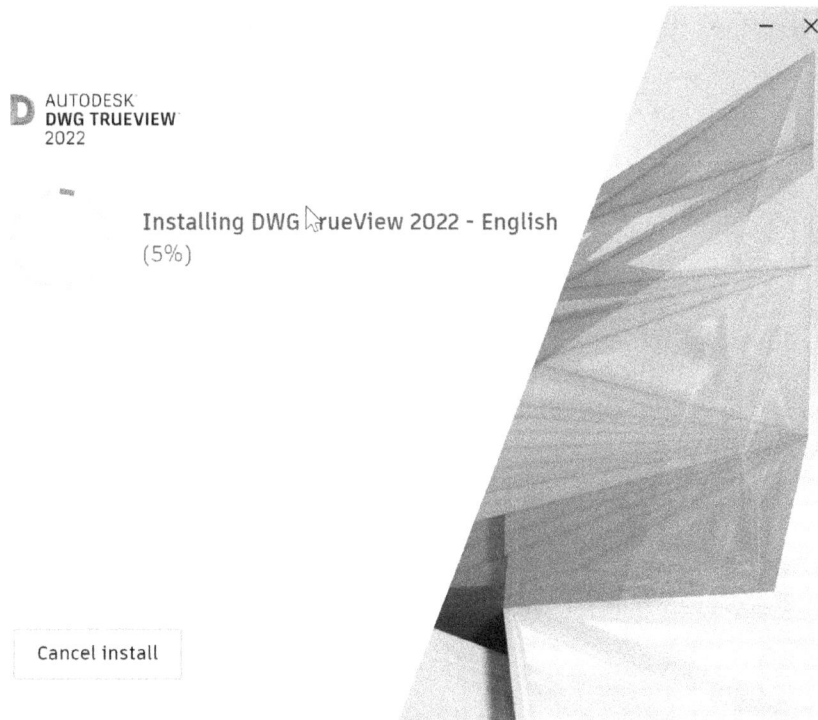

Figure 13.8 – Installing DWG TrueView

Start the application when it has finished installing. Select one of the DWG files from the package and view the result, as shown in the following screenshot:

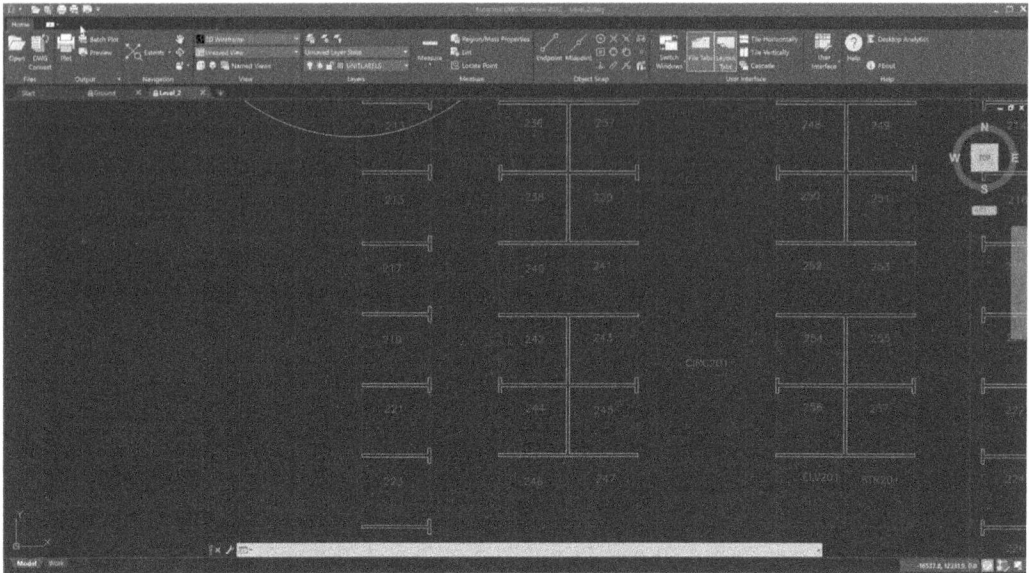

Figure 13.9 – An example of DWG TrueView with one of the floors and its units

Although we are using one of the Microsoft examples for an indoor map, it is up to you if you would like to create your own. You can learn more about creating your own indoor map here: `https://docs.microsoft.com/en-us/azure/azure-maps/tutorial-creator-indoor-maps` `https://www.autodesk.com/viewers` `https://www.autodesk.com/viewers`.

Our goal is to connect a digital twin from our Azure Digital Twins instance to a unit in the indoor map. We are going to implement an example that responds to the temperature change of a sensor and updates a unit in the indoor map with a specific color. For that, we need to specify a **feature stateset**. A feature stateset describes a property with rules that set the color for a unit. The process requires several steps before we can update the stateset. The steps are shown in *Figure 13.10* and explained in brief next. We will go deeper into each of the items in the upcoming sections of this chapter:

1. The drawing package is uploaded.

2. We need to check the upload status. An `{operationId}` value is provided in the previous call to check this.

3. The uploaded package is converted in an understandable format for the Azure Maps resource.

4. We need to check if the conversion has finished. An {operationId} value is provided in the previous call to check this. This call will give us a {conversionId} value.

5. A dataset is created using the {conversionId} value.

6. We need to validate that the dataset has been created by using the {operationId} value we got from the previous call. This call will give us a {datasetId} value.

7. A tileset is created using the {datasetId} value.

8. We need to validate that the tileset has been created by using the {operationId} value we got from the previous call. This call will give us a {tilesetId} value.

9. A feature stateset is created. The call will return us a {featurestatesetId} value.

10. We can update a feature stateset using the {featurestatesetId} value and specifying the unit in the URL.

You can see an overview of this process in the following screenshot:

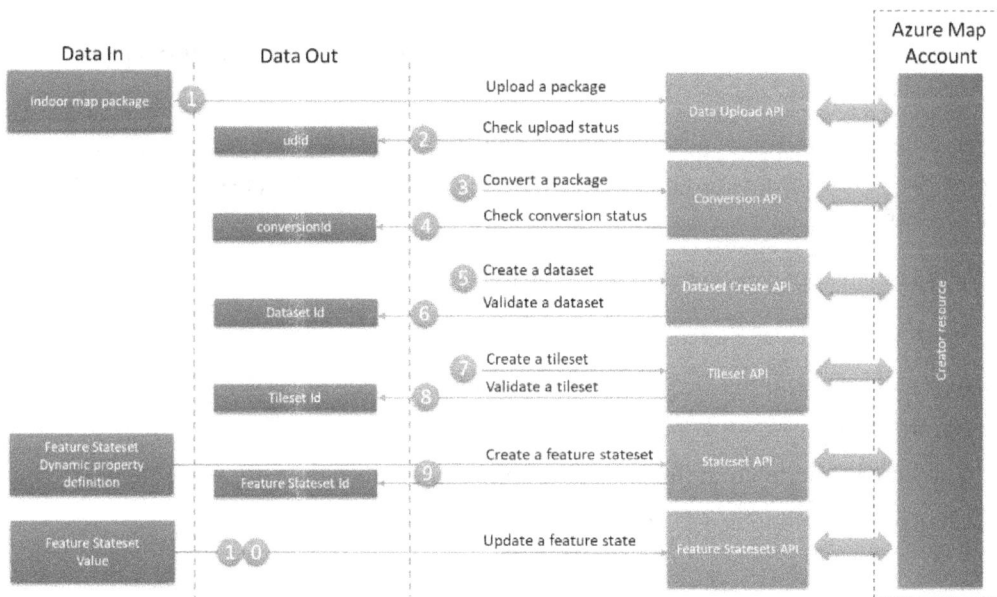

Figure 13.10 – An overview of all API calls to configure an indoor map with dynamic feature statesets

As we can see, several APIs are available to perform these calls. Each API call needs to have the api-version=2.0 and subscription-key={authentication key from Azure Map} properties defined. Some API calls have additional parameters and additional information in the body of the call.

We have learned on a high level what is required to create and configure an indoor map using the Azure Maps account resource and the Azure Creator resource. This will become much clearer when we start executing each of the API calls in the next sections of this chapter. We will be using Postman as an application to execute each of the queries. In the next section, we start by uploading a map.

Uploading a map

We start by uploading an indoor map as a drawing package to the Azure Maps resources. We need to create a POST URL for uploading the data. The URL definition is shown here:

```
https://eu.atlas.microsoft.com/mapData?
api-version=2.0&
dataFormat=dwgzippackage&
subscription-key={authenticationId}
```

Our {authenticationId} value is ce84cef4-95de-4cab-8499-a72430e965df. This is the primary or secondary key found under **Authentication** within the service. The URL becomes this:

```
https://eu.atlas.microsoft.com/mapData?api-version=2.0&dataF
ormat=dwgzippackage&subscription-key=ce84cef4-95de-4cab-8499-
a72430e965df
```

Start the Postman application and execute the following steps:

1. Select the DigitalTwinsBook environment.
2. Add a new collection called AzureMaps and add a new request.
3. Name the request DataUpload.
4. Copy the URL shown in the preceding code snippet into the URL field.
5. Change the method to POST.

The following screenshot highlights this process:

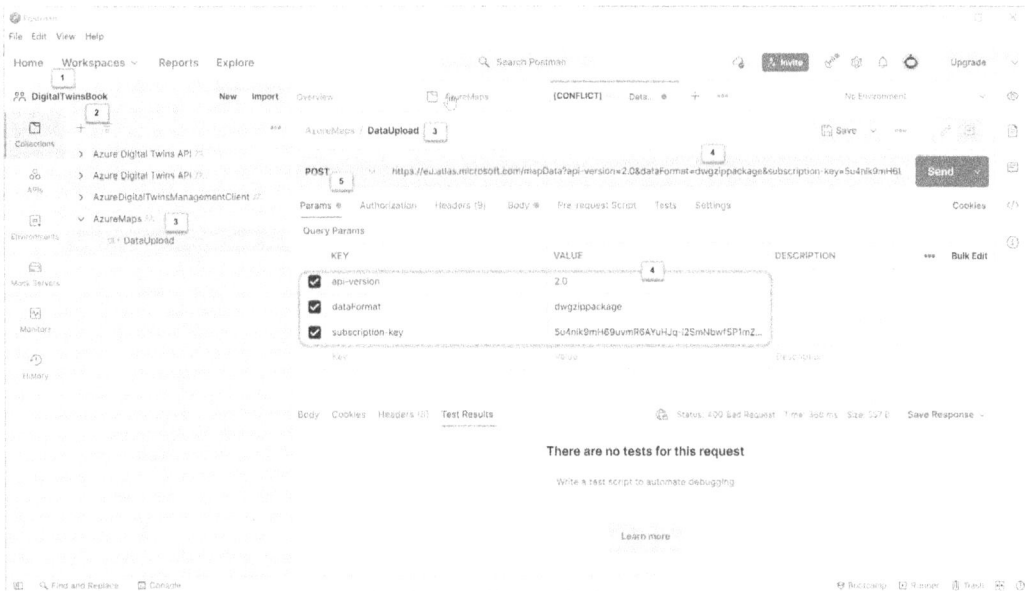

Figure 13.11 – Creating a DataUpload request with Postman

We will need to add an extra header. Execute the following steps:

1. Click on the **Headers** tab.

2. Add a `Content-Type` header with the value `application/octet-stream`.

The following screenshot highlights this process:

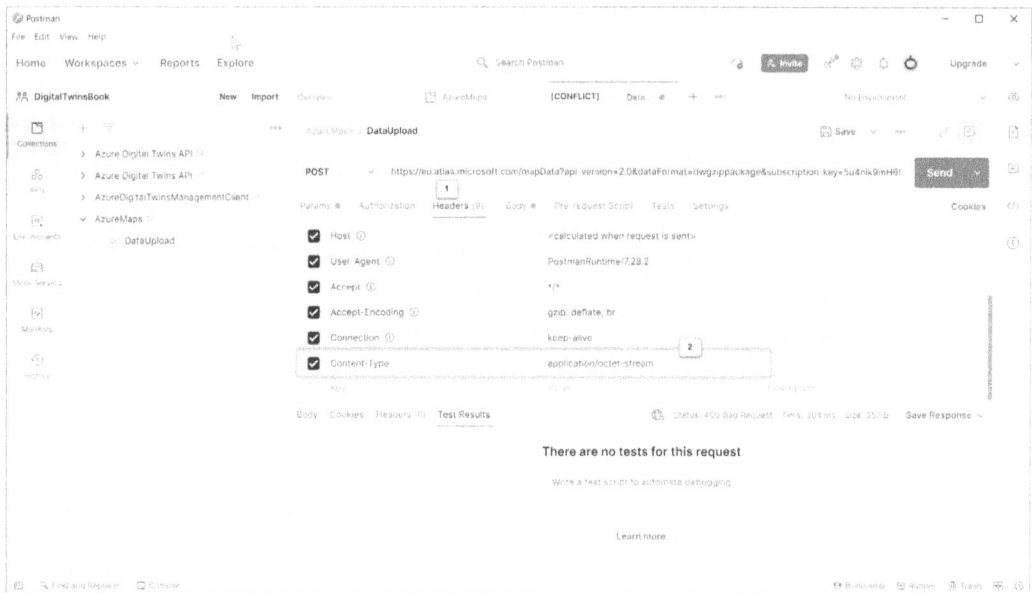

Figure 13.12 – Configuring the DataUpload request with Postman

We will need to add the drawing package file to the body and execute the request. Execute the following steps:

1. Click on the **Body** tab.

2. Select the **binary** option.

3. Browse for the `Sample - Contoso Drawing Package.zip` file.

4. Click on the **Send** button.

The following screenshot highlights this process:

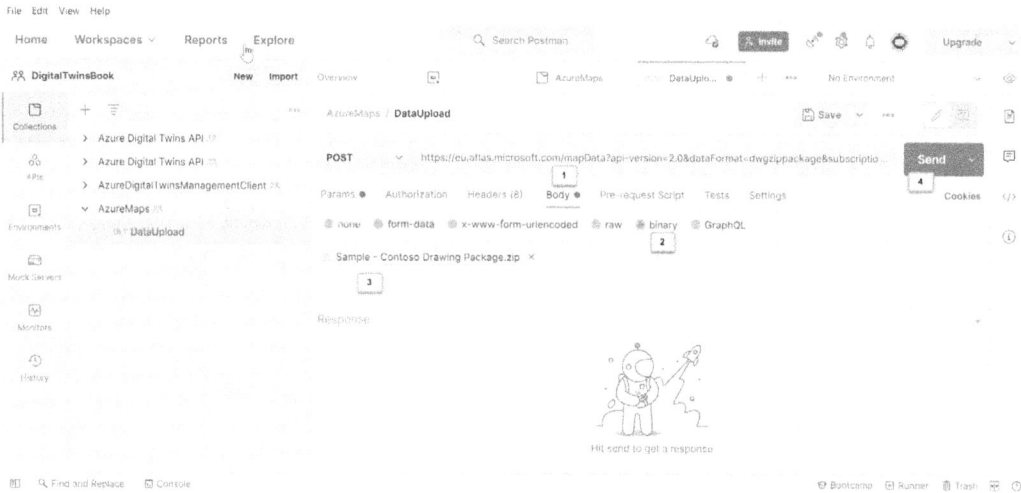

Figure 13.13 – Executing the DataUpload request with Postman

Uploading the drawing package can take some time. We need to check the status of the upload. To do that, we need to have the operation location URL. Execute the following steps:

1. Select the **Headers** tab of the result of the request.

2. Copy the URL from the Operation-Location field. It will look like this: https://eu.atlas.microsoft.com/mapData/operations/ fbc7ef5f-8910-4e5c-ad26-aea347542472?api-version=2.0.

   The following screenshot highlights this process:

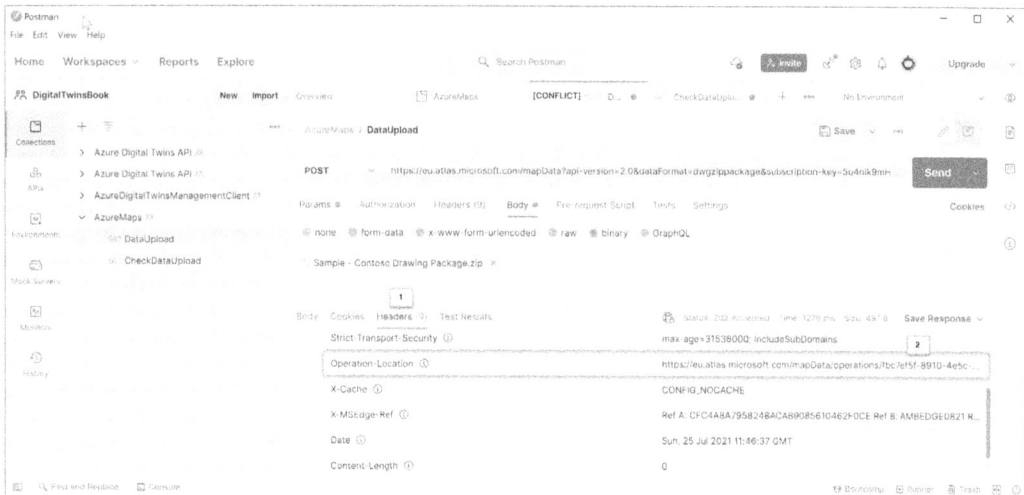

Figure 13.14 – Retrieving the operation location from the request result

We need to check if the upload succeeded. We will create another request in Postman. Execute the following steps:

1. Select the `DigitalTwinsBook` environment.

2. Add a new request and name it `CheckDataUpload`.

3. Copy the URL shown in the preceding screenshot into the URL field.

4. Add the `subscription-key` value to the list of parameters with the `{authenticationId}` value.

5. Click on the **Send** button to execute the request.

6. When the upload is successful, the `CheckDataUpload` request will return a `Succeeded` status in the return body.

    The following screenshot highlights this process:

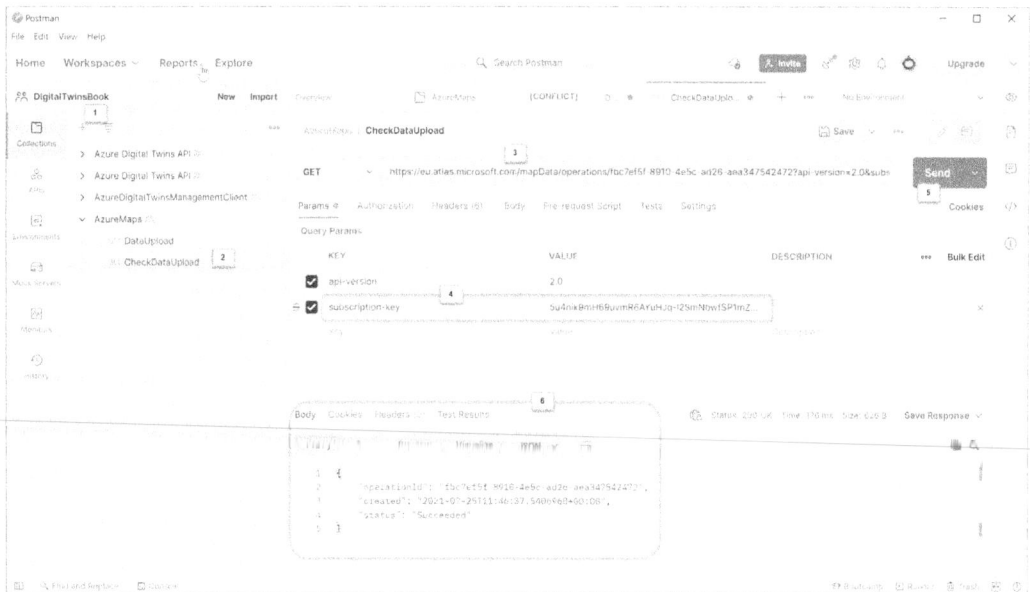

Figure 13.15 – Checking the status of the upload

We need to retrieve the resource location of the upload. Execute the following steps:

1. Select the **Headers** tab of the result of the request.

2. Copy the URL from the `Resource-Location` field. This contains a URL to the location of the uploaded package.

The following screenshot highlights this process:

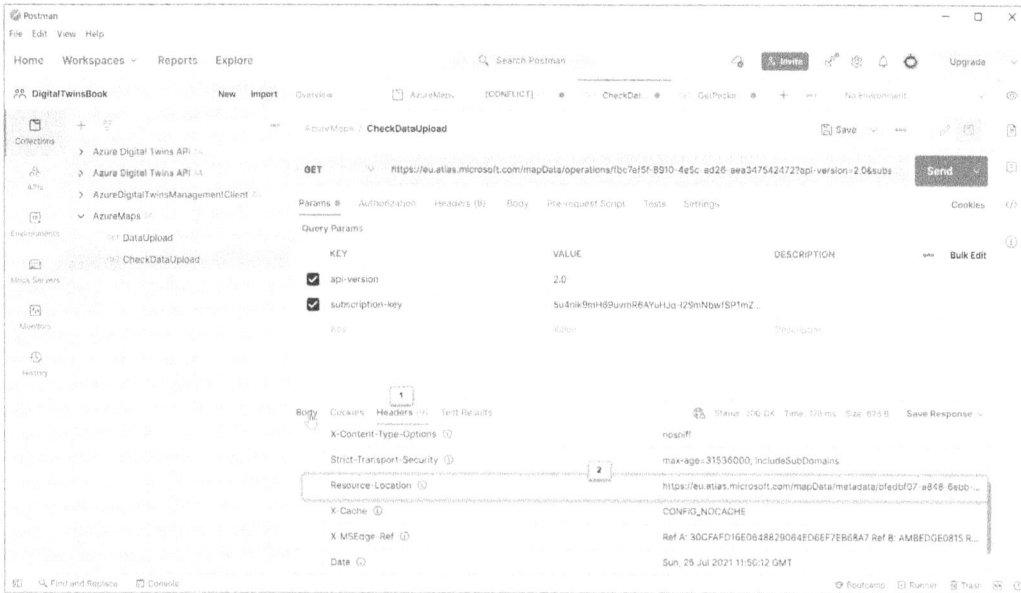

Figure 13.16 – Getting the resource location when the upload has succeeded

The final step is retrieving the metadata—in particular, the {udid} field, which is needed to convert the package—from the resource. Execute the following steps:

1. Select the DigitalTwinsBook environment.

2. Add a new request and name it GetPackageMetadata.

3. Copy the Resource-Location value we got from the previous call into the URL field.

4. Add the subscription-key value to the list of parameters with the {authenticationId} value.

5. Click on the **Send** button to execute the request.

The following screenshot highlights this process:

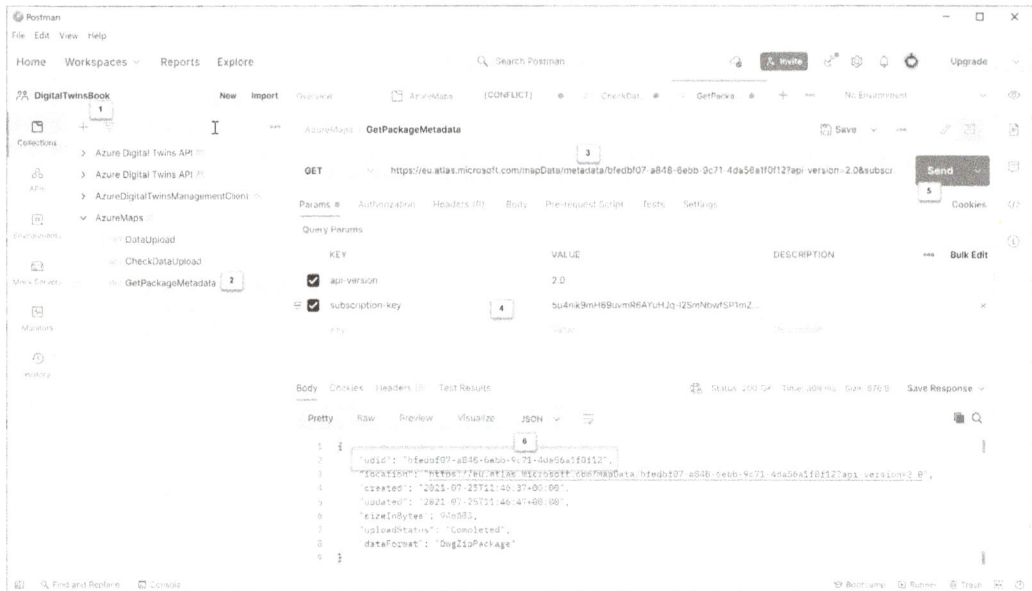

Figure 13.17 – Retrieving the udid value from the resource location

The result will contain a body with several metadata from the uploaded package. We will find the {udid} value in there.

We have now uploaded a drawing package into our Azure Maps account. We will be converting the package in the next section.

Converting a map

Each drawing package uploaded to the Azure Maps account needs to be converted. The conversion will convert the package into understandable map data. We will be using the Conversion API for this task. We need to create a POST URL for converting the data. Make sure to change the method to POST. The URL definition is shown here:

```
https://eu.atlas.microsoft.com/conversions?
subscription-key={AuthenticationId}&
api-version=2.0&
udid={udid}&
inputType=DWG&
outputOntology=facility-2.0
```

Our {authenticationId} value is ce84cef4-95de-4cab-8499-a72430e965df and the {udid} value is bfedbf07-a848-6ebb-9c71-4da56a1f0f12, which is the result from the previous call. The URL becomes this:

```
https://eu.atlas.microsoft.com/conversions?subscription-
key=5u4nik9mH69uvmR6AYuHJq-12SmNbwfSP1mZGxLfAkQ&api-
version=2.0&udid=bfedbf07-a848-6ebb-9c71-
4da56a1f0f12&inputType=DWG&outputOntology=facility-2.0
```

Execute the following steps:

1. Select the `DigitalTwinsBook` environment.

2. Add a new request and name it `ConvertPackage`.

3. Copy the URL shown in the preceding code snippet into the URL field and change the method to `POST`.

4. Press the **Send** button to execute the request.

5. Select the **Headers** tab in the result.

6. Copy the `Resource-Location` URL.

    The following screenshot highlights this process:

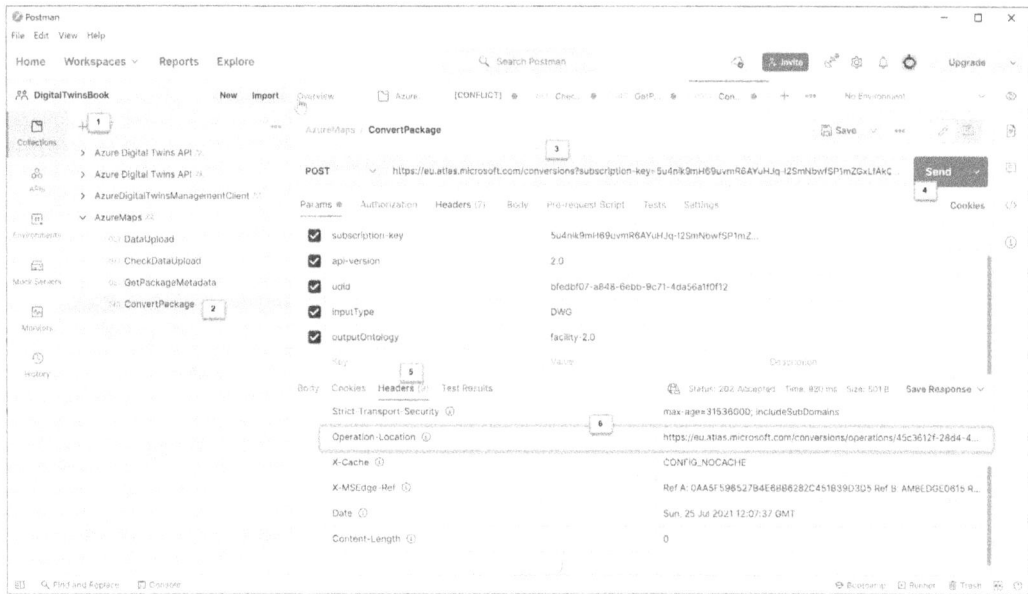

Figure 13.18 – Converting the drawing package

The `Resource-Location` URL allows us the check the conversion status. Execute the following steps:

1. Select the `DigitalTwinsBook` environment.

2. Add a new request and name it `CheckConversion`.

3. Copy the URL in the preceding screenshot into the URL field.

4. Press the **Send** button to execute the request.

5. When the conversion has finished, the body of the result will show a status of `Succeeded`.

6. The following screenshot highlights this process:

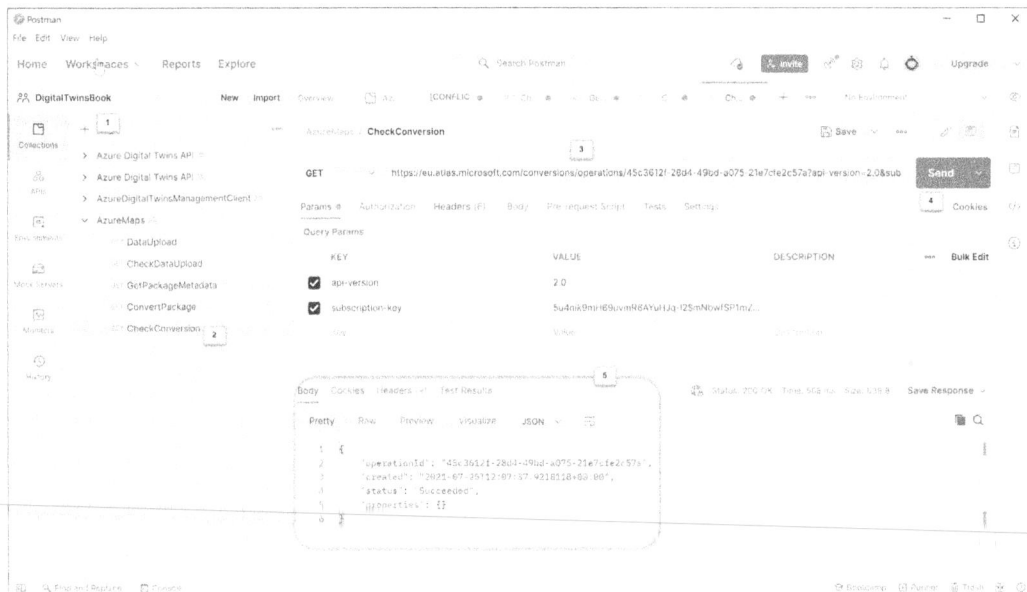

Figure 13.19 – Creating a request for checking the conversion status

We need to get the conversion **identifier (ID)** to perform the upcoming calls.

1. Click on the **Headers** tab of the result, as shown in *Figure 13.20*. The `Resource-Location` field contains a URL similar to this:

   ```
   https://eu.atlas.microsoft.com/conversions/b1e3dc72-2fdf-
   ebe9-1512-ee0a31887662?api-version=2.0
   ```

2. The bold part of the URL is the `{conversionId}` value. This will be used in the next call to create a dataset.

The following screenshot highlights this process:

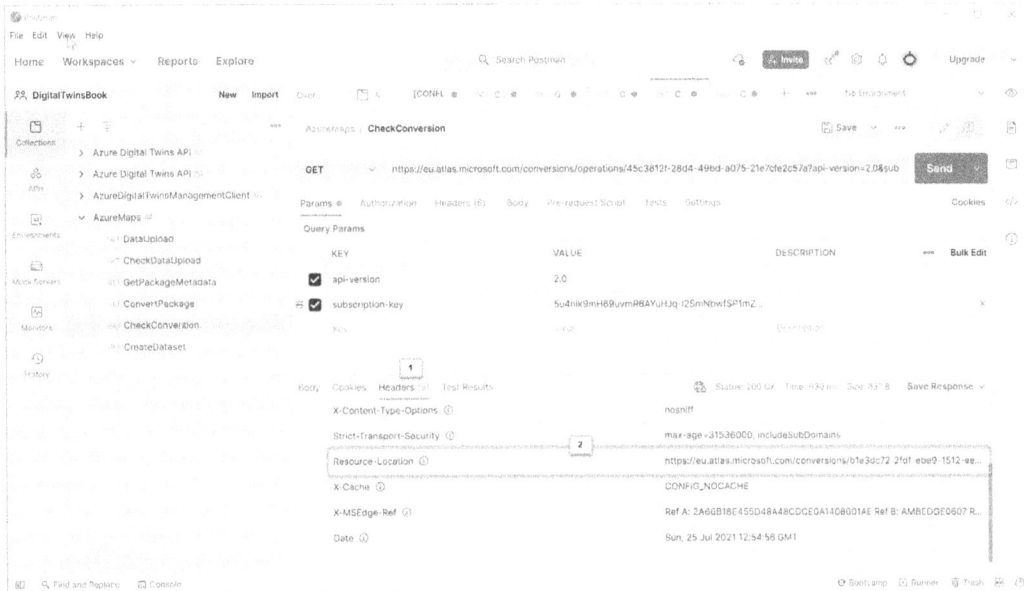

Figure 13.20 – Retrieving the conversion ID from the resource location URL

We have now converted the drawing package to understandable map data. In the next section, we will be creating a dataset.

Creating and validating a dataset

A **dataset** represents a collection of map features. Map features are areas such as the building, levels, and rooms. For that, we will need the `Dataset Create` API. We need to create a `POST` URL for creating a dataset. The URL definition is shown here:

```
https://eu.atlas.microsoft.com/datasets?
api-version=2.0&
conversionId={conversionId}&
subscription-key={AuthenticationId}
```

Our {`authenticationId`} value is `ce84cef4-95de-4cab-8499-a72430e965df` and the {`conversionId`} value is `b1e3dc72-2fdf-ebe9-1512-ee0a31887662`, which is the result from the previous call. The URL becomes this:

```
https://eu.atlas.microsoft.com/datasets?api-version=2.0&con
versionId=b1e3dc72-2fdf-ebe9-1512-ee0a31887662&subscription-
key=5u4nik9mH69uvmR6AYuHJq-l2SmNbwfSP1mZGxLfAkQ
```

Execute the following steps:

1. Select the `DigitalTwinsBook` environment.

2. Add a new request and name it `CreateDataset`.

3. Copy the URL shown in the preceding code snippet into the URL field and change the method to `POST`.

4. Press the **Send** button to execute the request.

5. Select the **Headers** tab in the result.

6. Copy the `Operation-Location` URL.

    The following screenshot highlights this process:

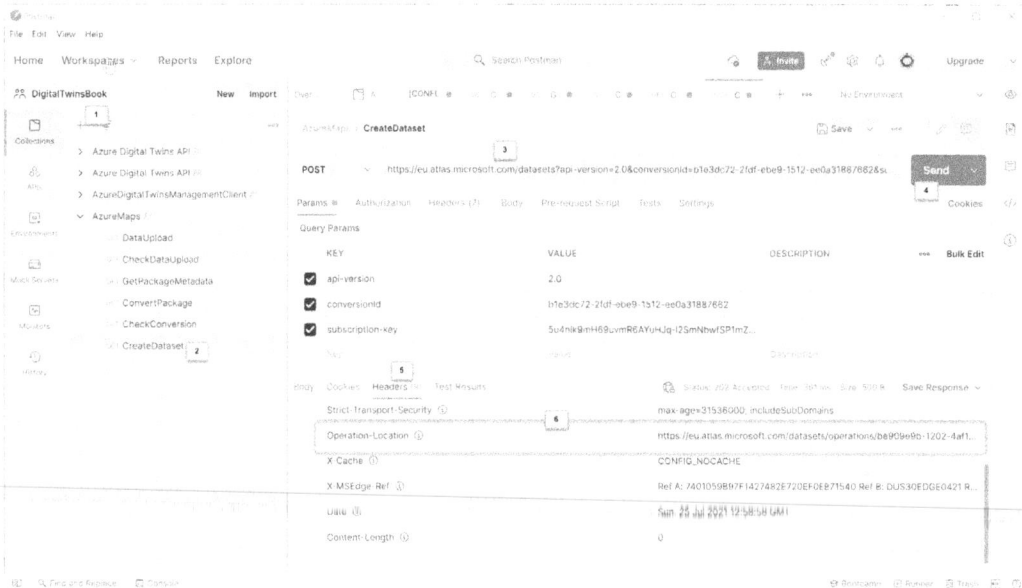

Figure 13.21 – Creating a dataset

The `Operation-Location` URL allows us the validate the creation of the dataset. Execute the following steps:

1. Select the `DigitalTwinsBook` environment.

2. Add a new request and name it `ValidateDataset`.

3. Copy the URL in the preceding screenshot into the URL field. Add the authentication key to the `subscription-key` value.

4. Press the **Send** button to execute the request.

5. When the dataset is created, the body of the result will show a status of `Succeeded`.

The following screenshot highlights this process:

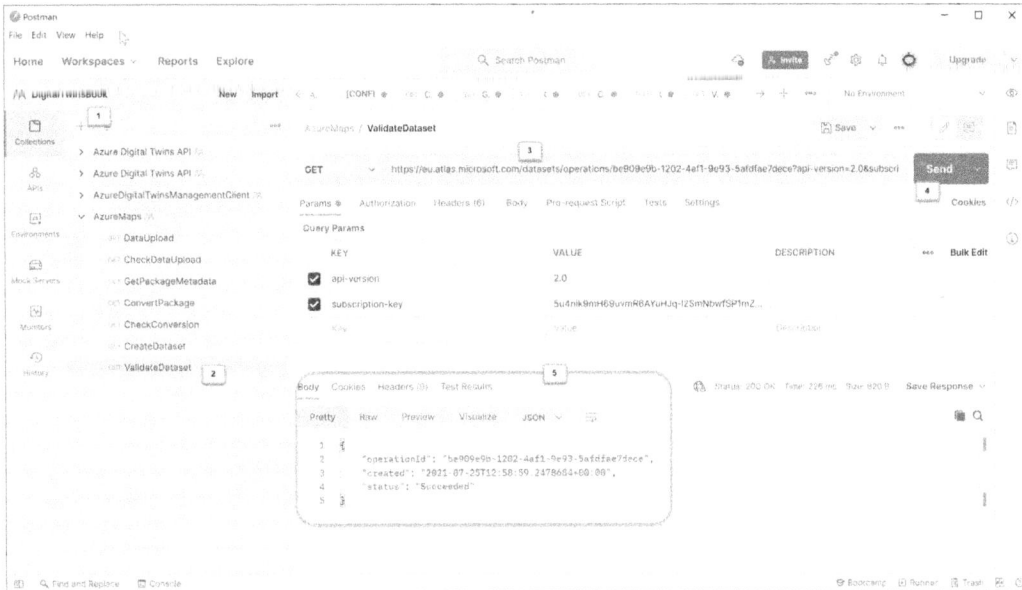

Figure 13.22 – Validating the creation of the dataset

We now need to get the dataset ID to perform the upcoming calls. Execute the following steps:

1. Click on the **Headers** tab of the result.

2. The `Resource-Location` value contains a URL similar to this:

```
https://eu.atlas.microsoft.com/datasets/bd187289-0b1e-
6b4e-6738-0e64f3b4cf9e?api-version=2.0
```

The bold part of the URL is the `{datasetId}` value. This will be used in the next call to create a tileset.

The following screenshot highlights this process:

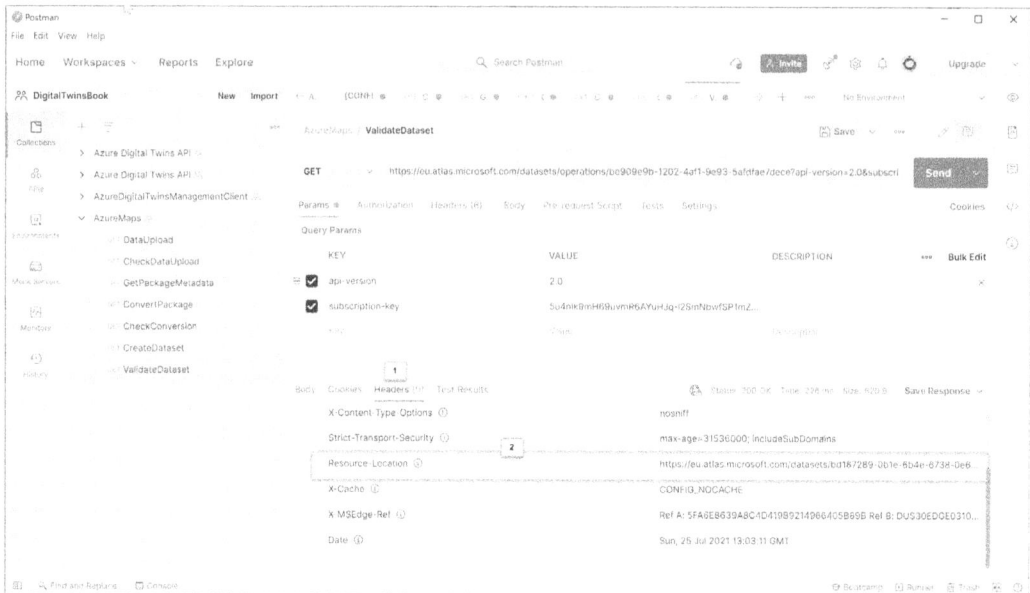

Figure 13.23 – Getting the dataset ID

We have now created a dataset. In the next section, we will be creating a tileset.

Creating and validating a tileset

A **tileset** represents a set of vector tiles that are rendered onto the indoor map. These vector tiles represent certain areas and are based on the content of the dataset. Even when the dataset is removed, the tileset will continue to live. For that, we need to use the Tileset API. We need to create a POST URL for creating a tileset. The URL definition is shown here:

```
https://eu.atlas.microsoft.com/tilesets?
api-version=2.0&
datasetID={datasetId}&
subscription-key={AuthenticationId}
```

Our {authenticationId} value is ce84cef4-95de-4cab-8499-a72430e965df and the {datasetId} value is bd187289-0b1e-6b4e-6738-0e64f3b4cf9e, which is the result from the previous call. The URL becomes this:

```
https://eu.atlas.microsoft.com/tilesets?api-
version=2.0&datasetID=bd187289-0b1e-6b4e-6738-
0e64f3b4cf9e&subscription-key=5u4nik9mH69uvmR6AYuHJq-
12SmNbwfSP1mZGxLfAkQ
```

Execute the following steps:

1. Select the DigitalTwinsBook environment.

2. Add a new request and name it CreateTileset.

3. Copy the URL shown in the preceding code snippet into the URL field.

4. Change the method to POST.

5. Press the **Send** button to execute the request.

6. Select the **Headers** tab in the result.

7. Copy the Operation-Location URL.

The following screenshot highlights this process:

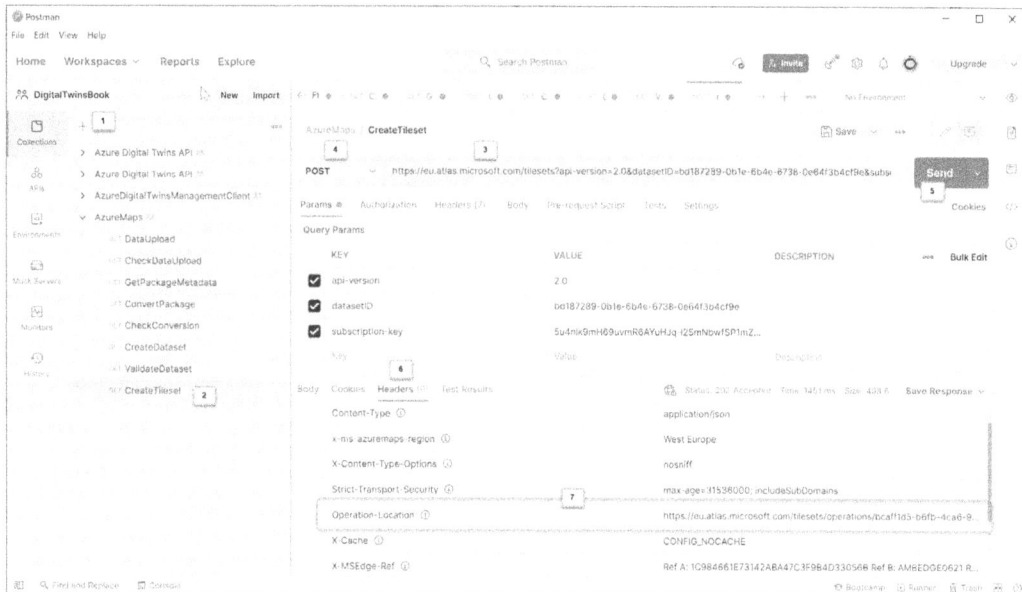

Figure 13.24 – Creating a tileset

The `Operation-Location` URL allows us the validate the creation of a tileset. Execute the following steps:

1. Select the `DigitalTwinsBook` environment.

2. Add a new request and name it `ValidateTileset`.

3. Copy the `Operation-Location` value shown in the preceding screenshot into the URL field. Add the authentication key to the `subscription-key` value.

4. Press the **Send** button to execute the request.

5. When the tileset is created, the body of the result will display a status of `Running`.

The following screenshot highlights this process:

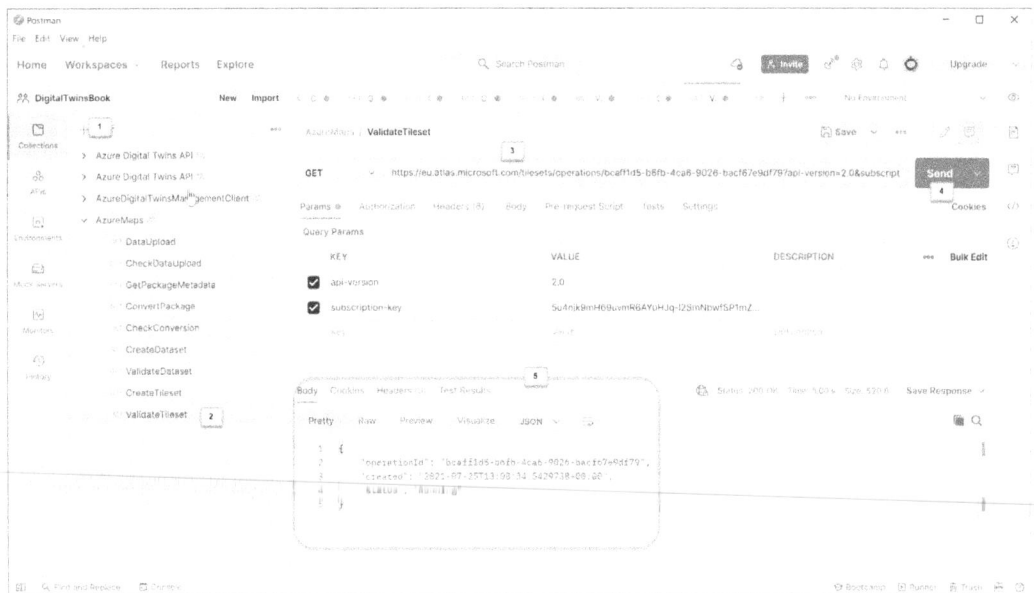

Figure 13.25 – Validating and checking the tileset creation

We now need to get the tileset ID to perform the upcoming calls. Execute the following steps:

1. Click on the **Headers** tab of the result.

2. The `Resource-Location` field contains a URL similar to this:

    ```
    https://eu.atlas.microsoft.com/tilesets/b9434ccd-a66e-
    9052-4b8c-948dc4579a33?api-version=2.0
    ```

The bold part of the URL is the `{tilesetId}` value.

The following screenshot highlights this process:

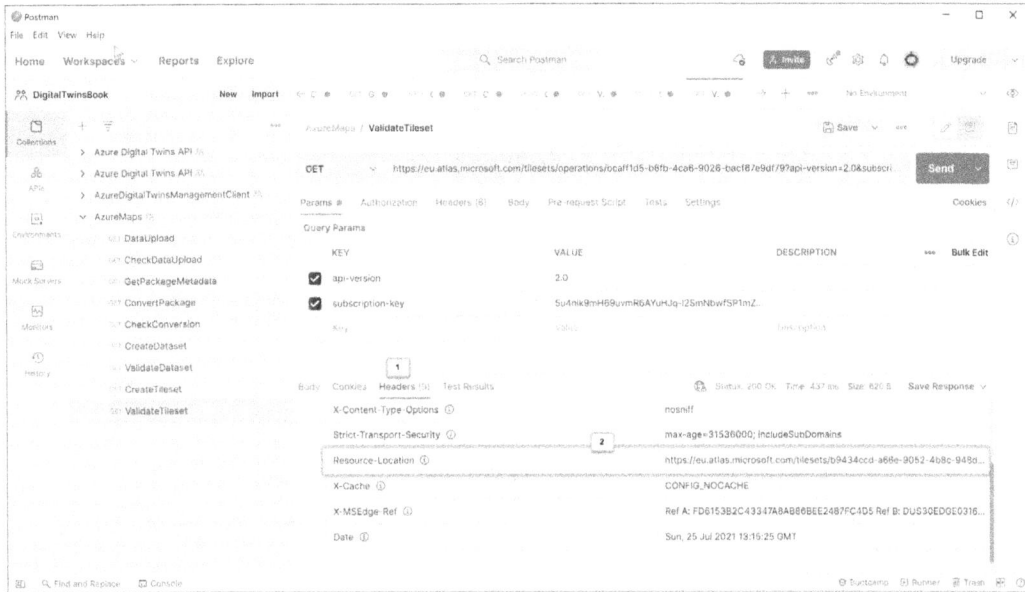

Figure 13.26 – Getting the tileset ID

We have now created a tileset. In the final section. we will be creating a feature stateset.

Creating and validating a feature stateset

A **feature stateset** represents the dynamics properties and values on features of the units in an indoor map. It allows, for example, to change the color of such a unit based on a value. For that, we need to use the `Stateset` API. We need to create a `POST` URL for creating a feature stateset. The URL definition is shown here:

```
https://eu.atlas.microsoft.com/featurestatesets?
api-version=2.0&
datasetId={datasetId}&
subscription-key={authenticationId}
```

Our `{authenticationId}` value is `ce84cef4-95de-4cab-8499-a72430e965df` and the `{datasetId}` value is `bd187289-0b1e-6b4e-6738-0e64f3b4cf9e`, which is the result from one of the previous calls. The URL becomes this:

```
https://eu.atlas.microsoft.com/featurestatesets?api-
version=2.0&datasetId=bd187289-0b1e-6b4e-6738-
```

```
0e64f3b4cf9e&subscription-key=5u4nik9mH69uvmR6AYuHJq-
12SmNbwfSP1mZGxLfAkQ
```

Execute the following steps:

1. Select the `DigitalTwinsBook` environment.
2. Add a new request and name it `CreateStateset`.
3. Copy the URL shown in the preceding code snippet into the URL field.
4. Change the method to `POST`.
5. Select the **Headers** tab and add `Content-Type` with the value `application/json`.
6. Select the **Body** tab.
7. Change the option to **raw**.
8. Copy the **JavaScript Object Notation (JSON)** payload in the body field. The JSON payload can be found in the screenshot shown next and is explained in more detail shortly.
9. Press the **Send** button to execute the request.
10. When the call is finished, the `{tilesetId}` value is returned in the body of the result.

    The following screenshot highlights this process:

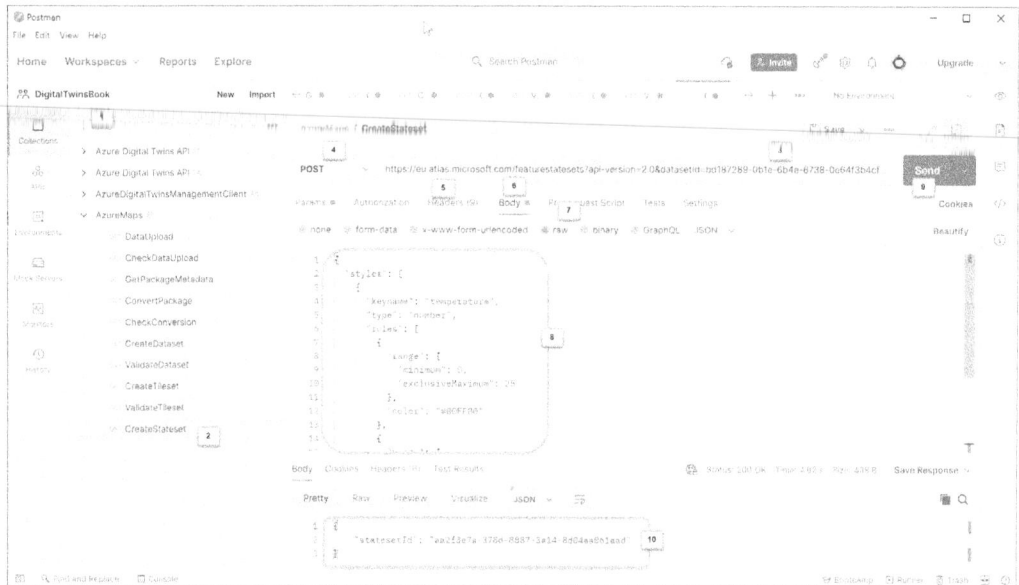

Figure 13.27 – Creating a feature stateset

The JSON payload for a feature stateset describes a specific feature and additional rules, as you can see in the following code snippet. In this scenario, we specified a feature called `temperature`, which is a `number` value:

```
{
    "styles": [
        {
            "keyname": "temperature",
            "type": "number",
            "rules": [
                {
                    "range": {
                        "minimum": 0,
                        "exclusiveMaximum": 25
                    },
                    "color": "#00FF00"
                },
                {
                    "range": {
                        "minimum": 25
                    },
                    "color": "#FF0000"
                }
            ]
        }
    ]
}
```

The second part of the JSON payload contains the rules. In this example, we have specified two rules. The first rule is valid when the temperature value is from 0 and 24. The second rule is valid when the temperature is 25 or higher. The rule that validates determines the color used to color the area. If more than one rule applies, the first rule that validates is used.

There are several ways of creating rules. You can find more information on this here: https://docs.microsoft.com/en-us/azure/azure-maps/schema-stateset-stylesobject.

In this final section, we have learned how to create a feature stateset. We have also learned how the rules apply to a certain feature to dynamically update an indoor map.

Summary

In this chapter, we have learned how to set up and configure an Azure Maps account. We have also learned how to configure Azure Maps using a drawing package, a dataset, a tileset, and a feature stateset. This is required to move to the next step to start connecting to Azure Digital Twins and to visualize a digital twin.

In the next chapter, we will learn how to connect this Azure Maps account to an event grid using an Azure function. We will learn how to update the feature stateset based on a change of the temperature value from a sensor.

Questions

As we conclude, here is a list of questions for you to test your knowledge regarding this chapter's material. You will find the answers in the *Assessments* section of the *Appendix*:

1. Which Azure service is used to upload a drawing package and create elements such as a dataset, tileset, and feature stateset?

    a. Azure Maps account

    b. Azure Creator service

    c. Azure Functions

2. Can you delete a dataset without harming the tileset?

    a. No

    b. Yes

3. What should the API URLs start with when you have selected a European location for the Azure Maps account?

    a. `https://eu.atlas.microsoft.com`

    b. `https://us.atlas.microsoft.com`

    c. `https://common.atlas.microsoft.com`

Further reading

To learn more on this topic, visit the following links:

- *Get the primary key for your account:*

  `https://docs.microsoft.com/en-us/azure/azure-maps/quick-demo-map-app#get-the-primary-key-for-your-account`

- *Manage Azure Maps Creator:*

  `https://docs.microsoft.com/en-us/azure/azure-maps/how-to-manage-creator`

14
Integrating Azure Maps

The Azure Maps account has been created and configured, as covered in the previous chapter. In this chapter, we will be integrating the Azure Maps account with our Event Grid. This integration involves setting up and creating an Azure Function, subscribing it to Event Grid, and monitoring the results. We will also visualize the indoor map model. We will learn how a change in a `temperature` property of a digital twin sensor is visualized using a **feature stateset**.

In this chapter, we will be covering the following topics:

- Updating a feature stateset
- Setting up an update Azure Function
- Publishing the Azure Function
- Configuring application settings
- Subscribing to Event Grid
- Monitoring updates
- Visualizing the model

By the end of this chapter, we will be able to update an external service such as Azure Maps using an Azure Function that is triggered by Event Grid through a subscription.

Technical requirements

We will be using the Azure portal to create several Azure services to support an indoor map. To preconfigure the elements, we will be using **application programming interface (API)** calls via the Postman application.

Updating a feature stateset

We started with building, uploading, and creating a map in the previous chapter. Based on the map, we created a dataset, tileset, and a feature set. In this section, we will be updating the feature set.

We will be focusing on building the integration between Event Grid and the Azure Maps account, as shown in the following diagram. We will need to create an Azure Function that is subscribed to an event at one of the Event Grid endpoints:

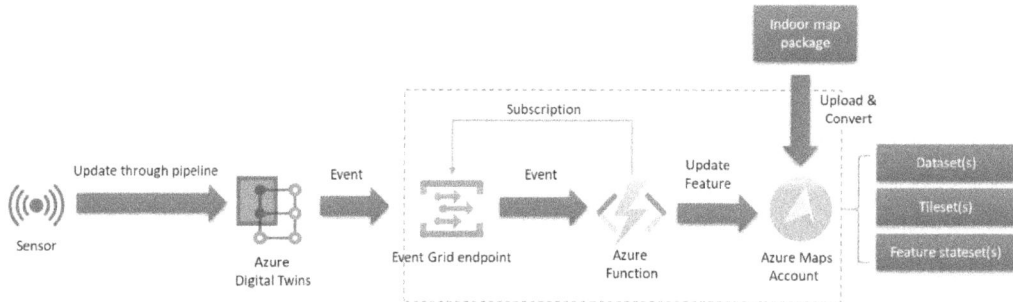

Figure 14.1 – High-level overview of the architecture of the mechanism of updating the Azure map

The Azure Function will call the `Feature Statesets` API to update a feature stateset for a particular unit. To do this, we will use the following API call:

```
https://eu.atlas.microsoft.com/featurestatesets/{statesetId}/
featureStates/{Area}?
api-version=2.0&
subscription-key={authenticationId}
```

Our {authenticationId} value is ce84cef4-95de-4cab-8499-a72430e965df and the {statesetId} value is aa2f3e7a-378d-8887-3a14-8d04aa061a, both being results of the API calls made in the previous chapter.

{Area} indicates the unit from which we want to update the feature stateset. We will be using UNIT111 as an example. The **Uniform Resource Locator (URL)** for it will look like this:

```
https://eu.atlas.microsoft.com/featurestatesets/aa2f3e7a-
378d-8887-3a14-8d04aa061aad/featureStates/UNIT111?api-
version=2.0&subscription-key=5u4nik9mH69uvmR6AYuHJq-
12SmNbwfSP1mZGxLfAkQ
```

We want to make the Azure Function run more dynamically. Therefore, we will be using the name of the digital twin as the name of the unit. This call requires to be executed as a PUT request that has a body containing the updated value. An example of the body is shown here:

```
{
    "states": [
        {
            "keyName": "temperature",
            "value": 26,
            "eventTimestamp": "2021-07-25T17:10:20"
        }
    ]
}
```

The **JavaScript Object Notation (JSON)** payload contains the keyName property of the property that needs to be set. The value is set through the value property. The eventTimestamp property needs to be set to a newer timestamp than the previous update for this unit; otherwise, the value is not updated.

We have learned how a feature stateset is updated to reflect the change of a digital twin. In the following section, we will be setting up the Azure Function to perform the update.

Setting up an update Azure Function

We start by setting up an Azure Function using Visual Studio. Open Visual Studio to start creating a new project, as shown in the following screenshot:

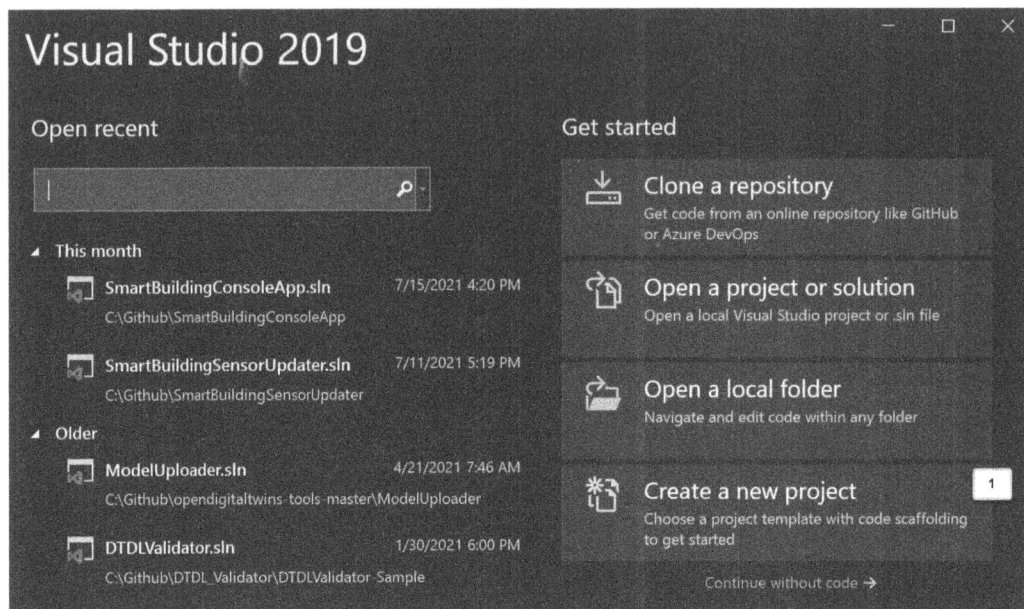

Figure 14.2 – Creating a new project with Visual Studio

We will need to select the right project template. Execute the following steps:

1. Search for Azure Function in the search box.

2. Select the **Azure Functions** template. Make sure you select a **C#** project.

3. Click on the **Next** button.

   The following screenshot highlights this process:

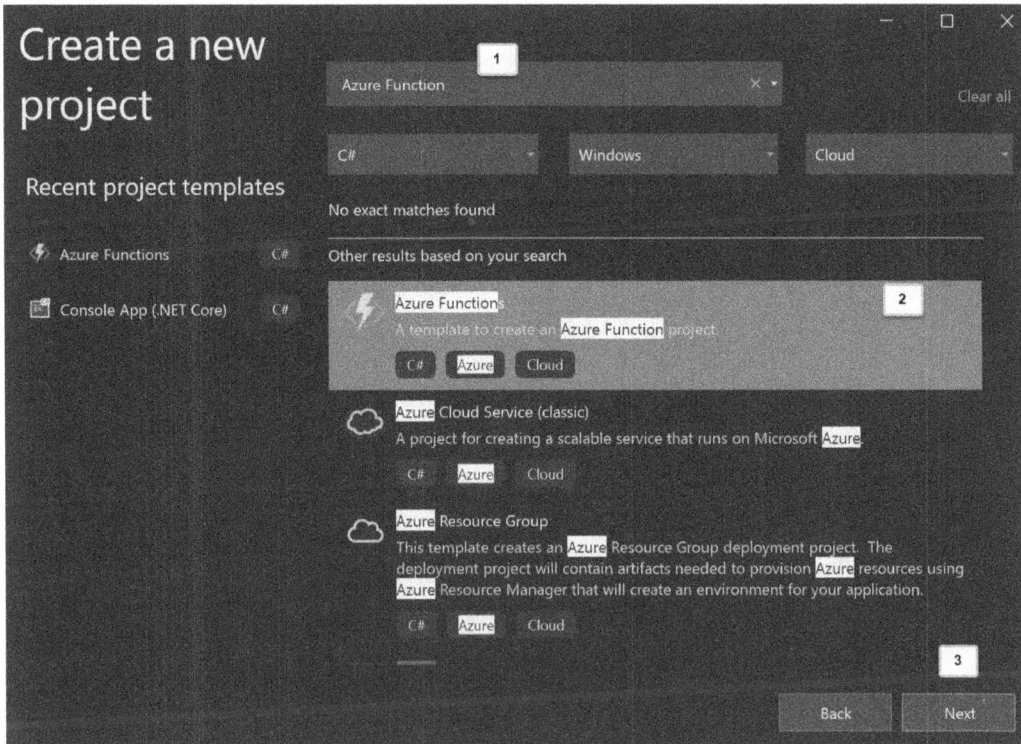

Figure 14.3 – Selecting the Azure Functions template

We will configure the project in the following dialog step. Execute the following steps:

1. Enter the name AzureMapUpdater.

2. Select the C:\Github\ location.

3. Make sure to check the **Place solution and project in the same directory** checkbox to *place the solution in the same folder as the project file*.

4. Click the **Create** button.

The following screenshot highlights this process:

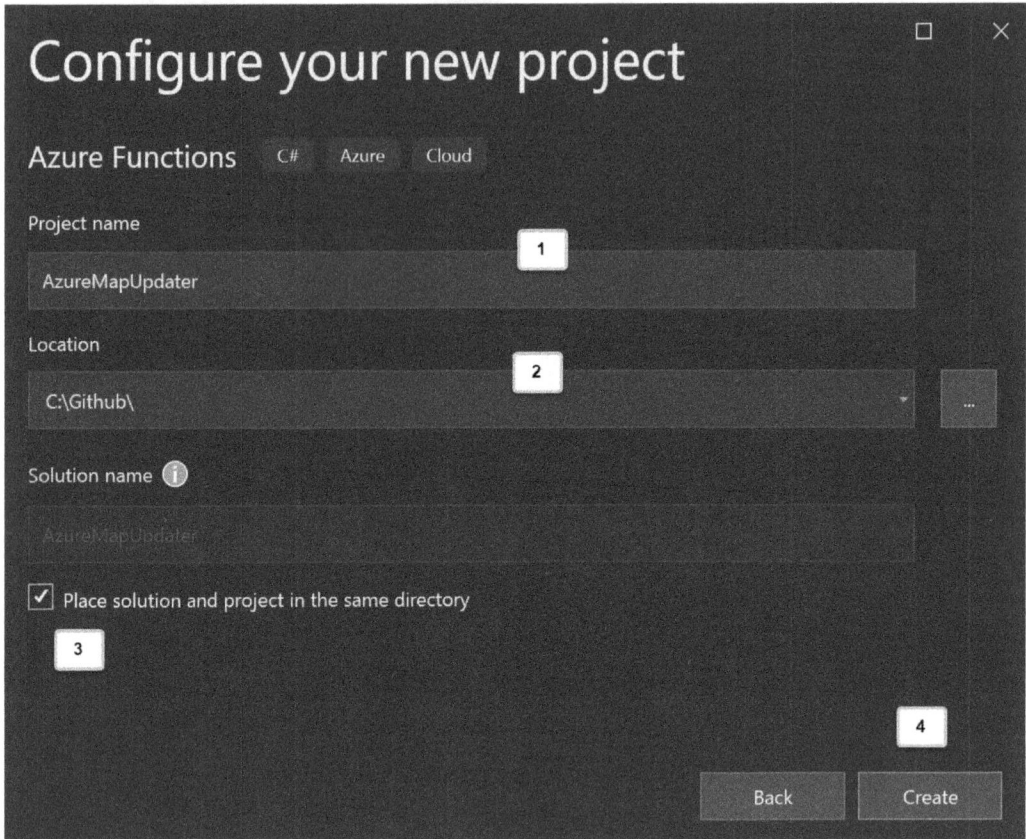

Figure 14.4 – Configuring our new project

In the next dialog step, we need to define which type of trigger we want. We also want to have a new storage account for this Azure Function. Execute the following steps:

1. Select **Event Grid trigger**.

2. Select the **Browse...** option for opening the dialog to select a storage account. This dialog will allow us to create a new storage account.

The following screenshot highlights this process:

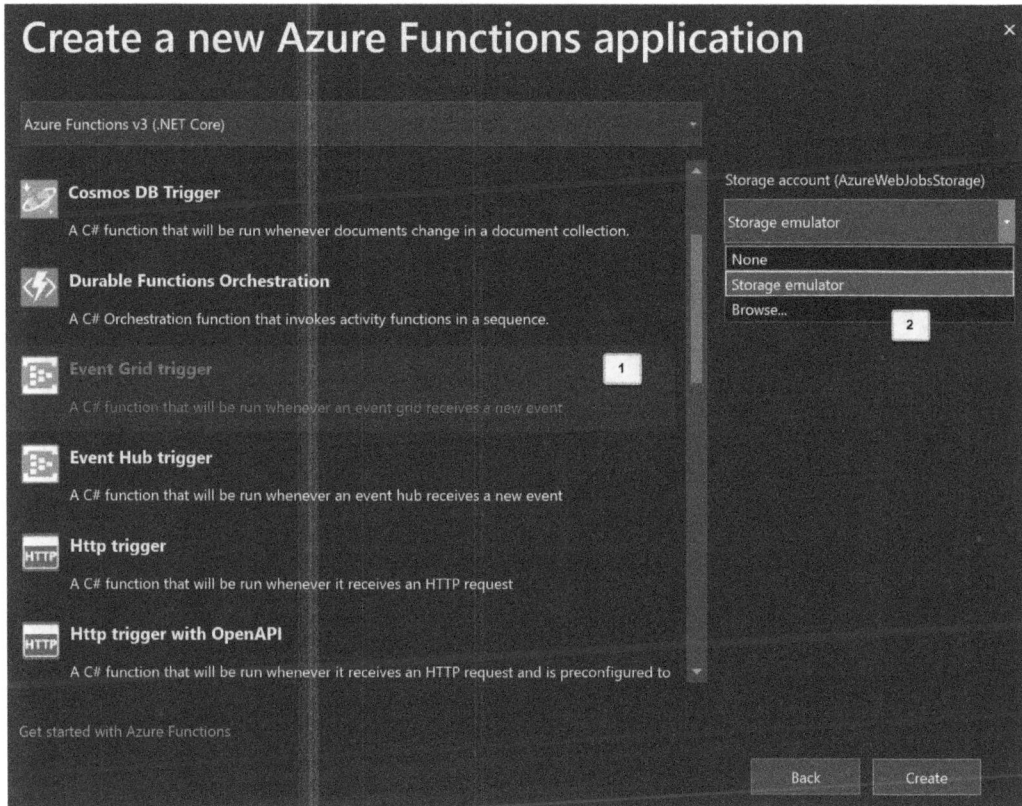

Figure 14.5 – Selecting the type of trigger and starting to create a storage account

We want to create a new storage account. Execute the following steps:

1. Select the right subscription.
2. Click the + sign to create a new storage account.

The following screenshot highlights this process:

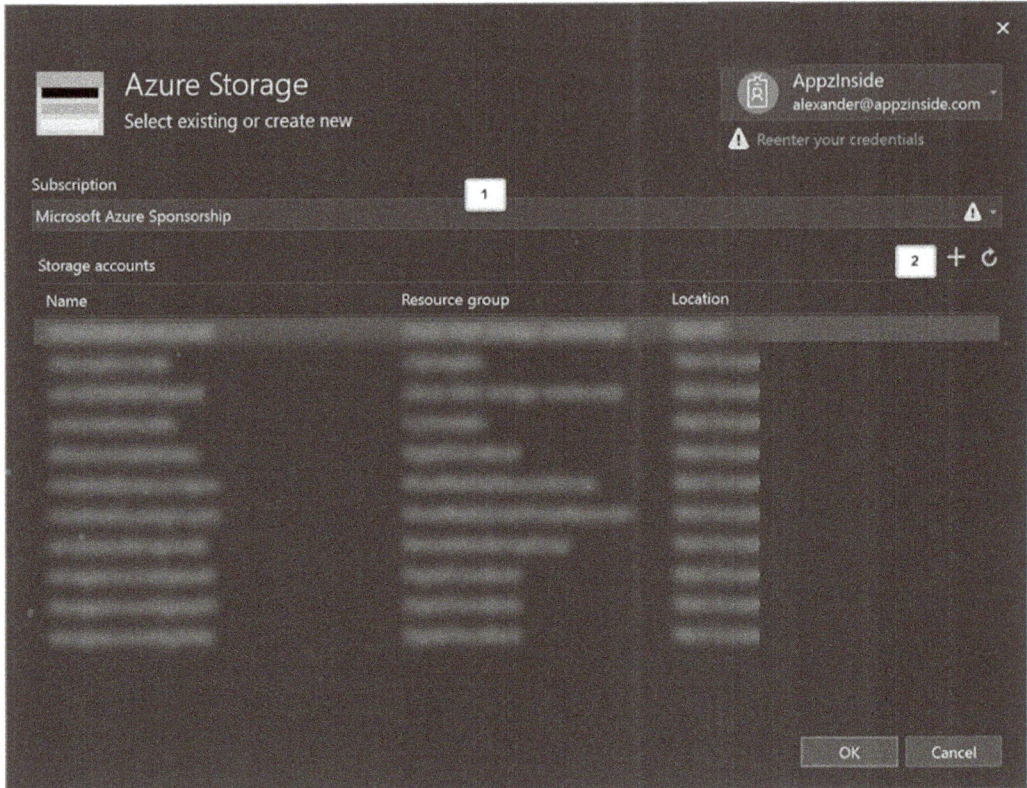

Figure 14.6 – Creating a new Azure Storage account

We are going to create a new storage account. Execute the following steps:

1. Enter `azuremapupdaterstorage` for the account name.

2. Select the right subscription.

3. Select the `DigitalTwinsbook` resource group.

4. Select the closest region. In my case, this is `West Europe`.

5. Select `Standard - Locally Redundant Storage` as the account type.

6. Click the **Create** button.

The following screenshot highlights this process:

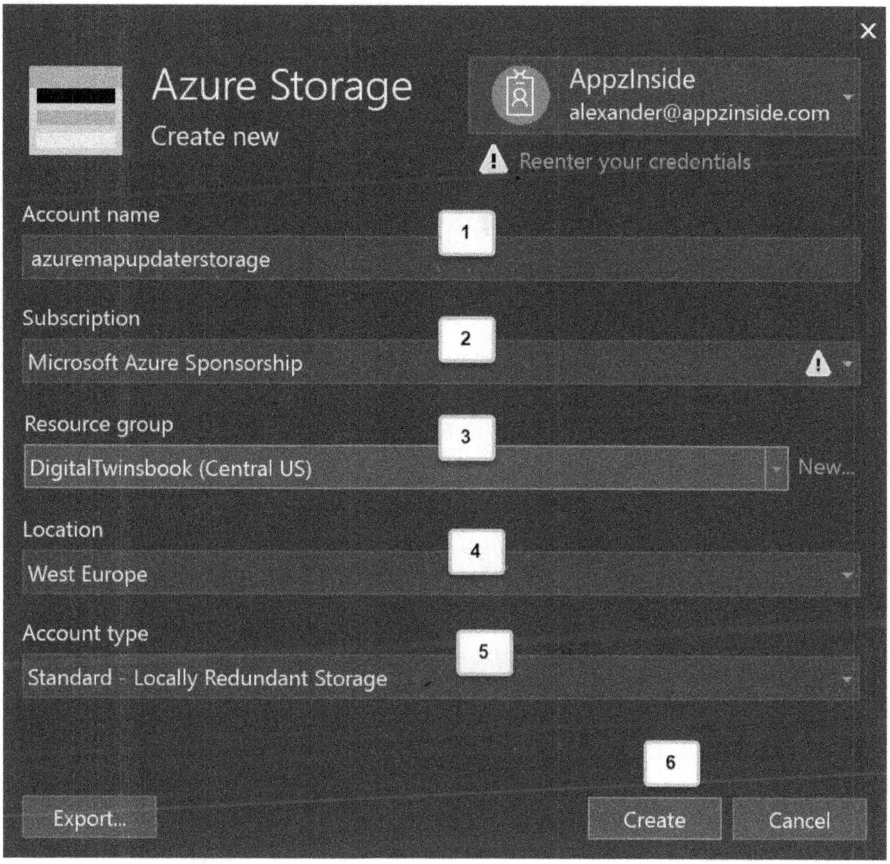

Figure 14.7 – Creating a new Azure Storage account (continued)

We need to select the created storage account for the Azure Function. Execute the following steps:

1. Select the `azuremapupdaterstorage` account from the list.
2. Click the **OK** button.

The following screenshot highlights this process:

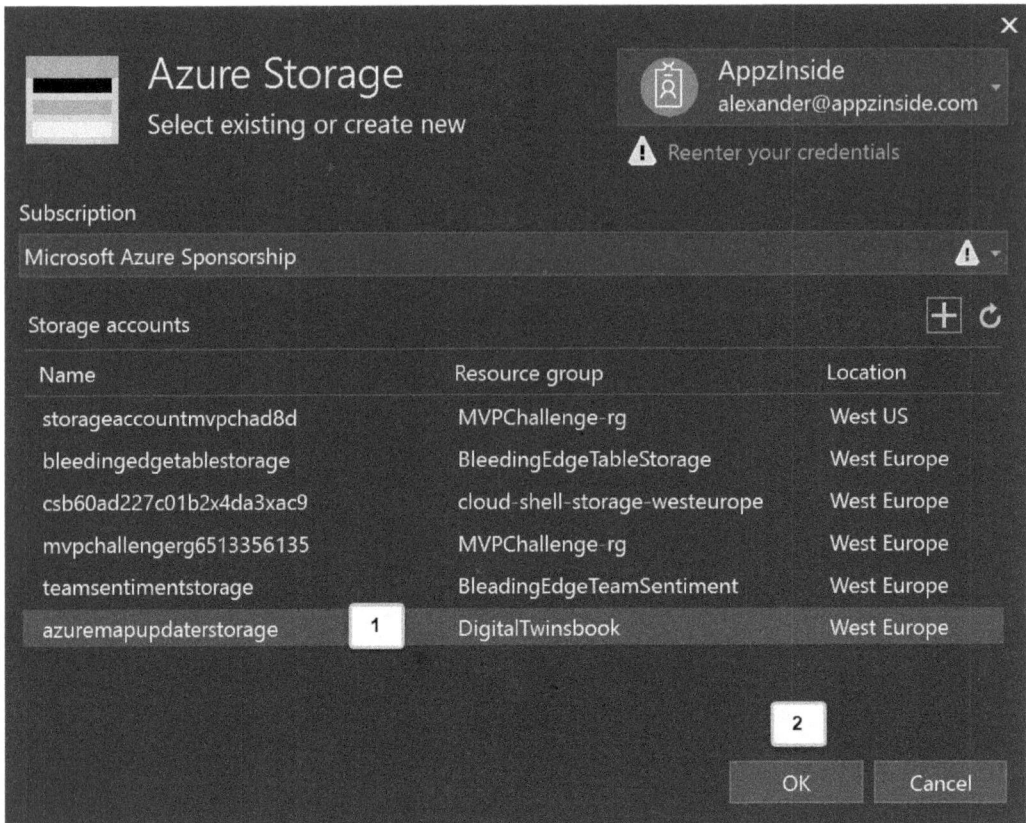

Figure 14.8 – Selecting the created Azure Storage account

We will return to the dialog step for creating an Azure Function. The newly created storage account is selected in the drop-down box. Click on the **Create** button to create an Azure Function.

We have created an Azure Function with a new storage account. In the next section, we will add code to the Azure Function to perform an update on the feature stateset.

The project will contain an empty function. Execute the following steps to rename and prepare the function:

1. Rename `Function1.cs` as `AzureMapUpdater.cs`.

2. Rename the class `AzureMapUpdater`.

3. Rename the function name `[FunctionName("AzureMapUpdater")]`.

The following screenshot highlights this process:

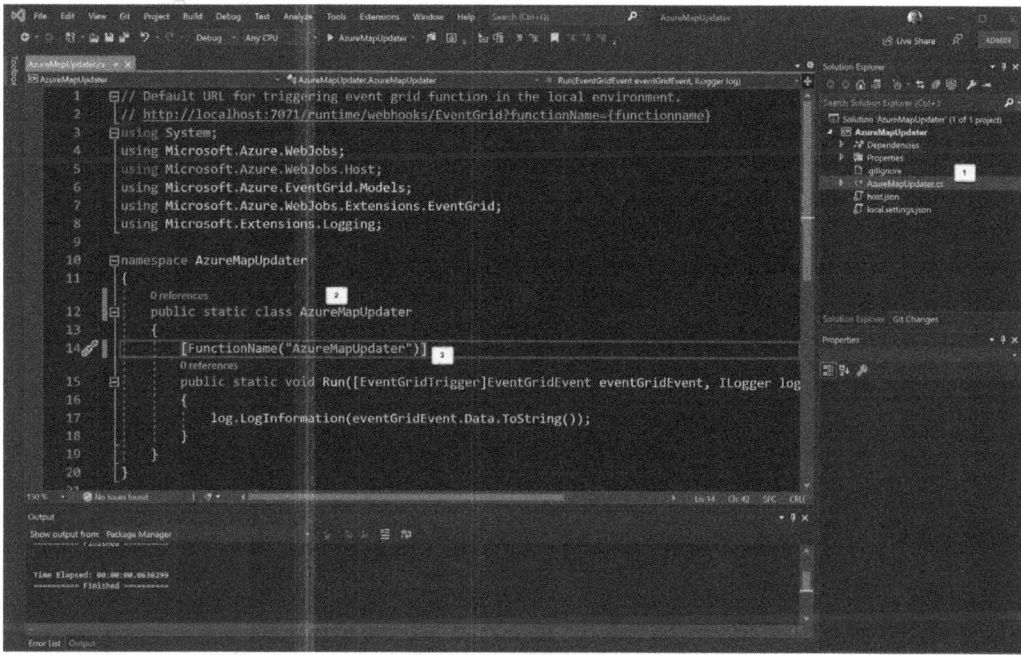

Figure 14.9 – Creating an update function

Let's make some changes to the `AzureMapUpdater.cs` file. Replace the `using` instances within the newly created file with the following code:

```
using System.Net.Http;
using Newtonsoft.Json.Linq;
using Newtonsoft.Json;
using Microsoft.Azure.WebJobs;
using Microsoft.Azure.WebJobs.Host;
using Microsoft.Azure.EventGrid.Models;
using Microsoft.Azure.WebJobs.Extensions.EventGrid;
using Microsoft.Extensions.Logging;
```

We need to have the stateset **identifier** (**ID**) and the subscription key to execute a call to the `Feature Statesets` API. For that, we will need to add two static properties to the class that retrieves both values as environment variables. The values for these variables will be defined later in the application settings of the Azure Function through the Azure portal.

Add the following code at the beginning of the `AzureMapUpdater` class:

```
private static string statesetID = Environment.
GetEnvironmentVariable("statesetid");
private static string subscriptionKey = Environment.
GetEnvironmentVariable("subscriptionkey");
```

We will implement a method that will update the feature stateset using the `Feature Statesets` API. This requires us to use an `HttpClient` post body. The `HttpClient` post body will contain the `keyName`, `value`, and `eventTimestamp` properties. The call will use the `HttpClient.PutAsync()` method. This method uses the URL, as mentioned at the beginning of this chapter, to update a feature stateset. This URL format is shown here:

```
https://eu.atlas.microsoft.com/featurestatesets/{statesetID}/
featureStates/{featureId}?
api-version=2.0&
subscription-key={subscriptionKey}
```

Copy the following `UpdateMapStateset` method into the `AzureMapUpdater` class:

```
static async void UpdateMapStateset(string featureId, string
value, ILogger log)
{
    HttpClient httpClient = new HttpClient();

    log.LogInformation($"AZUREMAP-START");

    try
    {

        var postcontent = new JObject(
            new JProperty(
                "States",
                new JArray(
                    new JObject(
                        new JProperty("keyName",
"temperature"),
```

```
                        new JProperty("value", value),
                        new JProperty("eventTimestamp",
DateTime.UtcNow.ToString("s")))))));


        log.LogInformation($"AZUREMAP-VALUES
FeatureId:{featureId} Value:{value} Subscription-
key:{subscriptionKey}");


        // think it need to be a putasync
        //var response = await httpClient.PostAsync(
        var response = await httpClient.PutAsync(

        $"https://eu.atlas.microsoft.com/featurestatesets/
{statesetID}/featureStates/{featureId}?api-
version=2.0&subscription-key={subscriptionKey}",
            new StringContent(postcontent.ToString()));

        string result = await response.Content.
ReadAsStringAsync();

        log.LogInformation($"AZUREMAP-RESULT:{result}");
    }
    catch(Exception ex)
    {
        log.LogInformation($"AZUREMAP-UPDATEERROR error:{ex.
Message}");
    }

    log.LogInformation($"AZUREMAP-END");
}
```

Replace the existing Run method in the AzureMapUpdater class with the following code:

```
public static void Run([EventGridTrigger] EventGridEvent
eventGridEvent, ILogger log)
{
```

```
    JObject message = (JObject)JsonConvert.
DeserializeObject(eventGridEvent.Data.ToString());

    string twinId = eventGridEvent.Subject;
    string modelId = message["data"]["modelId"].ToString();

    //Parse updates to "space" twins
    if (modelId == "dtmi:com:smartbuilding:Sensor;1")
    {
        // Iterate through the properties that have changed
        foreach (var operation in message["data"]["patch"])
        {
            if (operation["op"].ToString() == "replace" &&
operation["path"].ToString() == "/temperature")
            {
                string value = operation["value"].ToString();

                log.LogInformation($"AZUREMAP-RECEIVED
twinId:{twinId} modelId:{modelId} temperaturevalue:{value}");

                // Update the maps feature stateset
                UpdateMapStateset(twinId, value, log);
            }
        }
    }
}
```

The method starts by deserializing the data coming from the Event Grid event. The next step is determining the twinId and the modelId values. We need to check if the modelId value equals a digital twin of the model of type dtmi:com:smartbuilding:Sensor;1. If the digital twin is a sensor, the patch data is retrieved. The patch data is used to call the UpdateMapStateset() function to update the feature stateset of the unit with the name twinId.

We have now created an Azure Function to update a feature stateset of the indoor map specified by the Azure Maps account. We will be publishing the Azure Function in the next section of this chapter.

Publishing the Azure Function

We need to publish the Azure Function to Azure. In this section, we will be creating an Azure publishing profile. Execute the following steps:

1. Right mouse-click on the project and select the **Publish** menu option.
2. Select **Azure** as the target.
3. Click the **Next** button.

The following screenshot highlights this process:

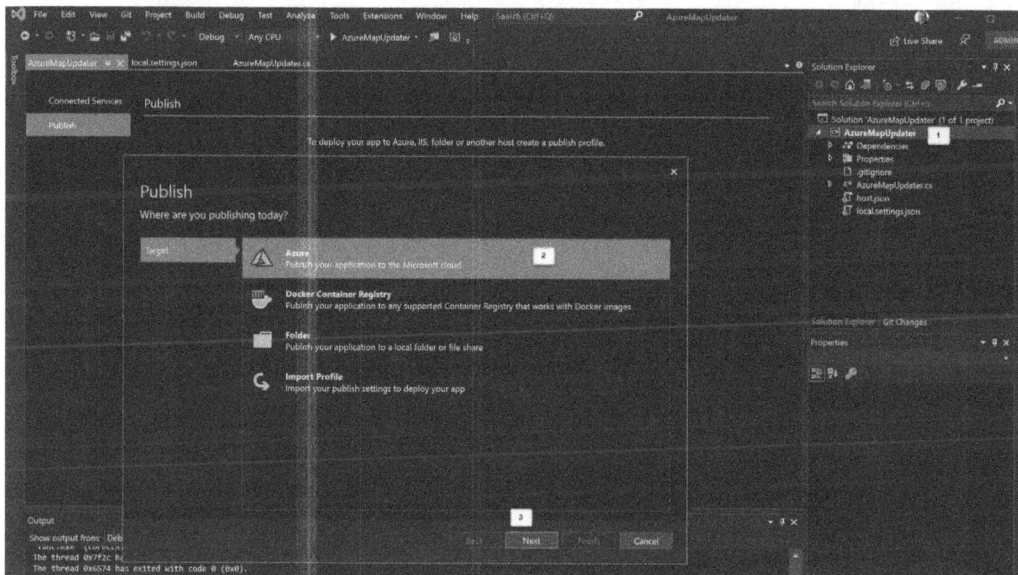

Figure 14.10 – Starting to create a publishing profile for the Azure Function

We need to specify which kind of target we want to use. Execute the following steps:

1. Select **Azure Function App** as the target.
2. Click the **Next** button to continue.

The following screenshot highlights this process:

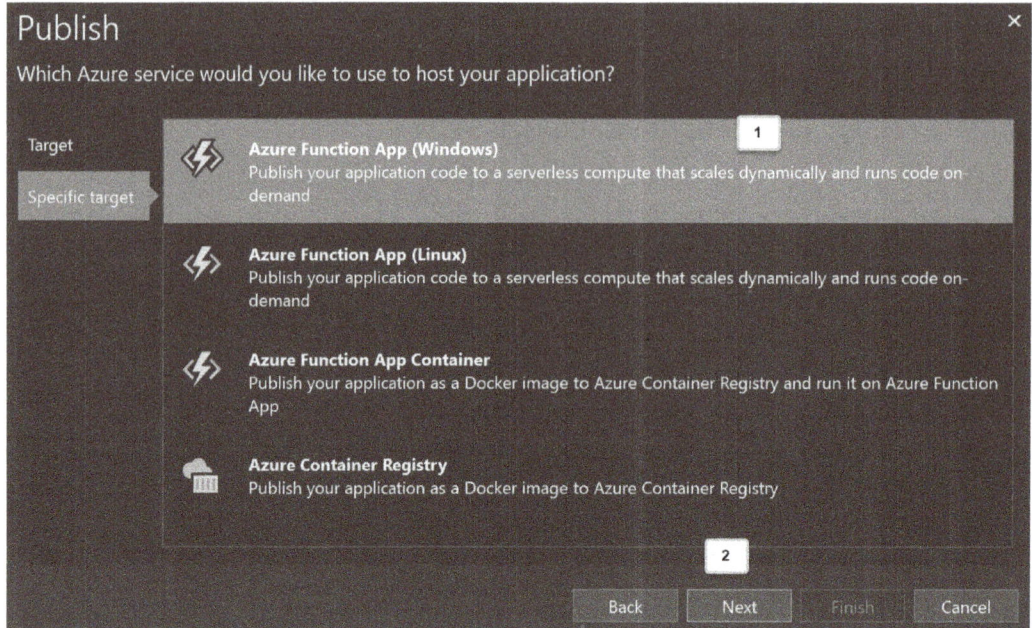

Figure 14.11 – Selecting a target for the publishing profile

We need to specify which instance in Azure we need to use as the destination for our Azure Function. Execute the following steps:

1. Set the **View** field to `Resource group`.
2. Click the + button to create a new function app.

The following screenshot highlights this process:

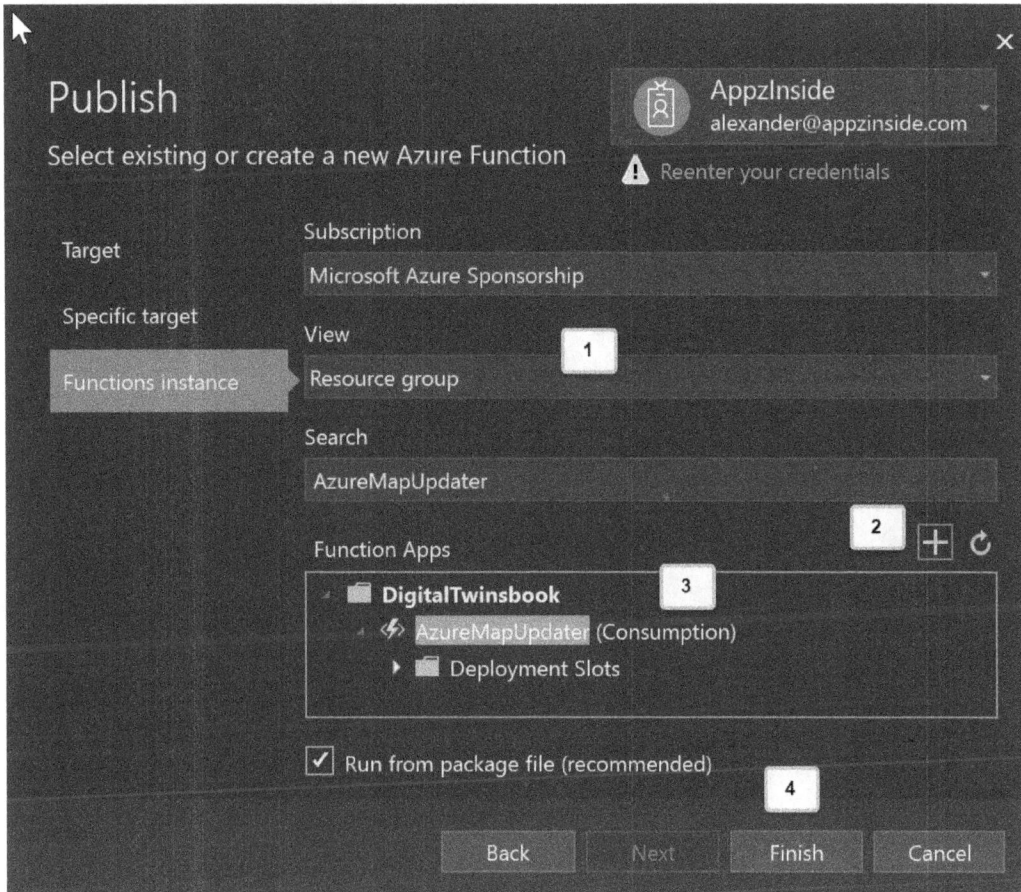

Figure 14.12 – Configuring the destination in the publishing profile

Before we continue with the other steps shown in *Figure 14.12*, we need to create an Azure function first by executing the following steps:

1. Enter the name AzureMapUpdater.
2. Select your subscription.
3. Select DigitalTwinsbook as the resource group.
4. Set the plan type to Consumption.
5. Choose the closest region. For me, this is West Europe.
6. Select azuremapupdaterstorage as the storage account.
7. Click on the **Create** button.

The following screenshot highlights this process:

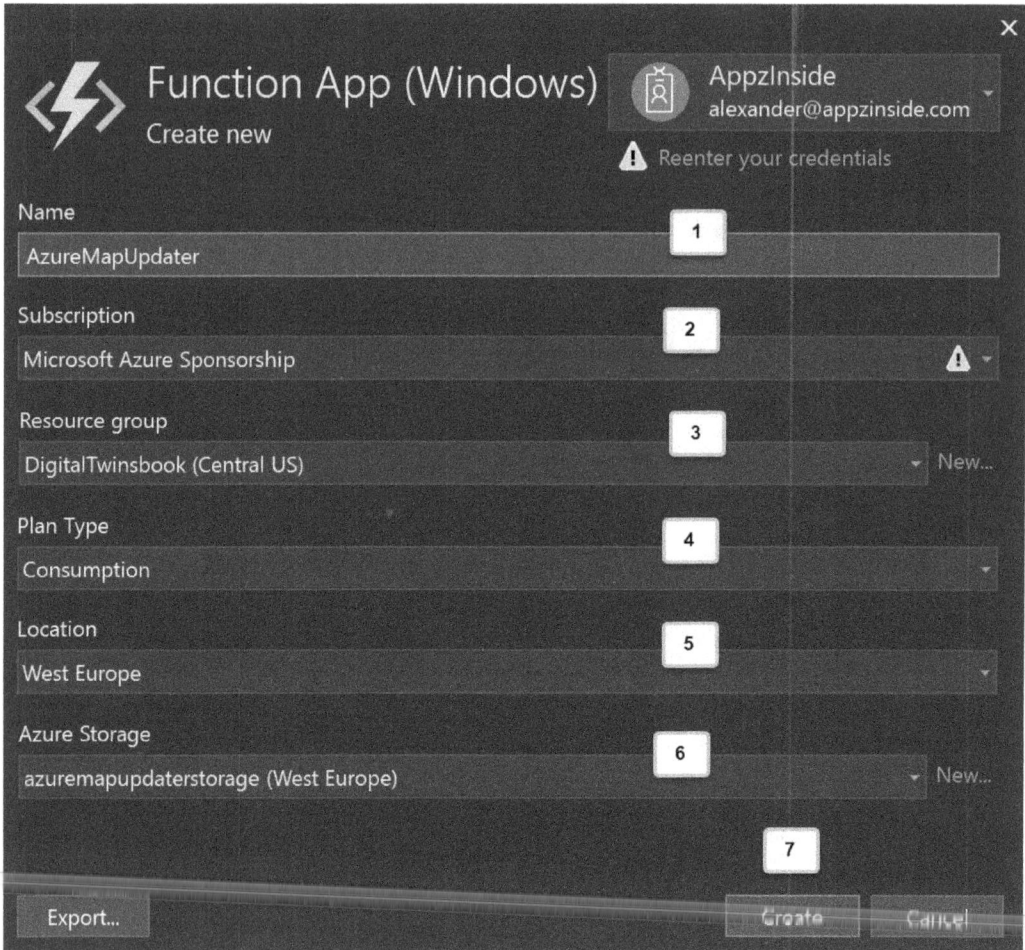

Figure 14.13 – Creating a publishing profile

We will return to the step shown in *Figure 14.12*. Execute the following steps:

1. Select the created Azure Function's destination. This is `AzureMapUpdater`.
2. Click on the **Finish** button to create a publishing profile.

The following screenshot highlights this process:

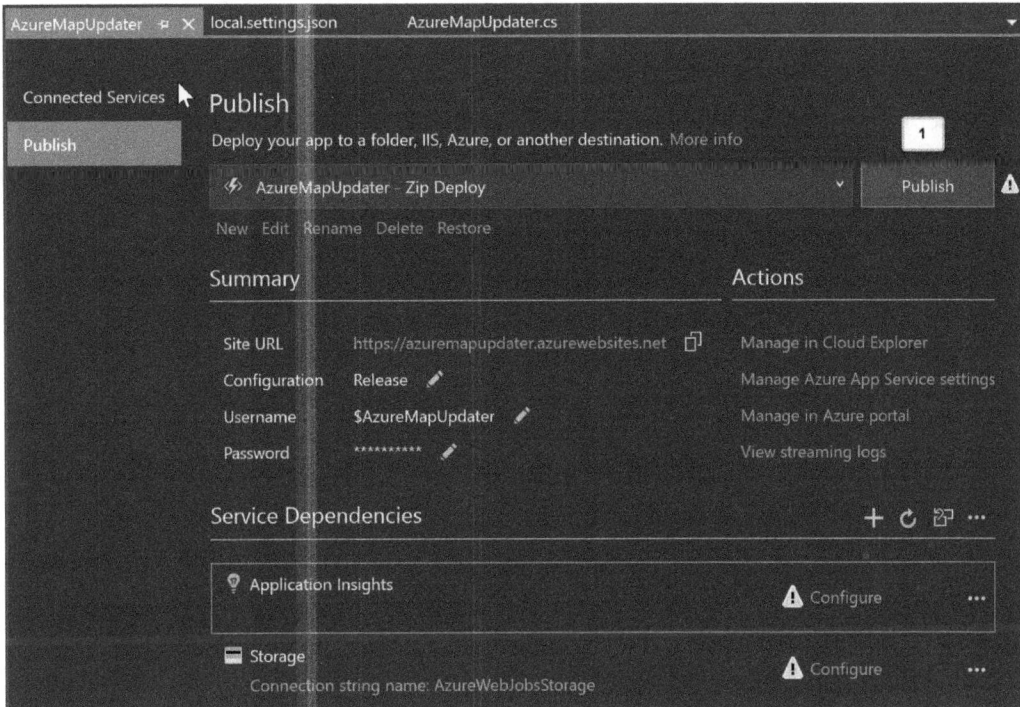

Figure 14.14 – Publishing the Azure Function using the publishing profile

An overview of the publishing profile is shown in *Figure 14.14*. Visual Studio will mention several service dependencies. These dependencies are not required and can be removed by clicking on the ellipsis (**…**) and removing them. Publish the profile by clicking on the **Publish** button.

We have created a publishing profile for the Azure Function and published it to Azure. In the next section, we will configure the application settings.

Configuring application settings

The Azure Function depends on two application settings. These application settings are read into two private static strings in the Azure Function, as shown in the following code snippet:

```
private static string statesetID = Environment.
GetEnvironmentVariable("statesetid");
private static string subscriptionKey = Environment.
GetEnvironmentVariable("subscriptionkey");
```

We will add both application settings through the Azure portal interface. Execute the following steps:

1. Select the `AzureMapUpdater` resource.

2. Select **Configuration** in the left menu.

3. Click on + **New application setting**. This allows us to specify a name and a value. Use this step to add the following two application settings:

Name	Description	Value
statesetid	{statesetId}	aa2f3e7a-378d-8887-3a14-8d04aa061aad
subscriptionkey	{authenticationId}	5u4nik9mH69uvmR6AYuHJq-12SmNbwfSP1mZGxLfAkQ

4. The application settings will appear in the list.

5. Click the **Save** button to save the changes.

The following screenshot highlights this process:

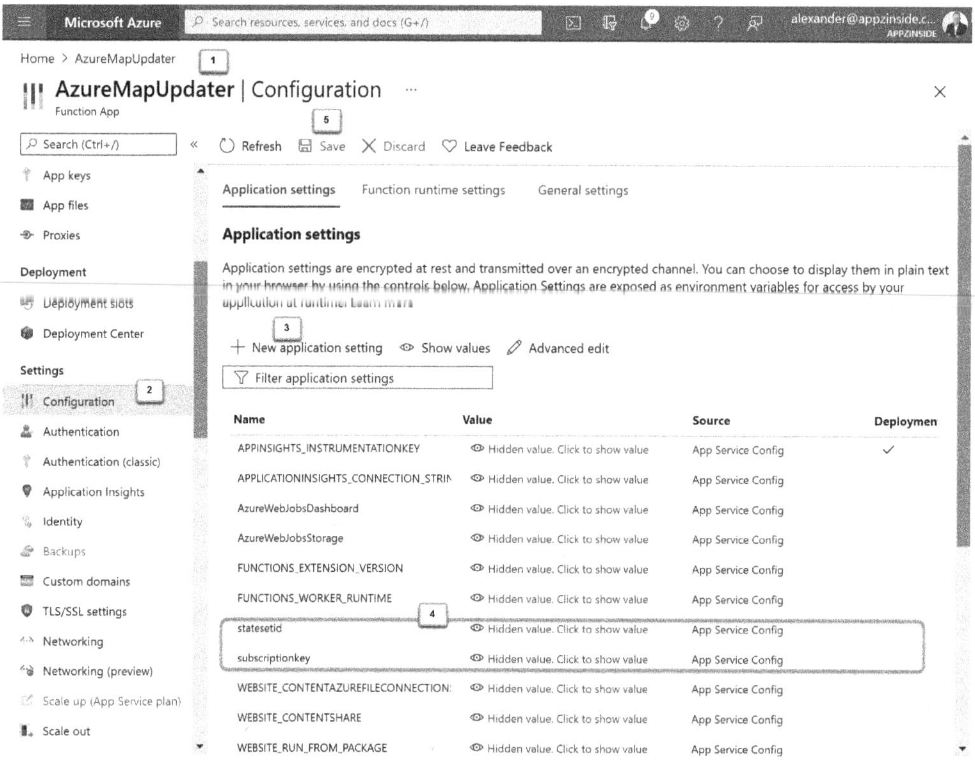

Figure 14.15 – Configuring application settings

We have now updated the `AzureMapUpdater` resource with two application settings that are used by the Azure Function. We will start subscribing to Event Grid in the next section.

Subscribing to Event Grid

The Azure Function is a trigger for Event Grid events. We need to subscribe to Event Grid to receive these events. This can be configured through the Azure portal. Execute the following steps:

1. Select the `AzureMapUpdater` resource.

2. Select **Functions** in the left menu.

3. Click on the function named `AzureMapUpdater`.

    The following screenshot highlights this process:

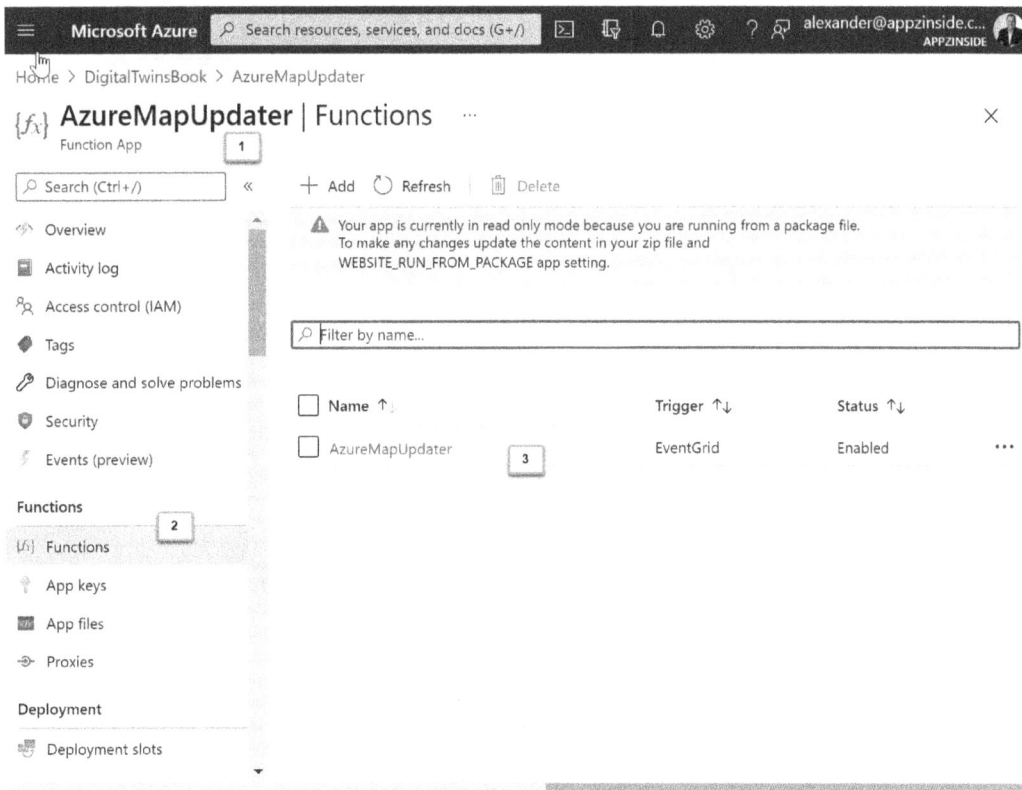

Figure 14.16 – Opening the Azure Function in the AzureMapUpdater resource

This will open the configuration of the `AzureMapUpdater` function. Execute the following steps to edit the trigger for this function:

1. Select **Integration** in the left menu.
2. Click on the **Event Grid Trigger** link.
3. Click on the **Create Event Grid subscription** link.

    The following screenshot highlights this process:

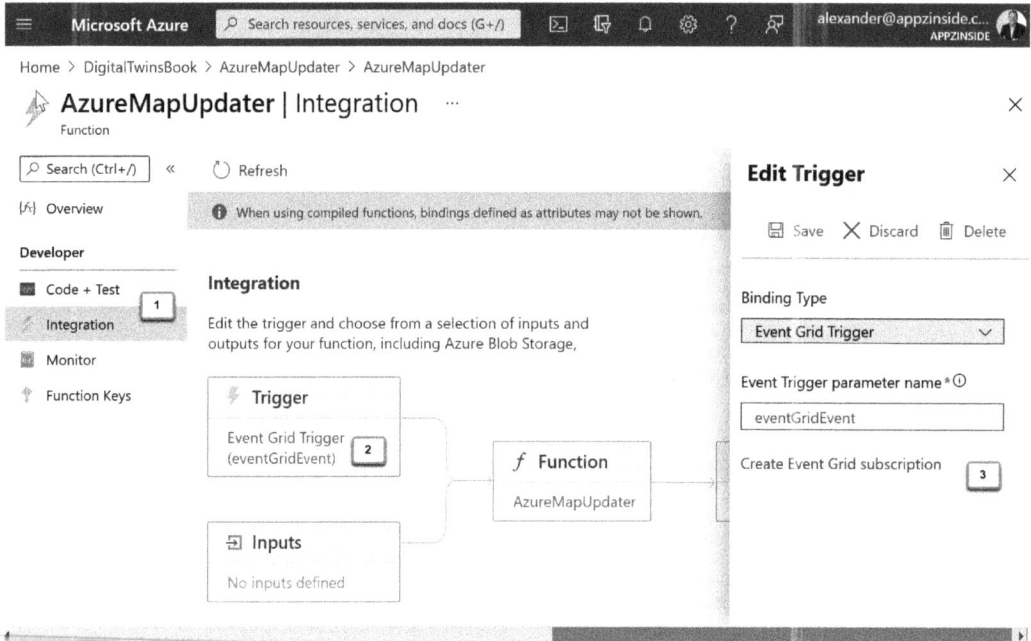

Figure 14.17 – Starting to edit the trigger of the Azure Function

We need to create a new event subscription to Event Grid. Execute the following steps:

1. Enter the name `GridEventTopicSubscription`.
2. Select `Event Grid Topics` from the **Topic Types** dropdown.
3. Select your subscription.
4. Select the `DigitalTwinsBook` resource group.
5. Select the `DTBEventGrid` event from the dropdown.
6. Click the **Create** button.

The following screenshot highlights this process:

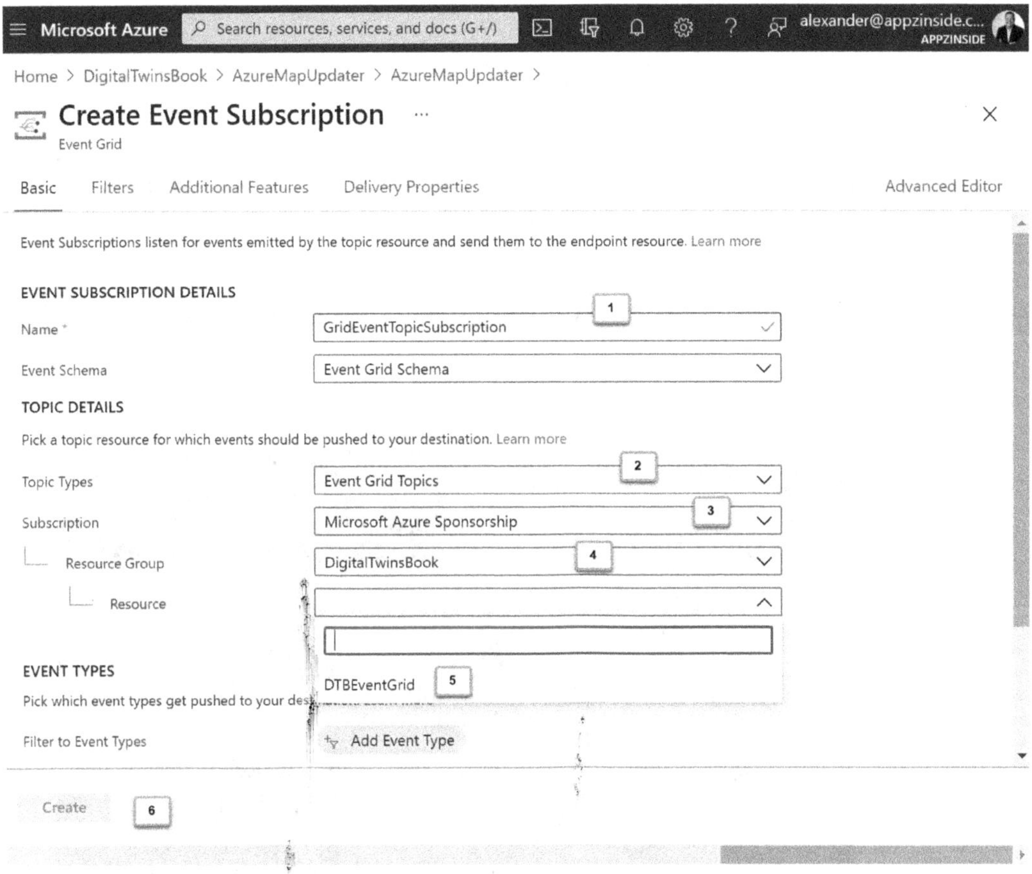

Figure 14.18 – Creating a new event subscription

We have now created a subscription from the `AzureMapUpdater` Azure Function to the `DTBEventGrid` event grid. The Azure Function will start receiving events from the event grid.

In the next section, we will monitor the events being handled by the Azure Function.

Monitoring updates

It is easy to monitor updates from the Azure Function. We have been using the `Ilogger.LogInformation` method to log information during the execution of the Azure Function. These log rows are viewable using the **Log stream** functionality in the Azure portal.

Execute the following steps to view logs in the Azure portal:

1. Select the `AzureMapUpdater` resource.

2. Select **Log stream** in the left menu. Sometimes, you will need to perform an additional step to have the **Log stream** functionality enabled. We have not described these steps here.

3. The log information will start floating in. Look for rows that describe the events coming from Event Grid produced by updates from the demo sensors.

   The following screenshot highlights this process:

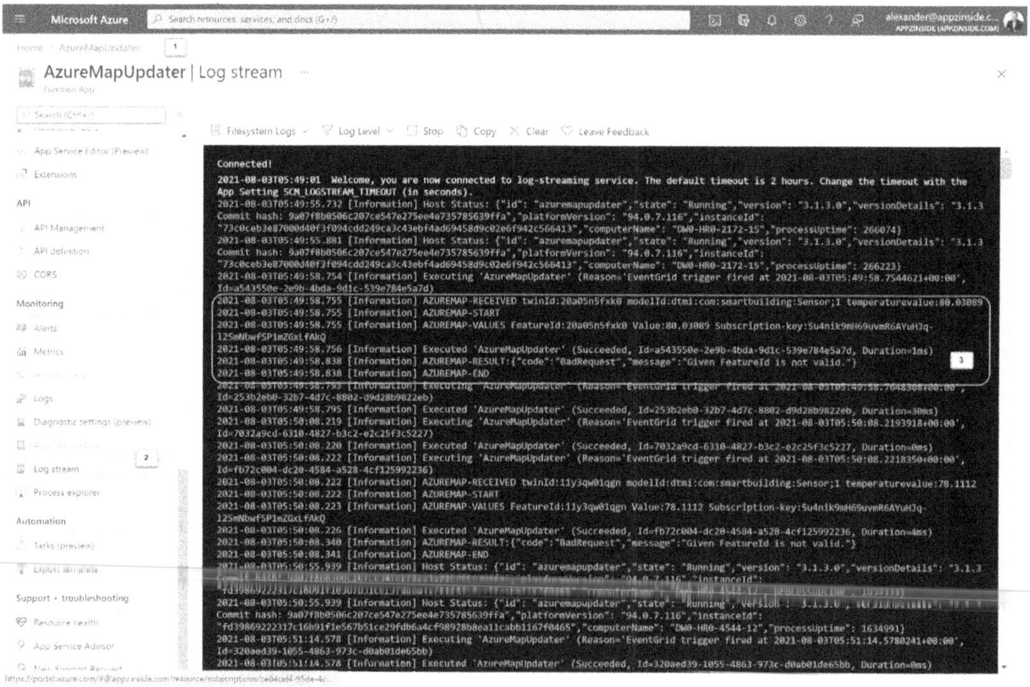

Figure 14.19 – Monitoring updates using Log stream

We can see that the event coming in is from the `20a05n5fxk0` demo sensor. Since there is no unit with that name in the indoor map, we receive an error. The error reports that the given `FeatureId` value is not valid. The reason for this is that we are not able to update a feature stateset for a non-existing unit.

Let's create a new digital twin based on the `Sensor` template and name it `UNIT111`. Then, update the value to—for example—25.8, as shown in the following screenshot:

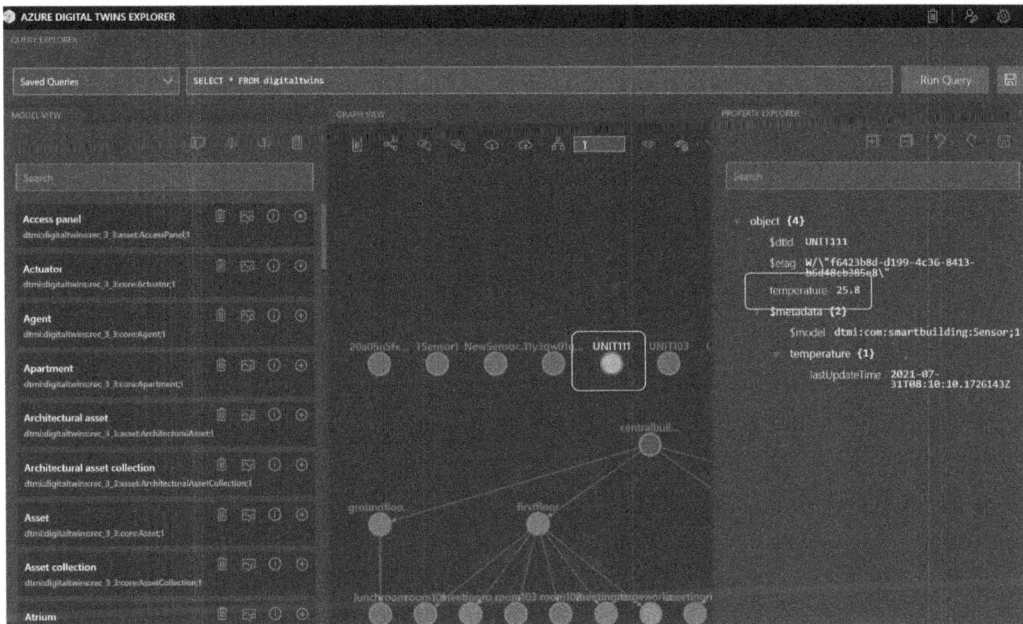

Figure 14.20 – Updating a new digital twin representing UNIT111

The result of this update can be seen in the following screenshot. The `FeatureId` property is set to `UNIT111` and the result is empty, which means the call has succeeded:

```
"3.1.3.0","versionDetails": "3.1.3 Commit hash: 9a07f8b0506c207ce547e275ee4e735785639ffa","platformVersion":
"94.0.7.116","instanceId": "73c0ceb3e87000d40f3f094cdd249ca3c43ebf4ad69458d9c02e6f942c566413","computerName":
"DW0-HR0-2172-15","processUptime": 3298791}
2021-08-03T06:41:07.081 [Information] Executing 'AzureMapUpdater' (Reason='EventGrid trigger fired at 2021-08-
03T06:41:07.0808729+00:00', Id=6d2cb457-1c12-40e7-b370-85ad27d133af)
2021-08-03T06:41:07.085 [Information] AZUREMAP-RECEIVED twinId:UNIT111 modelId:dtmi:com:smartbuilding:Sensor;1
temperaturevalue:25.8
2021-08-03T06:41:07.085 [Information] AZUREMAP-START
2021-08-03T06:41:07.085 [Information] AZUREMAP-VALUES FeatureId:UNIT111 Value:25.8 Subscription-
key:5u4nik9mH69uvmR6AYuHJq-12SmNbwfSP1mZGxLfAkQ
2021-08-03T06:41:07.088 [Information] Executed 'AzureMapUpdater' (Succeeded, Id=6d2cb457-1c12-40e7-b370-
85ad27d133af, Duration=7ms)
2021-08-03T06:41:07.243 [Information] AZUREMAP-RESULT:
2021-08-03T06:41:07.243 [Information] AZUREMAP-END
2021-08-03T06:41:15.811 [Information] Host Status: {"id": "azuremapupdater","state": "Running","version":
"3.1.3.0","versionDetails": "3.1.3 Commit hash: 9a07f8b0506c207ce547e275ee4e735785639ffa","platformVersion":
"94.0.7.116","instanceId": "2e2fc97107165db60a61013dd081d3a799a54b1199251bd03d6453e54ed7cad8","computerName":
"DW0-HR0-1343-11","processUptime": 5901694}
2021-08-03T06:41:15.687 [Information] Host Status: {"id": "azuremapupdater","state": "Running","version":
"3.1.3.0","versionDetails": "3.1.3 Commit hash: 9a07f8b0506c207ce547e275ee4e735785639ffa","platformVersion":
```

Figure 14.21 – The output result of the update of UNIT111

We have learned how to monitor event updates in the Azure Function by using the **Log stream** functionality. In the next and final section of this chapter, we will see how to visualize the indoor map, including the updates.

Visualizing the model

Microsoft offers an easy way of visualizing the indoor map model. We have the **Azure Maps Indoor** module available as part of the Azure Maps **software development kit (SDK)**. This module allows us to render an indoor map that was created with the Azure Maps Creator service.

It is possible to use this module in two ways. We can download the **Azure Maps Indoor** module to disk, or we can use the **Azure Content Delivery Network (Azure CDN)** version of the **Azure Maps Indoor** module.

In this example, we will be using the Azure CDN version. We can easily reference the JavaScript and style sheet files needed to render the model. Microsoft has the following web page that describes how to set this up: `https://docs.microsoft.com/en-us/azure/azure-maps/how-to-use-indoor-module#embed-the-indoor-maps-module`. It also contains the code we will be using to render the model. Execute the following steps to set up a **HyperText Markup Language (HTML)** file for rendering the model:

1. Create a new folder named `C:\Github\indoormapsmodule`.

2. Create a new empty file named `indoormap.html` in this folder.

3. Open the `indoormap.html` file using Visual Studio.

4. Copy the following HTML block into this file:

```html
<!DOCTYPE html>
<html lang="en">
<head>
    <meta charset="utf-8" />
    <meta name="viewport" content="width=device-width, user-scalable=no" />
    <title>Indoor Maps App</title>
</head>
<body>
    <div id="map-id"></div>
    <script>
    </script>
```

```
</body>
</html>
```

Add the following block of code in the header between `<head>` and `</head>`:

```
<link rel="stylesheet" href="https://atlas.microsoft.com/
sdk/javascript/mapcontrol/2/atlas.min.css" type="text/css" />
<link rel="stylesheet" href="https://atlas.microsoft.com/
sdk/javascript/indoor/0.1/atlas-indoor.min.css" type="text/css"
/>

<script src="https://atlas.microsoft.com/sdk/javascript/
mapcontrol/2/atlas.min.js"></script>
<script src="https://atlas.microsoft.com/sdk/javascript/
indoor/0.1/atlas-indoor.min.js"></script>
```

Add the following block of code in the header between `<head>` and `</head>`, below the previously added block:

```
<style>
    html,
    body {
        width: 100%;
        height: 100%;
        padding: 0;
        margin: 0;
    }

    #map-id {
        width: 100%;
        height: 100%;
    }
</style>
```

The next blocks of scripts need to be placed between `<script>` and `</script>` within `<body></body>`. Each script block follows up the next one.

We need to make a few changes before we can start rendering the indoor map. The first step is to set all the required values for accessing the indoor map. We need to specify the `subscriptionKey`, `tilesetId`, and `statesetId` values. All these values are known from the previous chapter. Make sure that the HTML file contains these values. The code looks like this:

```
const subscriptionKey = "5u4nik9mH69uvmR6AYuHJq-
12SmNbwfSP1mZGxLfAkQ";
const tilesetId = "b9434ccd-a66e-9052-4b8c-948dc4579a33";
const statesetId = "aa2f3e7a-378d-8887-3a14-8d04aa061aad";
```

Add the following script block, which will initialize the map:

```
const map = new atlas.Map("map-id", {
    center: [-122.13315, 47.63637],
    style: "blank",
    view: 'Auto',
    authOptions: {
        authType: 'subscriptionKey',
        subscriptionKey: subscriptionKey
    },
    zoom: 19,
});
```

Add the following script block. This script block will create a separate menu in the top right of the browser, allowing us to switch between different layers in the map. This control is specifically for this map, provided by Microsoft:

```
const levelControl = new atlas.control.LevelControl({
    position: "top-right",
});
```

Add the following script block. This script block will create a map based on the `tilesetId` and `statesetId` values:

```
const indoorManager = new atlas.indoor.IndoorManager(map,
{
    levelControl: levelControl, //level picker
    tilesetId: tilesetId,
```

```
        statesetId: statesetId // Optional
    });
```

Add the following script block. This script will take care of the styling:

```
    if (statesetId.length > 0) {
        indoorManager.setDynamicStyling(true);
    }
```

Since we have created our Azure Maps service in West Europe, we need to add the following line to the script to make sure that the correct region is used:

```
    indoorManager.setOptions({ geography: 'eu' });
```

Add the last script block. This script block contains methods to handle several events. While we are not using it directly, it will show some messages in the log console of the browser. The code looks like this:

```
    map.events.add("levelchanged", indoorManager, (eventData)
=> {
        console.log("The level has changed:", eventData);
    });
    map.events.add("facilitychanged", indoorManager,
(eventData) => {
        console.log("The facility has changed:", eventData);
    });
```

Run the `indoormap.html` file in a browser. The indoor map is displayed on screen. Changes to units through updates from digital twins with the same name are reflected in the map, using the colors specified in the feature stateset.

In this final section, we have visualized the indoor map and updates created by the digital twins.

Summary

In this chapter, you have learned how to integrate an Azure map using an Azure Function. The Azure Function is triggered and will update the feature stateset of the Azure map. We will learn more about how to monitor and troubleshoot all Azure Digital Twins-related services in the next chapter.

Questions

As we conclude, here is a list of questions for you to test your knowledge regarding this chapter's material. You will find the answers in the *Assessments* section of the *Appendix*:

1. What kind of Azure Function trigger is required to receive events from Event Grid?

    a. Event Grid trigger

    b. Event Hub trigger

    c. **HyperText Transfer Protocol (HTTP)** trigger

2. Can you update the feature stateset of a non-existing unit?

    a. No

    b. Yes

3. What is an easy way of monitoring the logs from an Azure Function?

    a. Logs

    b. Process Explorer

    c. Log stream

Further reading

To learn more on the topic, refer to the following links:

- How to integrate Azure Maps

  https://docs.microsoft.com/en-us/azure/digital-twins/how-to-integrate-maps

- Use the Azure Maps Indoor module

  https://docs.microsoft.com/en-us/azure/azure-maps/how-to-use-indoor-module#embed-the-indoor-maps-module

15
Monitoring and Troubleshooting

In this chapter, we will learn everything you need to know to monitor your Azure Digital Twins service using the Azure Log Analytics workspace and the diagnostic settings. Log Analytics, metrics, and alerts are tools provided in the Microsoft Azure platform to support monitoring, logging, and solving problems.

It allows us to monitor and troubleshoot problems that occur around the Azure Digital Twins instance by identifying logs and metrics. We can create alerts that can help us to take action upon those logs and metrics to inform or warn administrators and to resolve issues around the Azure Digital Twins instance and related services used by our solution.

The following topics will be covered in this chapter:

- Setting up a log analytics workspace
- Setting up diagnostic settings
- Viewing logs
- Viewing metrics
- Using alerts

After this chapter, we will understand how we can monitor and troubleshoot our Azure Digital Twins instance using logs and metrics.

Technical requirements

We will be using the Azure portal to create and configure Azure services to support monitoring and troubleshooting.

Setting up a log analytics workspace

A log analytics workspace is an environment created in Microsoft Azure in which log data from Azure monitoring is stored. This environment contains a data repository and a set of configurations. These configurations allow us to configure how data is handled and stored in the data repository of the environment.

An analytics workspace is used to collect data from different data sources. Data sources can be anything from Azure resources, device collections and diagnostics, and log data from Azure storage.

We will use a log analytics workspace to collect log data from our Azure Digital Twins resource.

Let's start creating a log analytics workspace through the Azure portal. Execute the following steps, as shown in *Figure 15.1*:

1.   Go to **Create a resource** in the Azure portal at `https://portal.azure.com`.
2.   Search for *Log Analytics workspace*.

3. Select **Create**.

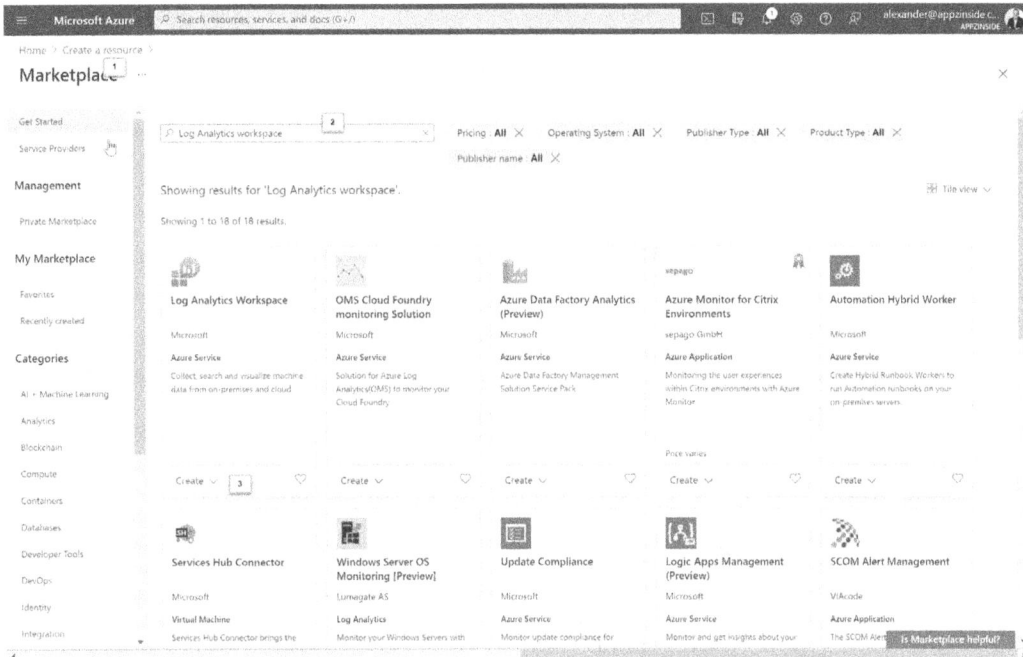

Figure 15.1 – Searching for Log Analytics workspace in the marketplace

Execute the following steps, as shown in *Figure 15.2*:

1. Select our resource group, **DigitalTwinsBook**.

2. Enter the name `dtbanalytics`.

3. Select your region. I have selected **West Europe** since that is my region.

4. Click on **Review + Create**:

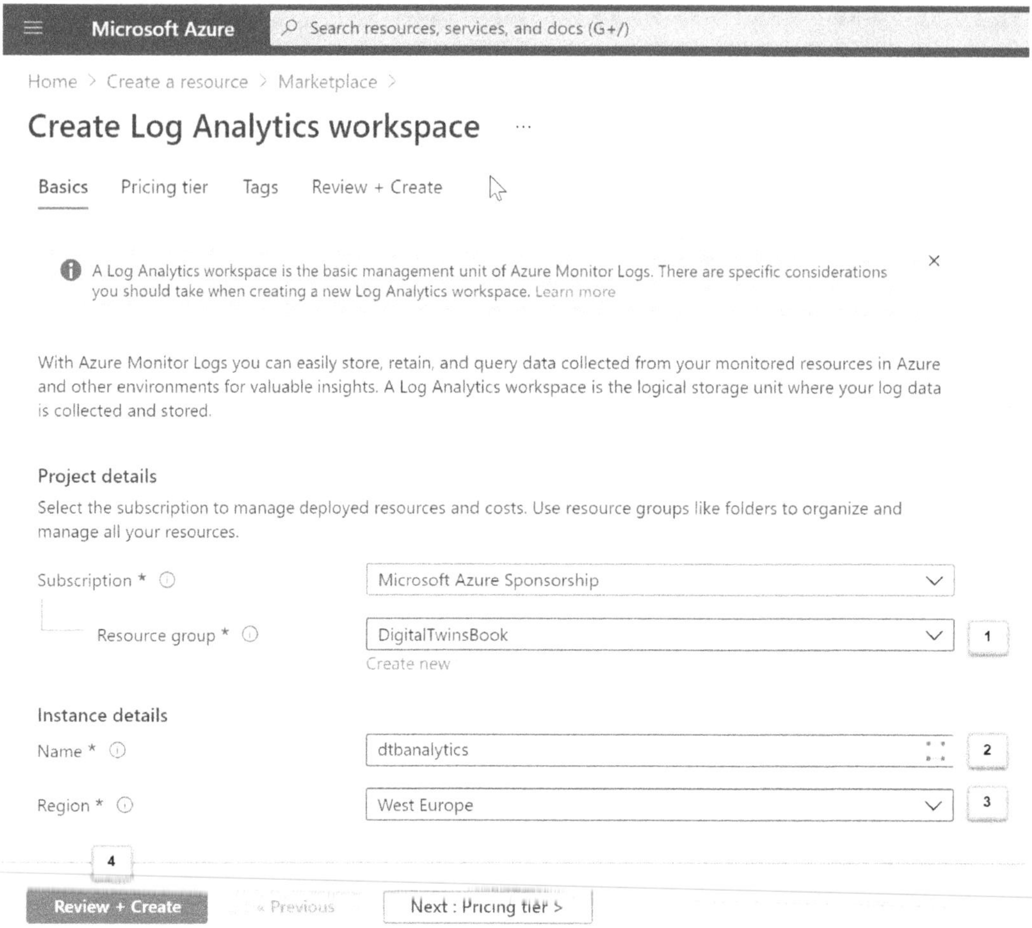

Figure 15.2 – Entering the basic configurations for our new Log Analytics workspace

In the next step, we will need to review the information we have filled in during the creation process. Press the **Create** button to create the Log Analytics workspace, as shown in *Figure 15.3*:

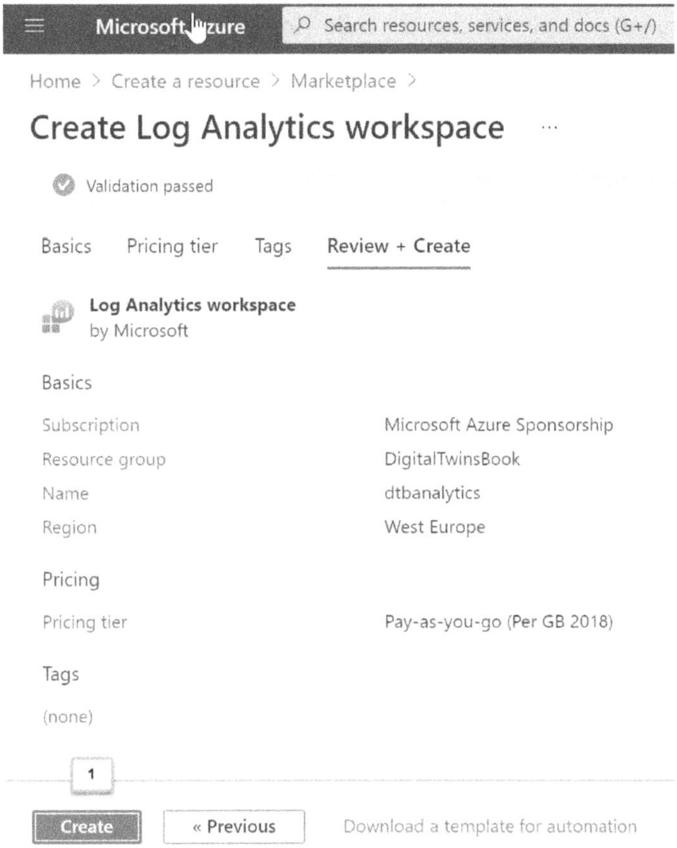

Figure 15.3 – Reviewing your filled-in information and starting to create the log analytics workspace

The creation of the Log Analytics workspace will take a bit of time. Open the service and have a look at the different tabs with configurations for the service. We will not make any changes since we will be using the default configurations.

In this section, we have set up the Log Analytics workspace service to collect the log and metrics data. In the next section of this chapter, we will be setting up the diagnostic settings using this service.

Setting up diagnostic settings

In this section, we will be setting up the diagnostic settings for the Azure Digital Twins service. These diagnostic settings need to be configured to view different logs and metrics. These logs and metrics will query the collected data in the Azure Log Analytics workspace data source. The diagnostic settings can have up to five different settings to allow you to collect different logs and metrics to different locations.

Let's create a diagnostic setting. Execute the following steps as shown in *Figure 15.4*:

1. Select the Azure Digital Twin **DTBDigitalTwins** resource.
2. Select **Diagnostic settings** in the left menu.
3. Select **+ Add diagnostic setting**.

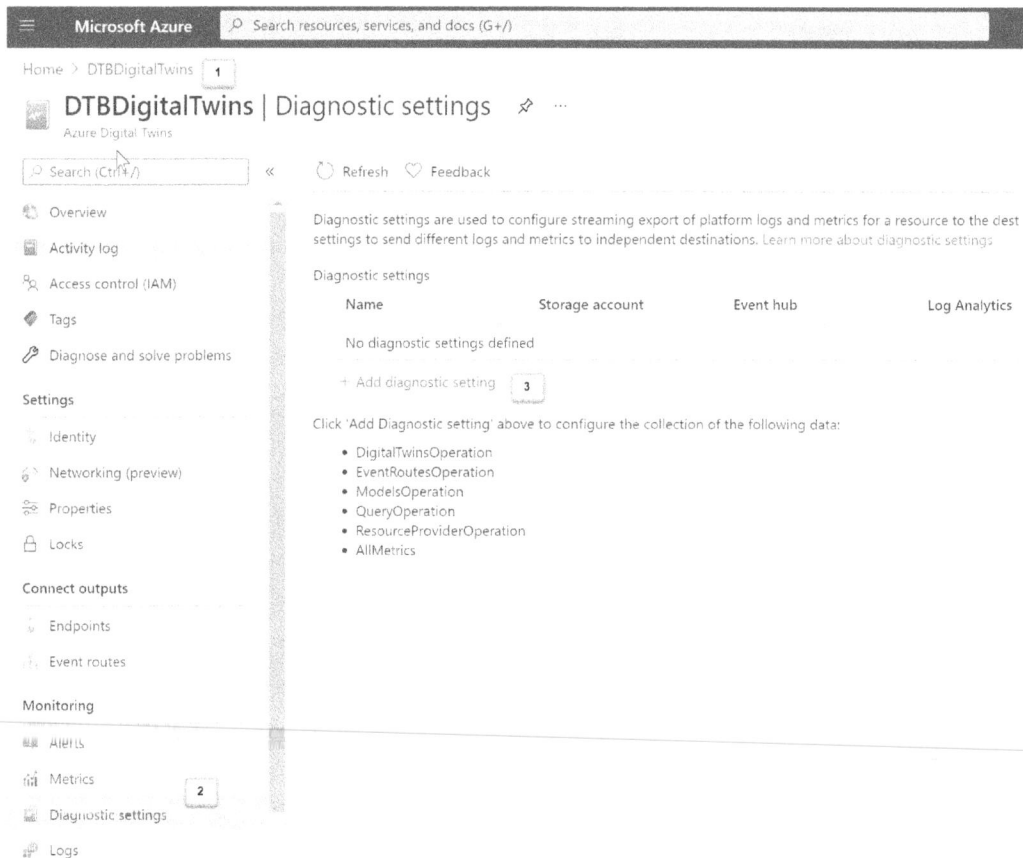

Figure 15.4 – Add a new diagnostic setting to the Azure Digital Twins service

Execute the following steps as shown in *Figure 15.5* to create our first diagnostic setting:

1. Enter a diagnostic settings name. We will use `logs and metrics`.
2. Select all the logs by checking all the checkboxes.
3. Check the **AllMetrics** checkbox.
4. Select the option **Send to Log Analytics workspace**.
5. Select **dtbanalytics** from the list as the Log Analytics workspace.

6. Press the **Save** button.

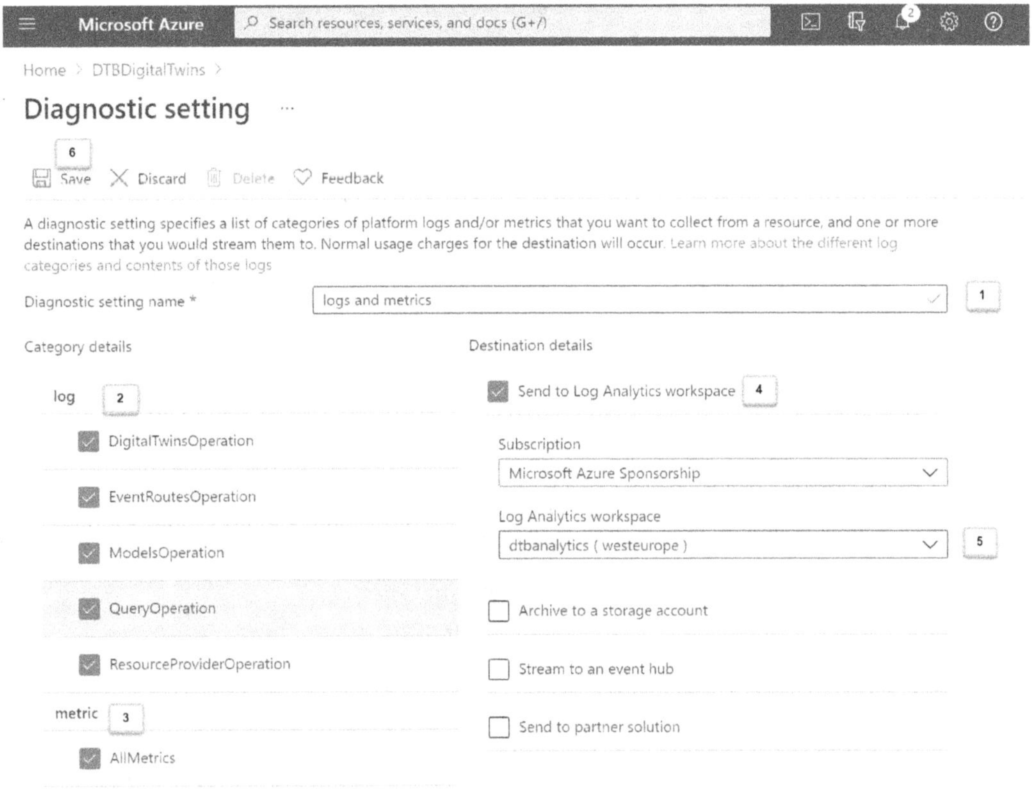

Figure 15.5 – Creating a diagnostic setting

This diagnostic setting has been configured to collect any type of log and metric. While this makes our next step to query data somewhat easier, normally you would specify only the logs that are required. This means that we need to think about what we exactly want to monitor to determine which categories need to be selected. Each log category consists of operations such as read, write, delete, and action. They allow us to collect data regarding the following:

- `DigitalTwinsOperation`: Logs all API calls that are related to digital twins. This includes changes to digital twins, such as changing the value of a property.

- `EventRoutesOperation`: Logs are operations that are related to event routes. This includes logs around ingress and egress actions.

- `ModelsOperation`: Logs all API calls related to models. This includes the create and delete models in our Digital Twins instance.

- `QueryOperation`: Logs all API calls related to executing queries. This includes queries such as those from the Azure Digital Twins Explorer.

- `ResourceProviderOperation`: Logs all operations on the resource itself, meaning changes to the settings of the Azure Digital Twins resource.

- `AllMetrics`: Logs all metrics available for the Azure Digital Twins resource. Metrics are different from log categories. Metrics focus mostly on how quickly an operation took place.

We have created a diagnostic setting to collect all possible logs and metrics. In the next section, we will start viewing the logs based on the collection of this data.

Viewing logs

Azure has a sophisticated engine to view and query log files. Execute the following steps as shown in *Figure 15.6.* to access the logs:

1. Select **Logs** in the left menu of our Azure Digital Twins resource.

2. Select **Usage** from the **All Queries** section within **Queries**.

3. Press the **Run** button for the **DigitalTwin API Usage** query.

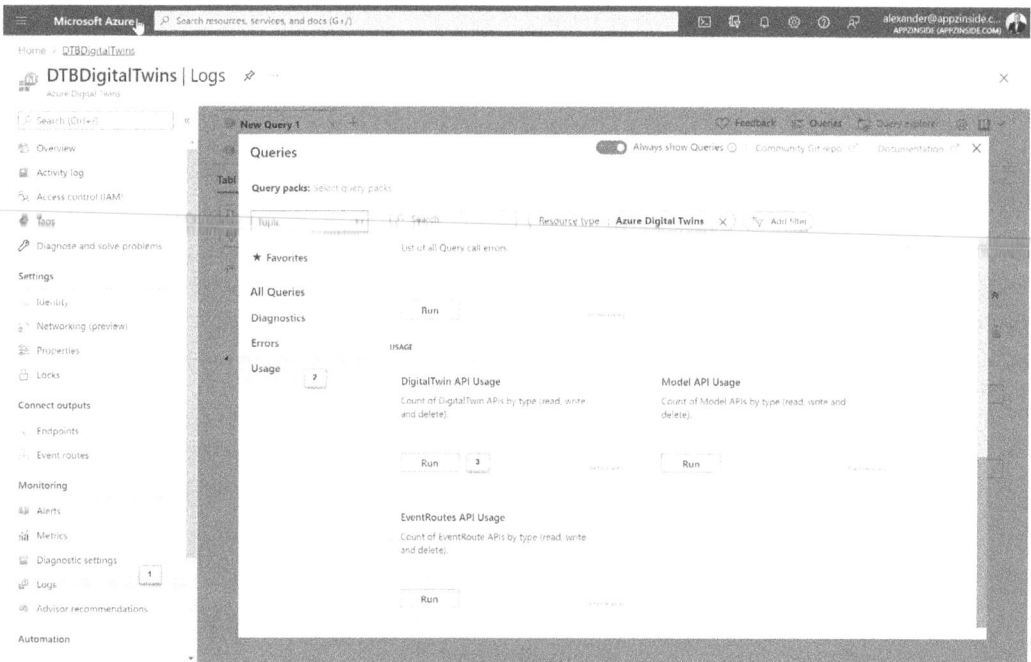

Figure 15.6 – Selecting one of the log queries to view the logs

After selecting the log query, the following view appears, as shown in *Figure 15.7*. This view has four important areas, which are explained here:

1. **Query area**: This area allows you to select one of the predefined queries available. These queries are divided into categories such as **Diagnostics**, **Errors**, and **Usage**. As soon as you double-click on one of the queries, the code required to execute the query is copied to the **Queries** editor.

2. **Queries editor**: This area contains query code that is executed to create the result for the log view. The editor allows us to make changes to what we query and how it should be represented in the **Results** view.

3. **Query bar**: This area shows all the queries that we are working on. It allows us to save a query by name and to define some filters. These filters determine the time range in which results need to be shown in the **Results** view. It even allows us to scope the result based on resources and to create alerts.

4. **Results view**: This area shows the result of the execution of the query code in the **Queries** editor. The code in the **Queries** editor determines what is shown. In this example, we have used the `render piechart` code, which shows us a pie chart of the summarized content per operation name.

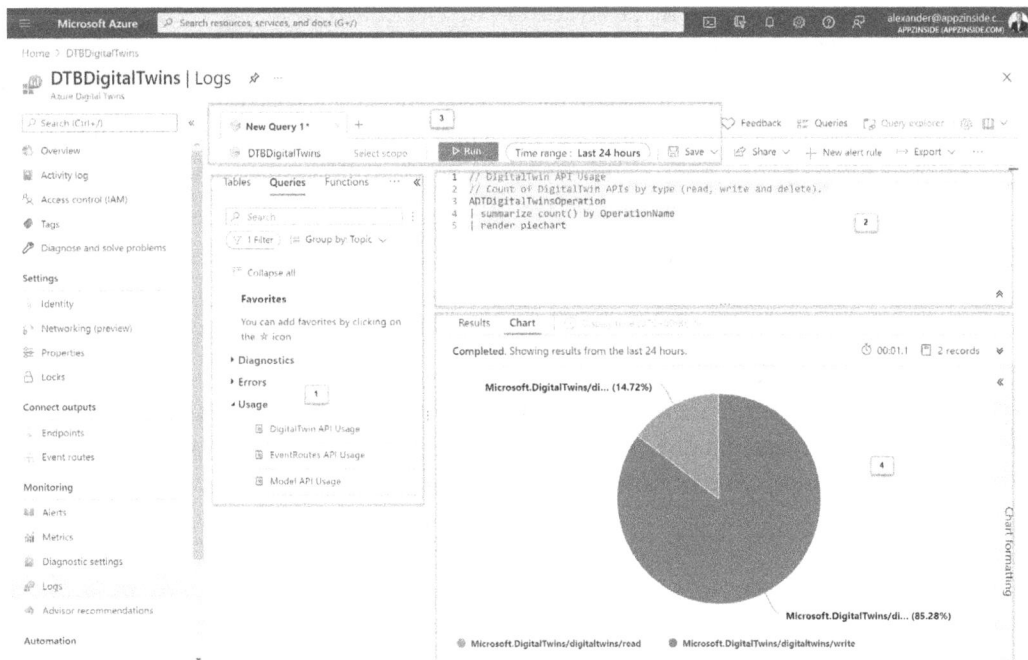

Figure 15.7 – View of a log query

In this section, we have seen how to view the log data. In the next section, we will view the metrics data.

Viewing metrics

Metrics give us more information about the state and health of a certain Azure resource. In this case, we want to view information about the health of the Azure Digital Twins instance. Metrics are enabled by default and do not *per se* require the setup of the Log Analytics workspace. They are often used as the first place to monitor, check, and try to resolve issues around your resource before contacting Azure support.

Our Azure Digital Twins instance provides several metrics that will provide us with an overview of the health of the resource. The following categories of metrics are available:

- **Service limits**: These are metrics that provide us with more information about the Digital Twins count and the Model count, also called `TwinCount` and `ModelCount`. At the time of writing this book, these metrics are still in preview.

- **API Request metrics**: These are metrics that provide us with information regarding the execution of API requests. These requests concern APIs, failure, and latency and are named `ApiRequests`, `ApiRequestsFailureRate`, and `ApiRequestLatency`.

- **Billing metrics**: These are metrics to do with the billing of the Azure Digital Twins instance. These involve operations, messages processed, and the querying of units. They are named `BillingApiOperations`, `BillingMessagesProcessed`, and `BillingQueryUnits`.

- **Ingress metrics**: These are metrics to do with data ingress. The metrics are divided between telemetry events, the failure rate, and latency. These requests are named `IngressEvents`, `IngressEventsFailureRate`, and `IngressEventsLatency`.

- **Routing metrics**: These are metrics to do with the routing of events. These requests concern routed messages, the failure rate, and latency. These requests are named `MessagesRouted`, `RoutingFailureRate`, and `RoutingLatency`.

To explain it in a bit more detail, we will compare two metrics to see what we can learn from them. Execute the following steps as shown in *Figure 15.8*:

1. Select **Metrics** from the left menu.

2. Define a metric with **Scope** set to **DTBDigitalTwins**, **Metric Namespace** set to **Standard metrics**, **Metric** set to **API Requests**, and **Aggregation** set to **Sum**.

3. The graphics will automatically update. It will show the sum of API requests made for our Azure Digital Twins instance. The number of requests is specified on the vertical bar, while the time is shown on the horizontal bar.

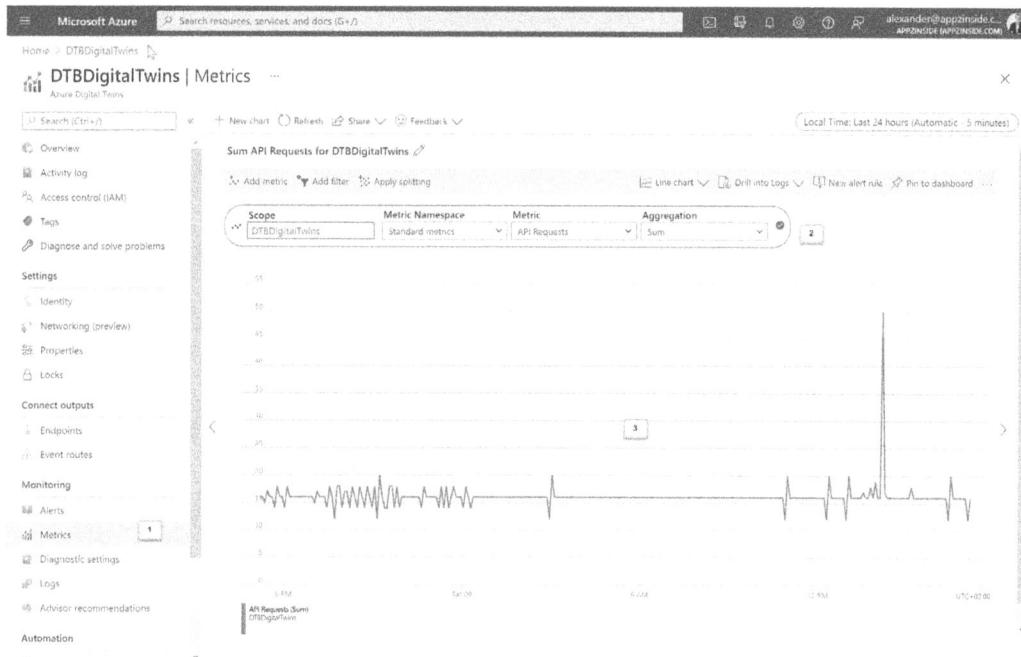

Figure 15.8 – API Request metrics of our Azure Digital Twins instance

Let's compare this metric to **Billing Query Units**. **Billing Query Units** defines the sum of query units for which we will be billed. Execute the following steps as shown in *Figure 15.9*:

1. Press the **Add metric** button to add a new metric to the graphics view.

2. Define a metric with **Scope** set to **DTBDigitalTwins**, **Metric Namespace** set to **Standard metrics**, **Metric** set to **Billing Query Units**, and **Aggregation** set to **Sum**.

3. The graphics will update automatically. It will add the sum of billing query units made for our Azure Digital Twins instance to the graphics. The number of requests is specified on the vertical bar, while the time is shown on the horizontal bar.

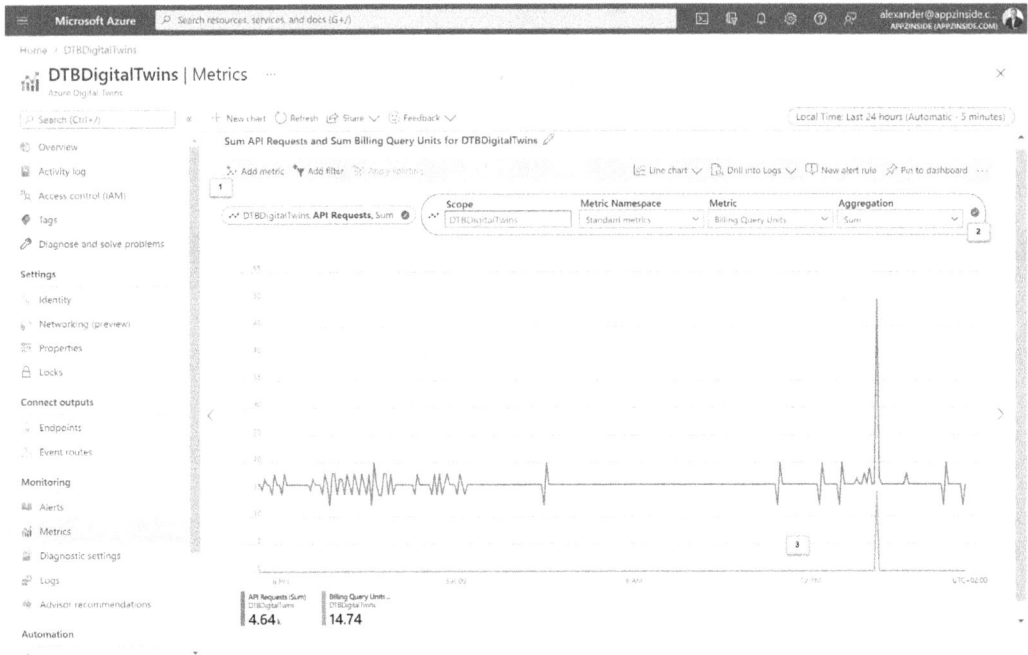

Figure 15.9 – Comparing API requests with Billing Query Units

We can see that there is a match between both metrics around **2 PM**. At that time, I did some queries through the Azure Digital Twins Explorer, which caused a larger hit of API requests. Since I was querying, the sum of query units that need to be billed had the same spike.

As you can imagine, using different metrics together will give you additional information regarding use of the resource. Try to experiment yourself by combining metrics after doing different types of API calls using the Azure Digital Twins Explorer.

We have learned how to view the metrics data. In the next and final section, we will learn how to use alerts to notify administrators and take action in the event of an alert.

Using alerts

Alerts allow us to respond to logs and metrics and act on them. There are different ways of creating alerts. Alerts are always created by defining rules. We can create alerts using the following:

- **Alert rules**: Create an alert rule to identify and address issues based on conditions regarding the monitoring of data. These alerts are focusing on metrics.

- **Log alert**: Create an alert rule around the resource's log. This is still in preview at the time of writing this book. The button is shown in *Figure 15.10*:

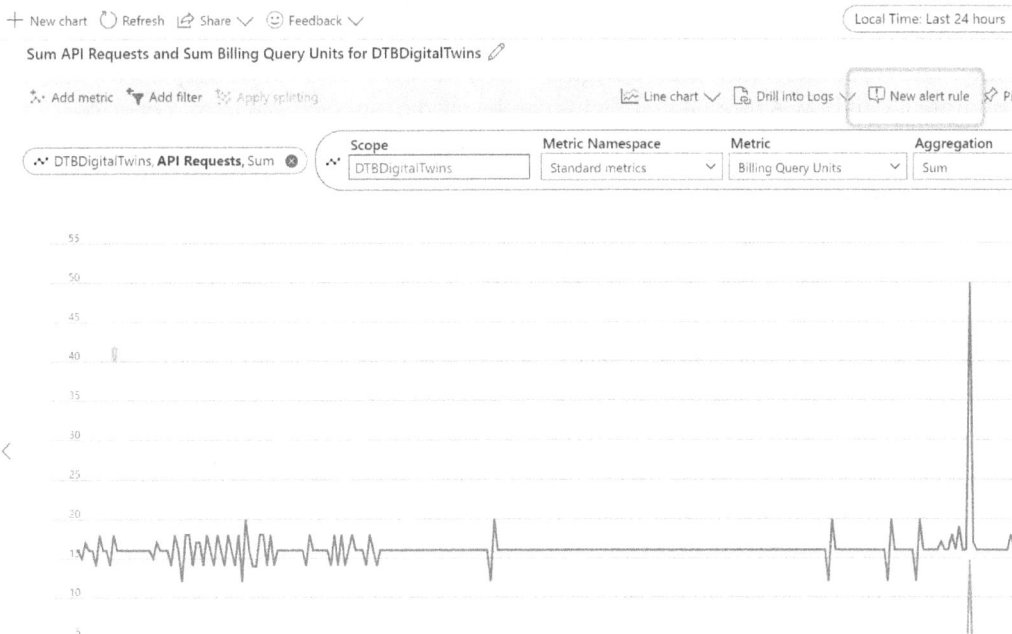

Figure 15.10 – Creating a new alert through the metrics page

Creating a rule consists of several steps:

- **Basic**: This determines how an alert is sent out. Think of just informational, warning, or even an error.

- **Scope**: The scope defines which resource is used to create an alert on.

- **Condition**: This allows us to specify around which log query the alert needs to be specified. It allows the definition of the measurement, dimensions, and alert logic. The alert logic allows us to define the final condition of the alert, for example, when a certain value is greater than value x.

- **Action**: This determines what kind of action needs to be executed when the alert is raised. We can create action groups. And each action group allows us to define how a notification takes place. There are several notification actions, including calling an Automation runbook, Logic app, Webhook, and even an Azure function.

This allows us to specify alert rules that just notify us as an administrator. But it also allows us to start an Azure resource that acts on the alert, which could, at some point, solve the alert. You can imagine that solutions are endless.

Summary

In this chapter, you have learned how to set up and configure the services required to monitor and troubleshoot the Azure Digital Twins resource. In the next chapter, we will learn more about how Azure Digital Twins can be used to support different scenarios.

Questions

As we conclude, here is a list of questions for you to test your knowledge regarding this chapter's material. You will find the answers in the *Assessments* section of the *Appendix*:

1. Why would we set up a Log Analytics workspace?

    a. It is always needed to collect data from different resources.

    b. It is one of the resources that can be used to collect data from different resources.

    c. It is required to perform analyses.

2. What different types of analytics can be monitored with Azure Digital Twins?

    a. Log files

    b. Metrics

    c. Logs and metrics

3. Is it possible to take any action with an alert generated by a log or metric?

    a. Yes

    b. No

Further reading

To learn more on the topic, refer to the following links:

- Log Analytics tutorial:

  `https://docs.microsoft.com/en-us/azure/azure-monitor/logs/`
  `log-analytics-tutorial`

- Azure Monitor Logs overview:

  `https://docs.microsoft.com/en-us/azure/azure-monitor/logs/`
  `data-platform-logs`

- Troubleshooting Azure Digital Twins: Diagnostics logging:

  `https://docs.microsoft.com/en-us/azure/digital-twins/`
  `troubleshoot-diagnostics`

- Troubleshooting Azure Digital Twins: Metrics:

  `https://docs.microsoft.com/en-us/azure/digital-twins/`
  `troubleshoot-metrics`

Section 4: Digital Twin Implementations in Real-world Scenarios

This final section will conclude the book with various examples of digital twin solutions using Azure Digital Twins. For each digital twin solution, we will explain the scenario, a possible solution design, and architecture that will help you to better understand how solutions to real-world challenges can be implemented using this technology.

This part of the book comprises the following chapters:

- *Chapter 16, Facility of the Future*
- *Chapter 17, Creating Digital Twins for Smart Building*
- *Chapter 18, Simulations Using a Digital Twin*

16
Facility of the Future

Until now, this book has focussed on creating **digital twins** using **Azure Digital Twins**. We have learned about each of the elements of the Azure Digital Twins service and how they integrate with other **Microsoft Azure** services. The next three chapters will look at how we can leverage Azure Digital Twins for a range of use cases.

Having insights into the activities taking place in a facility can accommodate different roles in an organization. Imagine having a virtual representation of the assets and equipment in our organization by using Azure Digital Twins. This will allow us to view real-time data in different ways, which can optimize business processes and support firstline workers.

Understanding how insights can contribute to different processes and roles within the organization will help you to model your own architecture with an Azure Digital Twins instance.

This chapter contains the following sections:

- Understanding the scenario
- Designing the digital twin solution
- The solution architecture

Understanding the scenario

Over the past year, something I have been involved with at work is a concept called *the facility of the future*. This concept refers to the aim of combining several existing sources of information and business applications within an organization to build rich solutions that support firstline workers and business processes. Today's organizations are facing challenges. While they often have systems in place to collect sensor data, use customer relationship management systems to support business processes, and use several applications on both desktop and mobile to support their firstline workers, they frequently have no idea how to bring these things together. And that is what *the facility of the future* is all about, as you can see in *Figure 16.1*:

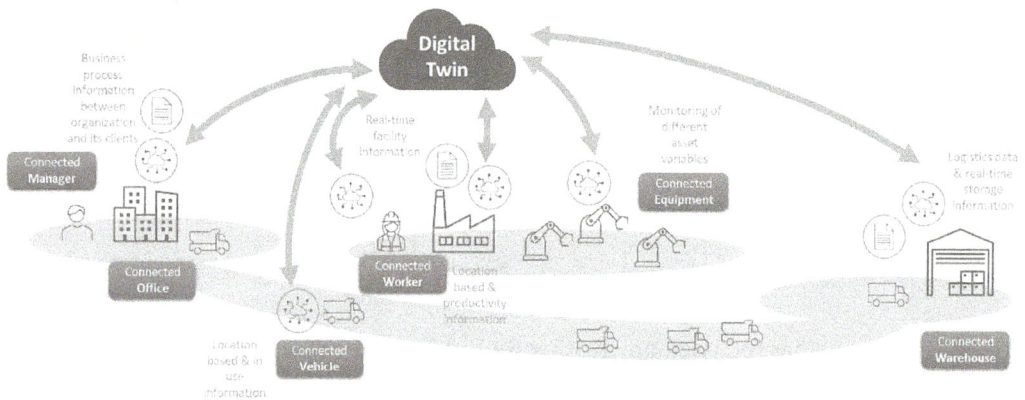

Figure 16.1 – The facility of the future concept

The facility of the future is all about being *connected*. It is about the connected vehicle, connected worker, connected equipment, connected warehouse, connected office, and more. All parts of a facility – including assets, equipment, vehicles, and firstline workers – are connected in some way. All of the connected parts have information stored for them based on their associated **internet of things** (**IoT**) sensors and their state. A digital twin is constructed from all of this real-time data and used to optimize processes.

In addition to this, *the facility of the future* focusses on the following principles:

- **Promoting safety**: We want to create a safe working environment for firstline workers. This involves training, redirecting workers to safe locations, and preventing accidents.

- **Enabling remote work**: Firstline workers will sometimes require support from remote experts. These remote experts can be firstline workers at other facilities, experts located in offices, and even experts from third-party companies that provide support for assets at the facility.

- **Increasing productivity**: It is important that the digital twin solution contributes to the productivity of processes so that work can be carried out more easily, quicker, or in better conditions.

- **Improving processes**: This is all about optimizing processes with automation so that the firstline workers can focus on the tasks they have been trained to do.

- **Having control**: This is about gaining control of the work and processes going on in a facility. More control can be achieved by generating better insights, using predictiveness, and supporting firstline workers.

- **Gaining insights**: This is about gaining better insights into the processes or assets affecting firstline workers by visualizing real-time data to gain a better understanding of the situation. This is often combined with the context of the work and the role of the person executing the work.

As we can see, each of these principles is primarily about supporting the daily work of firstline workers. But it's not only about firstline workers. The people working in management at the office can also benefit from *the facility of the future*.

> **The Ultimate Goal of the Facility of the Future**
> The ultimate goal of this concept is to equip any worker within an organization with the *right information* at the *right time*. This should take into account their location, the role they have within the organization, and the task they need to execute.

In this scenario, we will be focusing on the insights for a *facility*. We want these insights to support the daily work of different roles in the organization by leveraging existing data. Here, the existing data comes from IoT sensors that are positioned all over the facility and from existing backend systems. **Microsoft Dynamics 365 Field Service** is one of these backend systems, and it holds information about assets, firstline workers, equipment, facilities, work orders, cases, alerts, and much more. This backend system even supports **machine learning** (ML) in several ways, which allows us to leverage ML technology for predictiveness and scheduling firstline workers against work orders.

In this section, we discussed a scenario where Azure Digital Twins could help to improve business processes. In the next section, we will be going into more detail about how to implement this solution.

Designing the digital twin solution

The facility of the future is a perfect example of a scenario where you would use a digital twin. All of the data from the facility needs to be brought into a model that can clearly describe it with reference to entities with underlying relationships. In turn, this will allow us to get the right data to the end users.

To understand the architecture requirements we need to have a clear understanding of the end users. This will differ depending on the role, location, and tasks of the end user. In this example, we will be focusing on the following end users:

- **Operational manager**: The operational manager is responsible for maintaining all operations at the facilities of the organization. They are always involved when an issue occurs and are required to manage the work of the organization. Managing the work involves assigning work orders to field technicians.

- **Field technician**: Each field technician is specialized in several components within a facility. Their work involves solving issues on-site at different locations and performing inspections and audits.

- **Warehouse planner**: The warehouse planner is responsible for maintaining an organized warehouse and maintaining equipment. They are required to keep the warehouse structure organized, order any new parts necessary for the equipment, and understand how the parts are related to each other.

So, the information required by an end user will differ depending on their role and location. While an operational manager requires only high-level information, the field technician requires information about the asset they are working on. The warehouse planner also relies on algorithms to support them when organizing the structure of all the parts in the warehouse. This requires a different type of information and the support of smart services. Providing information in this way – which is tailored to the task, location, and role of the end user – is called **contextual information**.

Another important consideration is how this information is *presented* to the end users. As we can imagine, this will differ depending on their role and the task at hand. We could use standard devices, such as mobile devices or desktop computers. But nowadays, we also have augmented reality glasses! *The facility of the future* utilizes these modern techniques.

For example, take the operational manager. Normally, this role would rely on standard flat dashboards containing the relevant information. There are several examples out there that use tools like **Power BI** to create information dashboards containing real-time data from IoT sensors. In most cases, it is possible to drill down from a global view to more detailed views for specific assets. While these dashboards can be accessed from a desktop computer or a mobile device, they lack certain abilities that can give far better insights for people. One of these limitations is that a dashboard only provides flat data. Yes, there are dashboards that can show 3D presentations of assets, but these are still on a flat screen. Viewing 3D assets on a flat screen is far less intuitive for most people than seeing the actual asset represented in real space. It also does not allow the end user to pinpoint real-time data at a specific position on the asset. Another challenge is that the user will often lose the context in which they are viewing the information. Drilling down requires moving away from the initial view of the data to a new view of more detailed data. In some situations, an end user would want to maintain the initial view of the data.

What if you could view that same dashboard information from another perspective? And what if you could use different types of devices for that? I believe strongly in visualizing data in a visual field. Today, we have extended reality. **Extended reality** is a term to describe **virtual**, **augmented**, and **mixed reality**. Mixed reality allows us to combine digital information with the real world so that it can be bound to real-life objects. For example, imagine representing telemetry data from the IoT sensors of a machine directly at the specific parts of the machine. If this was a large machine, you would be able to walk around it and view different types of data at certain points. The virtual representation of the data would stay in place, as it would be connected to the asset.

Imagine if we could have a single application for any type of role. This application would require the end user to select the role and log in to the system.

Let's go through an example scenario. Imagine the operational manager gets a notification on their phone that reports an event at one of the facilities. They put on **Microsoft HoloLens 2** and log in as the operational manager user, as shown in *Figure 16.2*.

> **Microsoft HoloLens 2**
>
> Microsoft HoloLens 2 is a self-sustained device that can project digital information such as holograms on top of the real world. We call this *mixed reality*. It is often used to support workers in scenarios where they need their hands to be free to operate equipment but also require information to support their work.

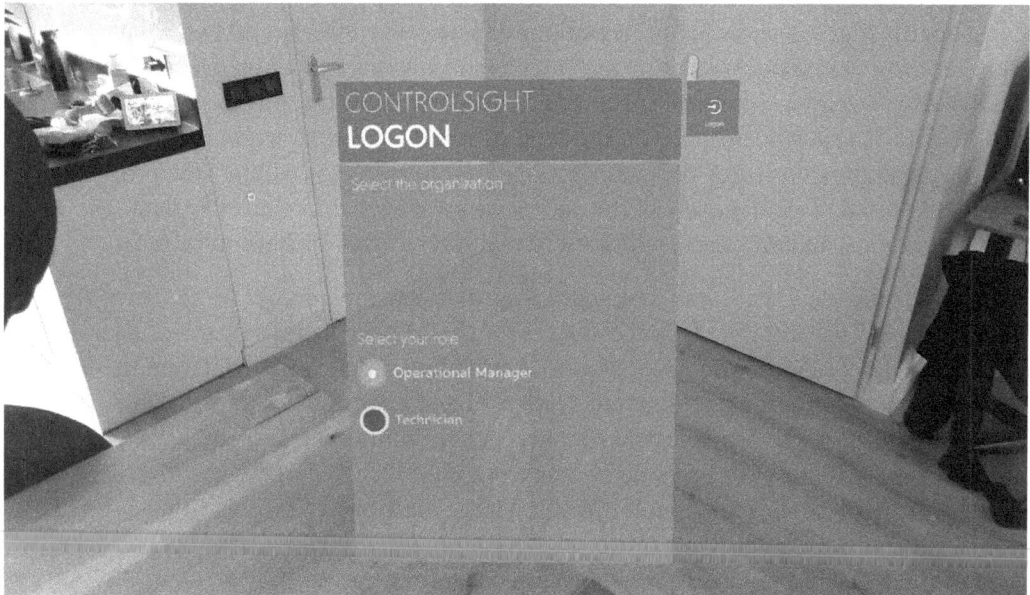

Figure 16.2 – An end user selecting a role and logging in to the system

The operational manager gets an overview of all of the facilities within the organization on a 3D map in space, as shown in *Figure 16.3*. This is the global map view for an operational manager. It shows a facility in Europe where an event has taken place. It seems that there is an issue with this facility.

Figure 16.3 – A global overview showing the organization's facilities

The operational manager selects the facility and drills down into the facility data itself. In this case, it shows a power station in detail. One of the power transformers seems to have an issue. Detailed information about the power, current, temperature, and frequency of the transformer is shown. The operational manager creates a work order and assigns it to a field technician, as shown in *Figure 16.4*. This is a simplified example that allows the operational manager to get a contextual view of the information they require. It also allows them to take certain actions when receiving an alert.

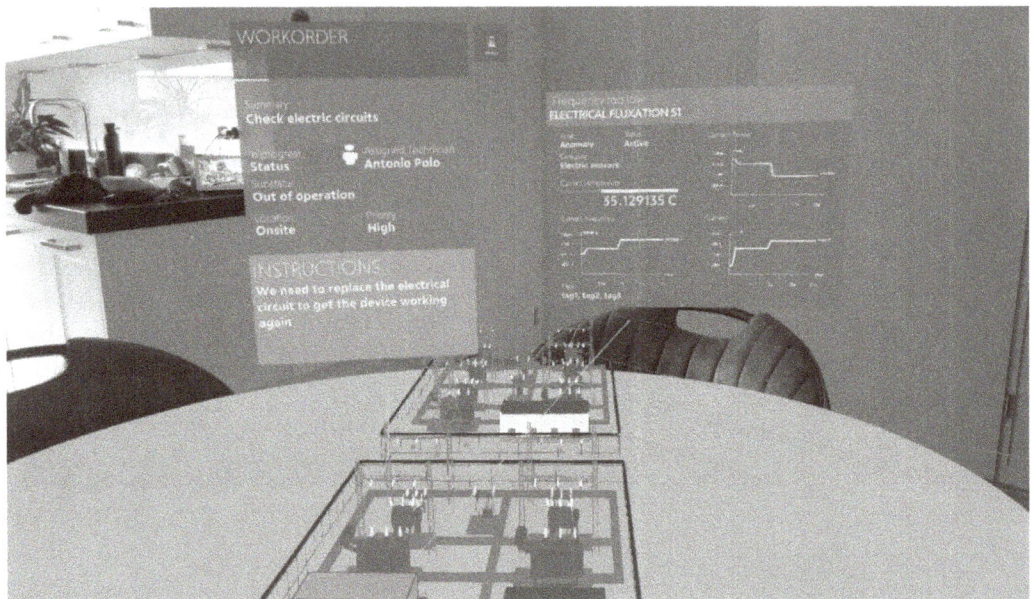

Figure 16.4 – Drilling down into the facility data to view more detailed information

At this moment, the field technician gets a notification that states that they need to go to the facility in question and work on a power transformer. The work order explains the work they need to do. After arriving at the power transformer, the field technician starts working. They require some additional information and so – just like the operational manager – they use the same application on a Microsoft HoloLens 2 device. But this time, they log in as the field technician. For a field technician, a large screen with information relevant to the power transformer is the optimum presentation of the data. Such screens can contain detailed information about the power transformer, as shown in *Figure 16.5*.

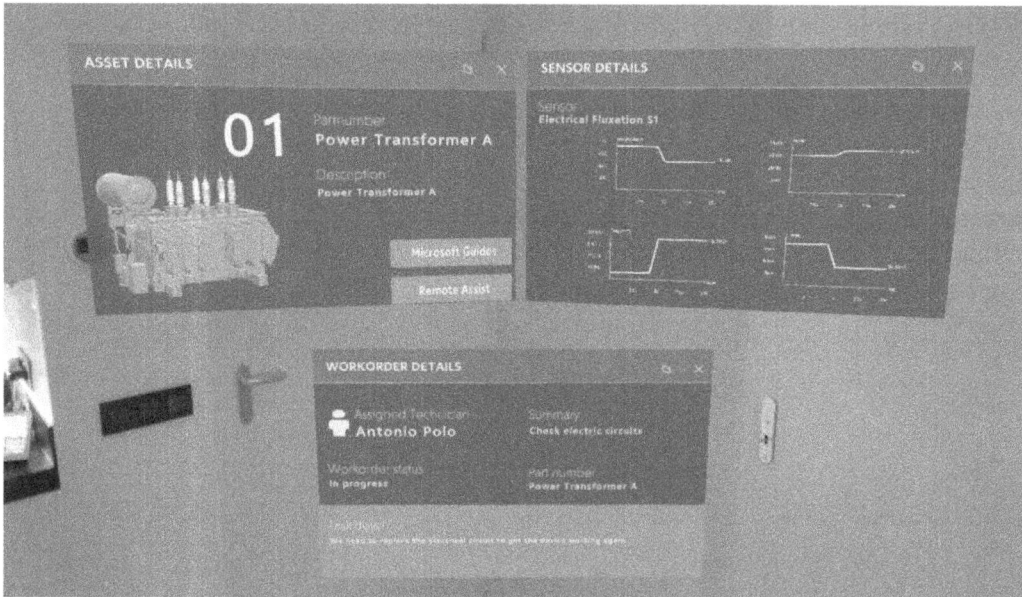

Figure 16.5 – Contextual information for a field technician working on a power transformer

While the big screens contain enough information for work to proceed, the field technician still has some issues. However, they can now use the help of an augmented training application to receive onsite guidance with step-by-step instructions, or maybe they can make a remote assistance call to an expert in the organization to help them further with over-the-shoulder support. These applications are **Microsoft Dynamics 365 Guides** and **Remote Assist**, and they are existing applications that are initiated from the application.

Microsoft Dynamics 365 Guides and Remote assist

These applications both run on a Microsoft HoloLens 2. Microsoft Guides is used for augmented training. Augmented training takes the form of step-by-step instructions, where the instructions are projected as holograms over an actual device such as a machine, in order to support a worker when executing a task. Microsoft Remote Assist allows a worker to set up a call to a remote expert. The remote expert uses **Microsoft Teams** to connect to the worker. Here, they can make annotations in the view seen by the worker and can share all kinds of information (such as documents and images) to help the worker resolve their issue.

We have now discussed how the digital twin solution will work for end users with different roles and views of the data. In the next section, we will detail a high-level design of the architecture required to implement the solution.

The solution architecture

Building a scalable digital solution will require some complex architecture, as shown in *Figure 16.6*. Let's explain the architecture from right to left. The facility has several IoT sensors that are feeding data to **IoT Central** and/or **Azure IoT Hub**. Azure IoT Hub is often used in enterprise organizations because it offers a good solution for scalable environments with large numbers of IoT sensors. Both services can send IoT data received from IoT sensors to **Azure Service Bus**. Both can also have connectivity with Microsoft Dynamics 365 Field Service to register IoT devices and IoT alerts.

Data from Field Service is synchronized with Azure Service Bus. This data could be information about assets, equipment, people, work orders, cases, and other entities within Field Service.

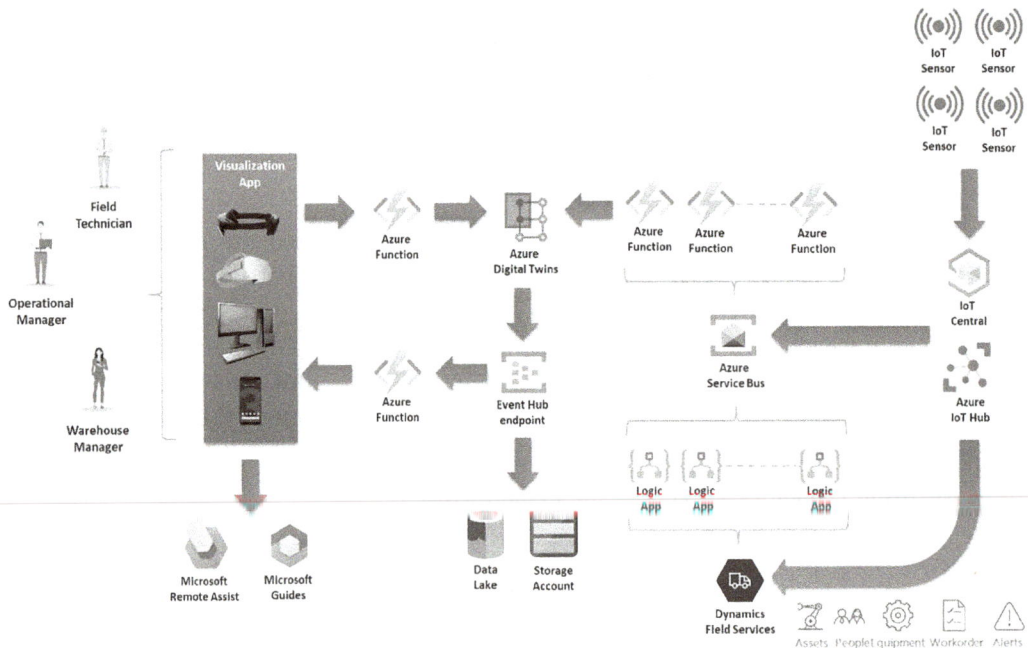

Figure 16.6 – The digital twin solution architecture

Azure Function is used to respond to messages appearing in Azure Service Bus. These messages are added, removed, and updated entities from Field Service and IoT sensors. These Azure Functions will update the digital twin depending on the available models.

Azure Event Hubs endpoints are used to expose the digital twin's data to data storage like storage accounts and data lakes. It is also used to expose the data through an API that is represented by an Azure Function.

A custom application, built for different platforms, is used to visualize the data (as shown in the previous section). Depending on the role of the user, the data is visualized in different ways. Existing applications, such as Microsoft Guides and Remote Assist, are integrated into the workflow.

Summary

In this chapter, you have learned what is required to build a digital twin solution supporting a *facility of the future* scenario. We have also learned how a *facility of the future* could be implemented using Azure Digital Twins, along with several other Azure services.

In the next chapter, we will learn about the implementation of Azure Digital Twins with smart building concepts.

Questions

As we conclude, here is a list of questions for you to test your knowledge regarding this chapter's material. You will find the answers in the *Assessments* section of the *Appendix*:

1. What is important to consider when you are designing a digital twin for a solution where you have to take into account the end users?

    a. The roles of the end users

    b. The location of the tasks

    c. The types of the tasks

    d. All the above

2. What is contextual information?

    a. Information tailored to the task, location, and role of the end user

    b. Information in the context of the work

    c. The idea that information everywhere is the same

3. Is Azure Event Bus a recommended way to send messages?

    a. Yes

    b. No

17
Creating Digital Twins for Smart Building

A **smart building** is a building where several integrated systems are used to provide vital information about it. This can then be used to automate and control a building's ecosystem, provide support services for workers, increase productivity, lower energy consumption, and more. We briefly discussed the concept of *smart buildings* in the first chapter. We also used smart building techniques in various examples throughout this book where we needed to monitor the temperature of rooms on different levels of a building. All of these exercises were examples of smart building techniques.

This chapter will go into more detail on what smart buildings are, how they can be used, and how a **digital twin** can support them.

We will cover the following main topics:

- Understanding the smart building ecosystem
- A smart building solution design
- The smart building architecture

Understanding the smart building ecosystem

Smart buildings can be used to support all kinds of office scenarios. The idea behind smart buildings is to use systems and sub-systems to collect important information that can be used to support all kinds of services. We call this the *smart building ecosystem*. But what do we mean by that?

Let's start by defining the basic components of smart buildings. There are four basic components in the smart building ecosystem:

- **Sensors**
- **Analytics**
- **User interfaces**
- **Automation**

We can illustrate these components and how they work together in *Figure 17.1*:

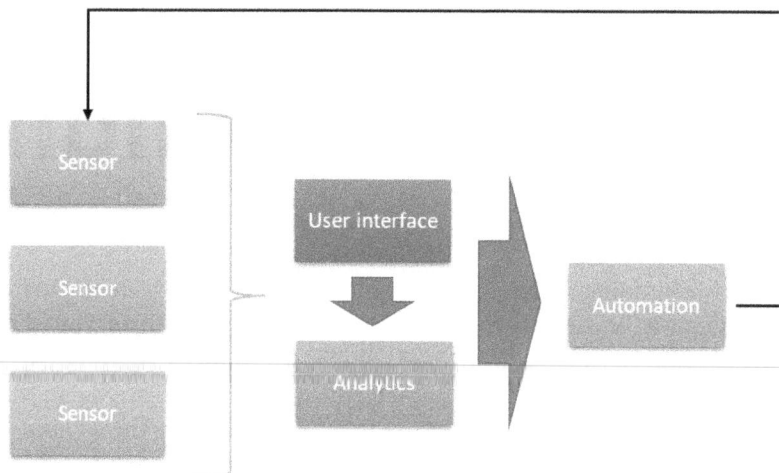

Figure 17.1 – An overview of the synergy between the basic components
of the smart building ecosystem

Each of these components is explained in more detail in the following sections.

Sensors

Sensors are devices that monitor a certain part of the environment of a building. Sensor devices can be connected to the internet or through a private network in the building. Today's sensor technology can monitor a range of physical elements, as shown in *Figure 17.2*:

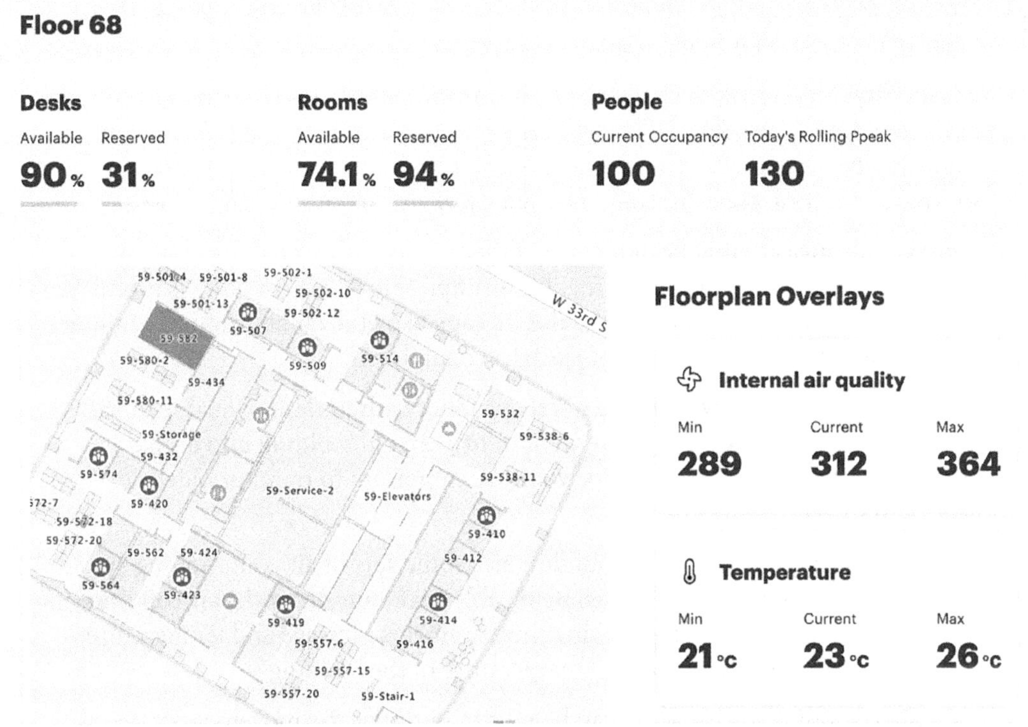
Floor 68
Desks
Available Reserved
90% 31%
Rooms
Available Reserved
74.1% 94%
People
Current Occupancy Today's Rolling Ppeak
100 130
Floorplan Overlays
Internal air quality
Min Current Max
289 312 364
Temperature
Min Current Max
21°C 23°C 26°C

Figure 17.2 – An example of different sensors providing sensor data and information

Let's look at some of these elements in detail. The following measurements are common for smart buildings:

- **Temperature**: This means measuring the temperature in rooms, floors, and other locations within the building. This allows us to identify what rooms are being used and to control heating systems or air conditioning units based on this.
- **Light**: This determines if lights are turned on or off. This information will help us decide whether to turn off lights to save energy.
- **Humidity**: This means measuring the humidity of the air. This information could inform certain actions, such as managing the air condition system or watering plants in a certain room.
- **Vibration**: This measurement can help us to identify if a certain area is being used.
- **Window shades**: Window shades protect the building from excess heat due to the sun. By bringing shades down or up based on weather conditions, the temperature within the building can be maintained or adjusted.

- **Motion**: Measuring the amount of movement in different locations allows us to identify if a location is being used by workers.

- **Air quality**: Measuring the air quality can contribute to the well-being of workers in the building. Nowadays, office windows are closed and buildings rely on air conditioning systems. This raises the risk of air pollution, so being able to open windows to refresh areas based on the quality of air can be a benefit.

- **Smoke**: This measurement can identify if there is an unexpected fire in an area. This could be caused by workers smoking, for example. However, smoking is prohibited in most areas nowadays. Smoke can also be caused by the combustion of a failing device, computer, or other piece of electrical equipment.

- **Carbon monoxide**: New buildings rarely produce carbon monoxide due to bad heating technology. However, if we want to turn an old building into a smart building, carbon monoxide sensors would be important to monitor older heating systems.

- **Air conditioning system**: Every building has some sort of air conditioning system that cools or heats rooms and areas. Being able to monitor how the air conditioning system is operating, including gaining measurements of usage and temperature, can help us to automate the overall temperature of the building.

- **Elevators**: Monitoring elevators can help us to optimize the movement of people over different floors of a building. Knowing where elevators are can allow us to move the right elevator to the right floor. This is of course influenced by the time of day. Think about how people entering the building in the morning and leaving the building in the evening could impact this.

- **Sound**: Modern buildings have quiet spaces and rooms. By monitoring sound, we can determine if these areas are providing the right service for workers. This can also be used to identify if a meeting room is being used or not.

- **Space**: This sensor operates like car park sensors that identify the available parking spots. In the case of a smart building, this can identify if a certain office desk is available to sit at. This allows the smart building to identify which office desk spaces are available for workers.

As we can see, most sensors generate data that can be combined with other data to support different types of smart building scenarios.

Several of these are standalone sensors and measure a certain ecosystem element. But some of them are related to assets or equipment. For example, we can track the use of the coffee machines on each floor to automatically create a schedule for the worker who is responsible for cleaning them.

Analytics

Analytics includes software that can analyze data from sensors to make certain decisions and/or provide vital information. This is an important part of the smart building ecosystem since it actually analyzes the data generated by the sensors. This could be real-time data or stored data from the past. The latter is called *historical data*.

Analytics is performed in two ways:

- **Business rules**: The easiest way to perform analytics is using business rules. This is a set of rules that is applied to the data coming from sensors to determine an outcome. This can be compared to running a workflow system. For example, you could have a business rule to turn off the lights in a room when there is no motion for more than 5 minutes.

- **Machine learning**: This allows us to use large sets of data to identify patterns of behavior over time to make assumptions and predict outcomes. These systems mainly use the historical data from the sensors. Based on this historical data, a predictive model is created and can make predictions based on the current sensor values. For example, think of cooling or heating a room based on a value that is predicted by a model that is built on the historical room usage data.

Analytics is an important part of the smart building ecosystem since it allows us to support one of the other basic components: *automation*.

User interfaces

User interfaces allow workers in the building to interact with the ecosystem by accessing vital information onscreen. These user interfaces help inform worker decision-making. For example, think of a screen outside each of the meeting rooms that specifies the availability and occupancy of the room. The screen allows a worker to reserve the room at a certain time for a certain period. If we start adding analytics to this screen, we would be able to identify scheduled rooms to be made free, as we would know that in the past, these scheduled rooms were not occupied at specific times. This often happens in large organizations where repeatable meetings are scheduled in meeting rooms. Often, these meeting rooms are unoccupied if a meeting was canceled but not adjusted in the reservation system.

Another example of a user interface is one that is used for elevators. These screens only allow you to enter the floor number that you want to go to. The elevator uses analytics to determine which elevator is required to move you from your current floor to the destination floor. In practice, these elevators are more efficient than ones where people enter first and then specify which floor they want to go to. In this scenario, the elevator knows the destination floor number upfront and can make the right decisions to optimize its use.

Automation

The *automation* component can make decisions based on the information from sensors and the analyzed data by using predefined business rules. Some complex ecosystems are even able to dynamically change their business rules based on historical data. While decisions are mostly made by the analytics component, the automation component uses the analytics outcomes to control the different elements of the smart building ecosystem. Automation allows us to control the windows shades, air conditioning, lights, and many more elements within the ecosystem. It is important to understand that automation can influence the system and its subsystems to get new measurements that will result in new outcomes.

An example of this is lowering the temperature in a room when it is occupied due to sensors that measure movement. The room being used will heat up so the system will start lowering the temperature in order to keep the room at the same temperature.

We have now learned about the smart building ecosystem and how this can help us to make a *smart building* by using *sensors*, *analytics*, and *automation*. In the next section, we will go through a solution design for an example of a smart building.

A smart building solution design

Let's use a **Contoso** building as our example. Imagine that this building was recently built and is state-of-the-art, containing several sensors for different purposes. The Contoso building is what we call a *smart building*. It can maintain and control itself by using sensors, analytics, and automation to support the workers that occupy it.

The Contoso building saves energy by turning off lights where workers are not active and changes the cooling system parameters for empty rooms to prevent over-cooling. It also controls the temperature in the building by bringing the window shades down when the outside temperature and the amount of sunlight is increasing. Although windows can be opened, workers rarely need to, as the building keeps the temperature constant.

The Contoso building is also a *green building*. Several areas contain a large number of plants that are maintained automatically. Humidity is measured to inform the amount of water needed for the plants. The green walls contribute to a safer and healthier environment for the workers.

Smart spaces are part of the Contoso building. This means that the building uses sensors to detect which rooms are occupied based on the schedules and the actual presence of workers. The building can also detect if office desks are available. A worker entering the building can easily use an app to check for available desk spaces.

The building uses several elevators. The elevators use a complex algorithm based on predictiveness to identify which elevator needs to move to which floor to minimize the waiting times for workers. This also saves energy by avoiding having elevators move around for no reason. The algorithm uses historical data to determine when the most crowded moments are, for example, the times when workers enter or leave the building. The algorithm can also use the meeting room schedules to decide when and where elevators should be available.

A lot of assets in the building are also controlled by the ecosystem. This includes the coffee machines on each floor, which need regular checkups and refills. The schedule of the worker responsible for their maintenance is managed by the ecosystem of the building, and this utilizes data from multiple sensors, including the occupancy of office desks, the meeting room schedules, and the machine itself. The same goes for the use of other assets, such as printers. By knowing how often they are used, we can make sure there is always enough paper available at the printer location and ink cartridges are replaced promptly.

The Contoso building is all about *making the lives of workers easier, becoming more sustainable by saving energy*, and *making the workplace safer, healthier, and more fun to be in*.

In this section, we described a smart building concept that uses a smart building ecosystem. In the final section, we will describe what the high-level architecture would look like for this concept.

The smart building architecture

This building scenario is a perfect use case for **Azure Digital Twins**. The building is using a large set of sensors that are connected to different levels, rooms, and desk spaces in the building. Also, sensors from assets across the building are part of the system. Buildings such as the Contoso building contain thousands and thousands of sensors that need to be connected. For that, **Azure IoT Hub** is the perfect Azure service. It is scalable enough for large sets of sensors to transport their data to several other locations within Azure using **Azure Service Bus**. Messages are posted on different queues within Azure Service Bus to move the data to an Azure Digital Twins instance and to **Azure Machine Learning**.

The Azure Digital Twins instance contains a comprehensive model and set of business rules to describe the whole ecosystem of the smart building. It contains a clear model that contains entities to describe each of the floors, rooms, desk spaces, elevators, plants, and other assets in the building, and it connects the data coming from the sensors to all of these entities. The digital twin gives us the ability to easily query the data that is required to feed Azure Machine Learning, and it allows us to automate several different processes in the building, for example, watering plants, changing a maintenance worker schedule, and more.

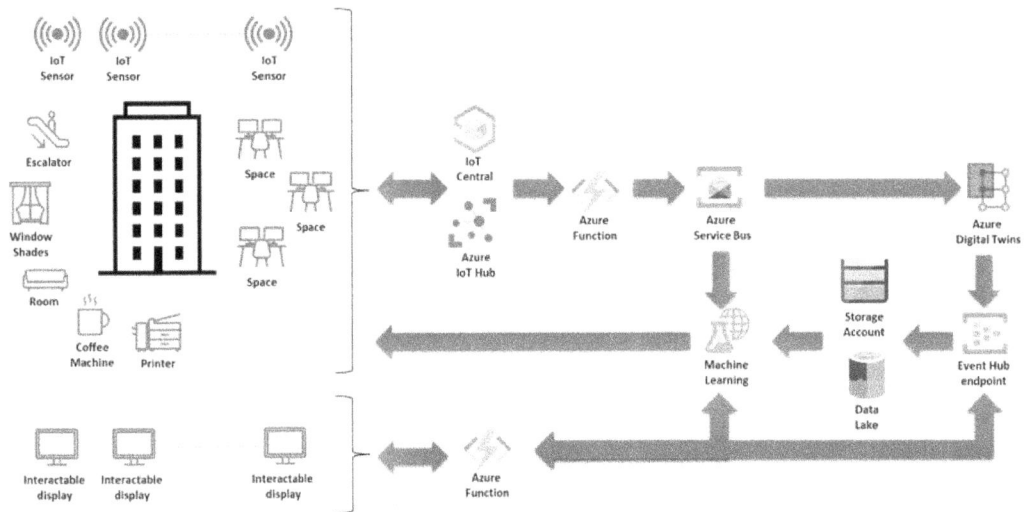

Figure 17.3 – An overview of the architecture of a smart building ecosystem using a digital twin

Azure Machine Learning allows us to build simple to complex artificial intelligence solutions based on models. These models contain algorithms to enable predictiveness, among other things. Large sets of data are used to train the models. A trained model can predict outcomes using input values from our ecosystem.

Azure Event Hubs endpoints are used to expose the Azure Digital Twins data to data storage like storage accounts and data lakes. It is also used to expose the data through an API that is represented by **Azure Functions**.

Data from the Azure Event Hubs endpoints and Azure Machine Learning is used to provide information for interactive displays throughout the building. These displays allow workers to control and influence some parts of the smart building ecosystem. Their actions feed back into the system to inform the predictiveness of the models in Azure Machine Learning and to update the digital twin through the endpoints of Azure Event Hubs.

In this section, we have learned what kind of high-level architecture would be required to set up the smart building ecosystem for our Contoso smart building example.

Summary

In this chapter, we learned about the smart building ecosystem. We learned about the basic components of the ecosystem and how they function in a smart building. We used an example to explore a smart building solution design and its high-level architecture.

In the next chapter, we will learn about *simulation*. Simulation often uses digital twins to support different scenarios with regard to work preparation and training, and it allows us to practice work before moving to an actual production line environment.

Questions

As we conclude, here is a list of questions for you to test your knowledge regarding this chapter's material. You will find the answers in the *Assessments* section of the *Appendix*:

1. What are the four basic components of the smart building ecosystem?

    a. Sensors, analytics, user interfaces, and automation

    b. Sensors, training, user interfaces, and data processing

    c. Sensors, user interfaces, automation, and learning

2. What would be the main reason to control lights in a building?

    a. To make sure that lights are turned off. Workers often forget to turn them off.

    b. To save energy and become more sustainable.

    c. No light is required when someone is not in the room.

3. What service will help us predict behavior in a building?

    a. Azure Machine Learning

    b. Azure Service Bus

    c. Azure Event Hubs

18
Simulations Using a Digital Twin

This chapter gives a better understanding of what simulation is and how simulation can benefit from using a digital twin. There are many forms of simulation. While we explain most of them at a concept level, we will deep dive into some of them by using real-life examples. In each of these examples, digital twins are used to support simulation in a certain way. This chapter will help you understand more about simulation in combination with a digital twin.

This chapter contains the following sections:

- Understanding simulation
- Solution design and architecture

Understanding simulation

Simulation is the imitation of some operation or process in the real world over time. Simulations are built based on two main core elements:

- **Model** – The model describes the operation, process, or system with its characteristics and behaviors. An example would be a machine as the system we want to simulate. Characteristics would be the parameters that can be adjusted on the machine. Behaviors describe how the machine acts when changing these parameters.

- **Simulation** – Simulation describes the evolution of the **model** over time. Taking the same example as above with a machine as a system, simulation would define how the outcome parameters of the machine would change during a certain period. These outcome parameters will specify how the machine is operating. Think of how the heating or rotation of a machine evolves over time when setting certain characteristics in the **model**.

There are so many examples of simulation in our world. We will try to organize them in a set of categories and explain each of them at a concept level. Examples of some of the categories will be given in the upcoming section in this chapter.

Category	Concept description
Testing	Testing is one of the most common scenarios when it comes to simulation. Simulation allows us to test new characteristics of a process or system over a certain period. The process or system is imitated using a model or other means. These characteristics will give us insight into what effect these characteristics have on that process or system. Such outcomes can help us tune and optimize performance before adding these characteristics to the actual process or machine in real life.
Training	In this case, we talk about training (updates) that support workers in performing uncommon tasks and in becoming compliant. Simulation can help training by imitating a certain process or system that is normally difficult to reach or dangerous. Processes and systems that are difficult to reach can be imitated at another more reachable location. Processes and systems that are normally dangerous can be imitated in a safe area to perform training.
Education	Workers are educated to gain expertise to be able to execute their daily work. Simulation allows creating controlled and safe environments where workers are educated before they move to the actual process or system. In some situations, simulation can partly replace the expertise required to educate the worker.
Work preparation	This is an interesting one since it comprehends the process of the preparation of work around a process or system. Simulation allows us to see all kinds of different outcomes when changing the characteristics to understand what the work will contain.

Work preparation	This is an interesting one since it comprehends the process of the preparation of work around a process or system. Simulation allows us to see all kinds of different outcomes when changing the characteristics to understand what the work will contain.
Engineering	Engineering is about the creation of a new process or system. It means that the process or system does not yet exist. Simulation can help to model the process or system to view its characteristics during the design. This allows us to make modifications to the design before the process or system is built.
Impact analysis	Simulation can be used to analyze the impact on a specific system, and is, in some situations, a dependent system, by changing the characteristics of the system causing the impact. Imagine a cascading system that measures the impact of each system or imagine systems that impact each other. These simulations can become extremely complex but can help you to understand the impact, which would normally be impossible to oversee.

We have learned about the two basic core components needed to do a simulation. We also went through several categories of types of simulation for a process or system. In the next section, we will go into more detail with several examples of these categories.

Solution design and architecture

In this section, we will be going into several solution designs using examples based on the categories mentioned in the previous section.

Work preparation

We have a government organization, Contoso Safety and Health, focusing on safety and health during large events. The organization needs to be sure to position safety and health posts accordingly to support different types of safety challenges during the event.

The organization is using a simulated model that visualizes the location of the event. Information from different sources is pulled in and layered on top of the visualized location to give the organization a better understanding of what they are against. Historical crowd data of the event is being used to get a better understanding of how the crowd moves during the day. All that information together will allow them to pinpoint the best locations during the day where safety and health posts need to be installed. While in normal situations, without simulation, locations are somewhat strategically selected. This simulation allows them to select strategic locations that can even differ over time during the day.

Digital twins are a great way of supporting such a simulation. The digital twin will collect all the data from the different sources and bring it together in an understandable entity model. This model will allow us to query the data, which is then visualized to the people who are involved in doing the work preparation.

And we can even move further. By allowing the people using the work preparation tool to make annotations over time, important information can be stored and made available during the day to support all the posts at the event. This could, for example, be the number of people needed at a certain post during the day or the expected health materials needed to keep the inventory at a post sufficient to support all issues.

An overview of a possible high-level architecture for this scenario is shown in *Figure 18.1*. An application for desktop and mobile shows the environment and its layers and allows the user to interact with the model and add annotations and vital information needed during the event. The application is using Azure Maps to visually generate the environment by using Bing Maps. The layer information is retrieved through the digital twin, which uses an entity model storing all the layers, annotations, and information. The layer data comes from different data sources and is stored in Azure storage using Azure Functions and Azure Service Bus.

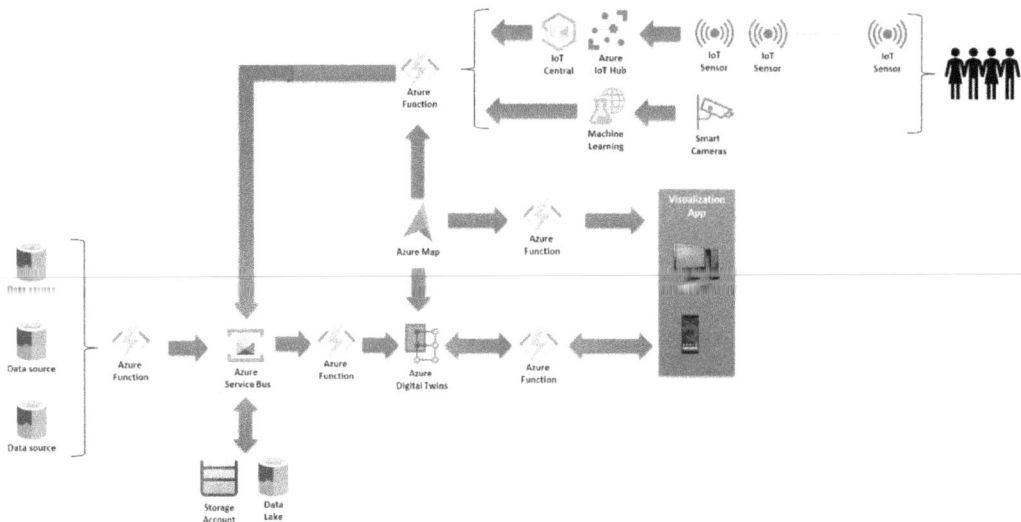

Figure 18.1 – Possible high-level architecture overview for work preparation

Crowd data is collected through IoT sensors and smart cameras that use machine learning to identify people and their movements. That information is synchronized with data from Azure Maps and stored in Azure storage using Azure Functions and Azure Service Bus.

Training

We have an organization called Contoso Hospital Training. Contoso Hospital Training delivers simulated training for safety staff in all hospitals. This involves areas that are often not available to perform the yearly safety training, which is required to be safety compliant. These areas are, for example, operating rooms. Using the actual room will have an enormous impact on a hospital. Besides the operating room not being available during training, it needs to be cleaned afterward to make it operational again.

Contoso Hospital Training has created a simulated environment of these operating rooms in virtual reality. It allows safety staff to train on different scenarios involving fire, explosions, and removing **hazmat (hazardous materials)** in virtual reality.

The process of simulating such a scenario and the environment itself is stored with all its characteristics in an entity model as part of a digital twin. The digital twin is used for the two main parts of the process. It is used to keep track of the environment and all the assets in there such as, for example, a fire extinguisher. And the digital twin is used to quickly identify whether the training has been performed successfully by the safety staff to create a valid report. This report is used by the hospital to prove its safety compliance and regulation.

A possible high-level architecture for this scenario is shown in *Figure 18.2*. The architecture in this case is not very complicated since we are not using sensor data. Azure Digital Twin is only used to store a copy of the environment and its assets and related characteristics and the scoring of safety staff executing the training.

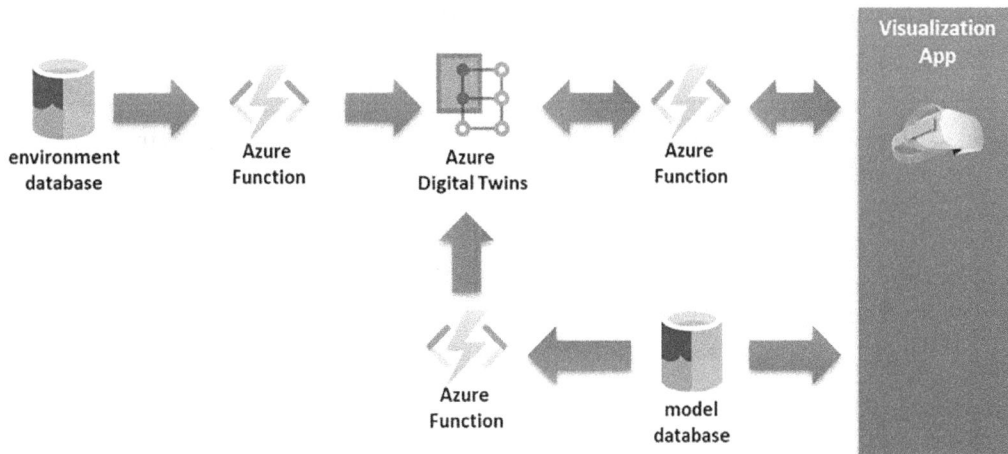

Figure 18.2 – Possible high-level architecture for the scenario

The visualization app is using the information from Azure Digital Twins and the model database to build and simulate the virtual environment.

Testing

Let's introduce the company Contoso Cars. Contoso Cars is using simulation to test and optimize their production lines. A production line works at its best based on how workstations are positioned with workers, materials, and machines. The architectural design of these workstations is of the utmost importance. The materials need to be positioned perfectly for a worker to pick them up and turn toward the production line to place the material onto the car. And all of that requires you to have a safe and healthy environment. This means preventing the worker from bending over, which could cause back injuries over time.

The simulation involves a person using a VR suit and a VR device in a room. The VR suit sends over all movements of that person to the system. The system projects the person into virtual reality together with all the materials and equipment of the workstation. The person can watch the workstation in VR through virtual reality glasses. Another person positions the materials and equipment in the workstation, allowing the person using the VR suit to try out whether they can move around easily and perform the work. The outcome of the simulation will define each exact position of the materials and equipment as part of the workstation aligned to the production line.

In this scenario, the digital twin is representing the digital workstation and continuously collecting all the data from the digital visualized workstation and the movement of the person wearing the VR suit. The entity model incorporates the ability to store the data based on a timestamp, which allows them to store data over time.

A possible high-level architecture is shown in *Figure 18.3*. This architecture shows that different sources of information are routed through Azure Service Bus to an Azure Digital Twin; sources such as the materials and the asset database, haptic feedback from the VR suit, and information about the models representing the materials and assets.

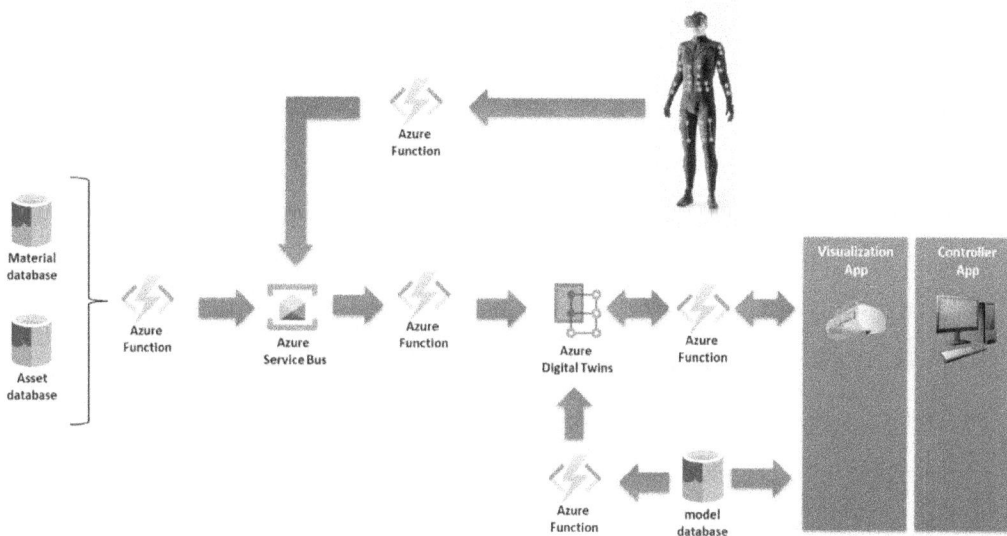

Figure 18.3 – Possible high-level architecture for the scenario

Both the app for the wearer of the suit and for the person who is making changes communicate with Azure Digital Twins. The Azure Digital Twin uses entities to store all the relevant information for the materials and assets, which contains the positions defined by the second person.

Summary

In this chapter, we have learned about the two core elements of a simulation and the different categories. We have several different forms of simulation and several of them were explained in more detail by using a real-life example and how it relates to the use of a digital twin.

This was the last chapter of the book. I want to congratulate you on finishing the book and learning everything you need to know to start working with Azure Digital Twins. I hope you enjoyed the book and that it will help you in creating your digital twin solutions.

Questions

As we conclude, here is a list of questions for you to test your knowledge regarding this chapter's material. You will find the answers in the *Assessments* section of the *Appendix*:

1. What are the two core elements of a simulation?

    a. Model and visualization

    b. Model and simulation

    c. Model and entities

2. What does behavior do within a simulation?

    a. It describes how a machine acts when characteristics are changing over time.

    b. It describes how the simulation visualizes the data.

    c. It describes the movement of a person during simulation.

3. Can any simulation benefit by using a digital twin?

    a. Yes

    b. No

Assessments

In the following pages, we will review all the practice questions from each of the chapters in this book and provide the correct answers.

Chapter 3 – Digital Twin Definition Model

1. c. Five levels
2. b. Semantic
3. c. The validation process creates an overview of your model

Chapter 4 – Understanding Models

1. b. `DigitalTwinsClient`
2. a. Setting the **Copy to Output** directory to **Always copy**
3. a. `DigitalTwinsClient.UpdateDigitalTwin<T>`

Chapter 5 – Model Elements

1. a. Property
2. c. The real-time object is part of another whole and cannot be seen as a self-sustained object.
3. a. C# event

Chapter 6 – Creating Relationships between Azure Digital Twin Models

1. b. `JsonPatchDocument`
2. c. Relationship names
3. a. Yes

Chapter 7 – Querying Digital Twins

1. c. `Pageable<>` class
2. b. `exact`
3. b. It allows me to implement paging and is more suitable for production.

Chapter 8 – Building Models Using Ontologies

1. a. Adopt, Conversion, Author
2. b. Azure Digital Twins Model Uploader
3. a. WebVOWL

Chapter 9 – APIs and SDKs

1. a. An API is used by an SDK to use a service.
2. a. Postman
3. c. To prevent getting results back from the digital twin.

Chapter 10 – Building a Digital Twin Pipeline

1. b. Azure IoT Central application
2. a. An IoT development kit
3. c. `iotcentral-device-id`

Chapter 11 – Updating the model

1. b. Logstream
2. a. `local.settings.json`
3. c. `EntityPath`

Chapter 12 – Event Routing

1. a. Upstream services
2. c. EventHub and Azure Function.
3. c. Subscribing with an Azure Service Bus queue.

Chapter 13 – Setting up Azure Maps

1. b. Azure Creator service

2. b. Yes

3. a. `https://eu.atlas.microsoft.com`

Chapter 14 – Integrating Azure Maps

1. a. Event Grid trigger

2. a. No

3. c. Log stream

Chapter 15 – Monitoring and Troubleshooting

1. b. It is one of the resources that can be used to collect data from different resources.

2. c. Logs and metrics

3. a. Yes

Chapter 16 – Facility of the Future

1. d. All the above

2. a. Information tailored to the task, location, and role of the end user

3. a. Yes

Chapter 17 – Creating Digital Twins for Smart Building

1. a. Sensors, analytics, user interface, and automation

2. b. To save energy and become more sustainable.

3. a. Azure Machine Learning

Chapter 18 – Simulations Using a Digital Twin

1. b. Model and simulation

2. a. It describes how a machine acts when characteristics are changing over time

3. a. Yes

Index

Symbols

.NET Console application
  creating 82-88
.NET SDK 198

A

access token
  configuring 204
alerts
  using 373, 374
analytics
  about 395
  business rules 395
  machine learning 395
API metrics
  monitoring 213-215
API permissions 184
application architecture 222
Application Programming
    Interface (API) 198
application settings, AzureMapUpdater
  configuring 349-351
architecture requirements, for
    digital twin solution
  field technician 382

  operational manager 382
  warehouse planner 382
Autodesk viewers
  reference link 307
automation 396
Azure account 26
Azure Active Directory (AAD) 26, 183
Azure Analytics 21
Azure CLI
  used, for managing Azure
      Digital Twins 216
Azure Content Delivery Network
    (Azure CDN) 356
Azure Digital Twins
  about 14, 397
  costs 17
  managing 19
  open modeling language 15
  service 15
  terminology 15
Azure Digital Twins, architecture
    components
  about 18
  Azure Analytics 21
  Azure Digital Twins, managing 19
  Azure Function 19

Azure IoT Hub 20
Azure IoT Hub DPS 20
Azure Logic Apps 21
Azure Service Bus 22
Azure Storage 21
Azure Digital Twins Data Owner 183
Azure Digital Twins Explorer
  about 41
  compiling 44-48
  digital twin models, uploading
    into 49-54
  downloading 41-43
  installing 43, 44
  URL 41
  used, for creating digital twin 48
Azure Digital Twins instance
  querying, with REST API 210, 211
  retrieving, with Postman 205
Azure Digital Twins REST API
  about 199
  exploring 22, 23
Azure Digital Twins service
  about 26
  access control, configuring 33, 35
  creating 29-33
  resource group 26-29
Azure Event Hubs 388, 398
Azure Function
  about 19, 223, 388, 398
  connection string, setting 257, 266
  creating, with Visual Studio 249-256
  publishing 262, 264, 266, 345-349
  update, monitoring 353-355
  update, setting up 334-344
Azure function permissions
  granting 261

Azure function placeholder
  creating 257-261
Azure IoT Central
  about 222
  using, to set up demo
    sensor 224-230, 232
Azure IoT Central application
  data export, creating in 238
  telemetry data, creating from 237
Azure IoT Hub 16, 20, 388, 397
Azure Logic Apps 21
Azure Machine Learning 397
Azure Maps
  about 300
  account, creating 301-303
Azure Maps software development
  kit (SDK) 356
Azure portal
  reference link 224
Azure Service Bus
  about 22, 388, 397
  sensor messages, obtaining on 232-240
Azure Service Bus queue 223
Azure Service Bus resource
  creating 234
Azure Storage 21
Azure Time Series Insights Gen2 21

C

CloudEvents standard 278
CNCP Serverless workgroup
  reference link 278
collection
  about 201
  access token, configuring 204
  importing, into Postman 203
complex properties 113-115

complex schemas
  about 68
  Array 71, 72
  Enum 72, 73
  fields 69
  Map 70
  Object data type 69
components 122, 124-126
computer-aided design (CAD) 306
contextual information 382
Contoso building 396
Contoso Cars 406
Contoso Hospital Training 405
control plane 201, 202
Creator resource
  creating 304-306

D

data egress 274, 277
data export
  creating, in Azure IoT Central
    application 238
data ingress 274, 277
data plane
  access token, obtaining 209
  base URL, configuring 210
dataset
  about 207, 301
  creating 319-321
  imported collection 208
  validating 319-321
delegate function 170
demo graph
  setting up 150-154
demo sensor
  setting up, with Azure IoT
    Central 224-230, 232

developer landscape 198
Device Provisioning Service (DPS) 20
diagnostic settings
  setting up 365-368
digital replica 4
Digital Twin
  creating 99-101
  creating, for sensor model 268
  creating, with Azure Digital
    Twins Explorer 48
  concept 4
  deleting 103, 104
  digital replica 4
  entity 5
  managing 99
  obtaining 104, 105
  reality 5
  relationships 5
  result, viewing 269, 270
  updating 101-103, 244
Digital Twin environment
  exploring 6-8
Digital Twin Graph 49
Digital Twin interface
  about 59-61
  primitive schemas 67, 68
  schema 67
Digital Twin interface, content
  about 62
  component 66, 67
  properties 62, 63
  relationship 64, 65
  telemetry 63, 64
digital twin model identifier (DTMI) 60
Digital Twin models
  uploading, into Azure Digital
    Twins Explorer 49-51, 53, 54

Digital Twins Definition
    Language (DTDL)
  about 15, 58, 80, 177
  JavaScript Object Notation for
    Linked Data (JSON-LD) 59
  versioning 59
digital twin solution
  architecture 388
  architecture, requirements 382
  designing 382, 383
  example scenario 384, 386, 387
digital twins relationship
  about 130-132
  creating 132-135
  deleting 141, 142
  list of relationships, obtaining 137-141
  obtaining 136
  properties 143, 144
  properties, using 145-147
  single relationship, obtaining 136, 137
downstream services 274
DWG format 306

E

endpoint
  creating 284-286
entity 5
Event Grid
  about 281
  subscribing 351-353
Event Grid topic
  creating 282-284
event messages
  subscribing 288-293
event notifications 277-279

event route
  about 280, 281
  creating 286-288
  messages, monitoring 294-296
extended reality 383

F

facility of the future concept
  about 380
  principles 380, 381
  scenario 380
  ultimate goal 381
fast Compound Spring Embedder
    (FCoSE) 155
feature stateset
  about 301, 308
  creating 325-327
  updating 332, 333
  validating 325-327
filter events
  reference link 287
functional limits 217

G

General Data Protection
    Regulation (GDPR) 300
Geospatial schemas 73, 74

H

hazardous materials (hazmat) 405

I

identity and access management
    (IAM) 183
IEEE Standard 68
indoor map model
    visualizing 356-359
industry-standard ontologies
    using 179-182
interface 198
Internet of Things (IoT) 14, 380
IoT Central 16, 388

J

JavaScript Object Notation for Linked
    Data (JSON-LD) 15, 59, 80
JavaScript Object Notation
    (JSON) 15, 58, 326, 333

L

limits
    preventing 218
    situation, handling 218
log analytics workspace
    setting up 362-365
logs
    viewing 369, 370

M

machine learning (ML) 381
map
    building 306-309
    converting 316-319
    uploading 310-316

mebibyte (MiB) 305
metrics
    viewing 370-372
Microsoft Dynamics 365 Field Service 381
Microsoft Dynamics 365 Guides 387
Microsoft HoloLens 2 384
Microsoft Teams 387
Microsoft Visual Studio
    installing 35-38
    URL 35
mixed reality 383
models
    creating 89, 91, 92
    deleting 96
    designing 80, 81
    inheriting from 94, 95
    managing 88
    multiple model, creating 92, 93
    obtaining 97, 98
    recommendations 81
    smart build, modeling 80
    validating 74-77
MXCHIP device
    about 227
    reference link 227

N

Node.js
    installing 40, 41
    URL 41
NuGet packages
    Azure.DigitalTwins.Core 84
    Azure.Identity 84

O

ontologies
  about 176, 177
  using, advantages 176, 177
ontology models
  uploading 183-191

P

platform as a service (PaaS) 15
Postman
  about 199
  collection, importing into 203
  URL 199
  used, for retrieving Azure Digital
    Twins instances 205
  workspace, creating 200
Power BI 383
primitive properties 108-112
primitive schemas 67, 68
projection 164

Q

query
  basic 154-159
  executing, with code 166-169
query asynchronous calls
  with code 170-172
query by model 160-162
query relationships 162, 164
query results
  filtering 164, 166

R

rate limits 217
reality 5
real-world applications
  about 9, 10
  education 11
  historical data 12
  insight and control 13, 14
  simulation 12
  smart building 10, 11
relationships 5
Remote Assist 387
Representational State Transfer
    (REST) 199
Resource Description
    Framework (RDF) 59
resource group
  creating 26-29
REST API
  using, to query Azure Digital
    Twins instance 210, 211
REST API call 199
RFC 3339 68
Role-based access control
    (RBAC) 33, 35, 300

S

schemas 67
semantic types 74
Sensor 278, 392-394
sensor messages
  obtaining, on Azure Service
    Bus 232-240

sensor model
  digital twin, creating for 268
  updating 244, 245
sensor elements
  air conditioning system 394
  air quality 394
  carbon monoxide 394
  elevators 394
  humidity 393
  light 393
  motion 394
  smoke 394
  sound 394
  space 394
  temperature 393
  vibration 393
  Window shades 393
Service Bus Explorer
  using, to view incoming messages 240
service limits 217
shared access policy 223
simulation
  about 401
  examples 402, 403
simulation, core elements
  model 401
  simulation 402
smart building
  architecture 397-399
  ecosystem 391
  solution design 396, 397
smart building, components
  about 392
  analytics 395
  automation 396
  sensors 392, 393
  user interfaces 395

smart spaces 397
Software Development Kit (SDK)
  about 198
  based on REST API, for Azure
    Digital Twins 212
solution design and architecture
  about 403
  testing 406, 407
  training 405
  work preparation 403, 404
source digital twin 130
stateset identifier (ID) 341
storage account
  creating 245, 246, 248, 249
strategy
  modeling 177, 178
StylesObject Schema
  reference link 327

T

target digital twin 130
telemetry 116, 117, 119-121
telemetry data
  creating, from Azure IoT
    Central application 237
three-dimensional (3D) 306
tileset
  about 301
  creating 322, 324, 325
  validating 322, 324, 325
two-dimensional (2D) 306

U

Uniform Resource Locator
    (URL) 305, 333
upstream services 274
user interfaces 395
UTF-8 68

V

Visual Studio
  used, for creating Azure
    Function 249-256

W

Web-based Visualization of
    Ontologies (WebVOWL) 180
Web Ontology Language (OWL) 180
Windows Azure CLI
  installing, with Windows
    PowerShell 39, 40
Windows PowerShell
  access token, obtaining for
    control plane 203
  used, for installing Windows
    Azure CLI 39, 40

Packt>

Packt.com

Subscribe to our online digital library for full access to over 7,000 books and videos, as well as industry leading tools to help you plan your personal development and advance your career. For more information, please visit our website.

Why subscribe?

- Spend less time learning and more time coding with practical eBooks and Videos from over 4,000 industry professionals

- Improve your learning with Skill Plans built especially for you

- Get a free eBook or video every month

- Fully searchable for easy access to vital information

- Copy and paste, print, and bookmark content

Did you know that Packt offers eBook versions of every book published, with PDF and ePub files available? You can upgrade to the eBook version at packt.com and as a print book customer, you are entitled to a discount on the eBook copy. Get in touch with us at customercare@packtpub.com for more details.

At www.packt.com, you can also read a collection of free technical articles, sign up for a range of free newsletters, and receive exclusive discounts and offers on Packt books and eBooks.

Other Books You May Enjoy

If you enjoyed this book, you may be interested in these other books by Packt:

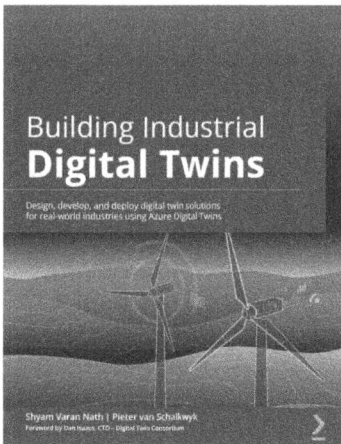

Building Industrial Digital Twins

Shyam Varan Nath | Pieter van Schalkwyk

ISBN: 978-1-83921-907-8

- Identify key criteria for the applicability of digital twins in your organization
- Explore the RACI matrix and rapid experimentation for choosing the right tech stack for your digital twin system
- Evaluate public cloud, industrial IoT, and enterprise platforms to set up your prototype
- Develop a digital twin prototype and validate it using a unit test, integration test, and functional test
- Perform an RoI analysis of your digital twin to determine its economic viability for the business
- Discover techniques to improve your digital twin for future enhancements

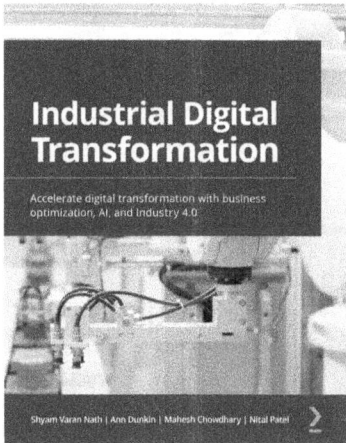

Industrial Digital Transformation

Shyam Varan Nath | Ann Dunkin | Mahesh Chowdhary | Nital Patel

ISBN: 978-1-80020-767-7

- Get up to speed with digital transformation and its important aspects
- Explore the skills that are needed to execute the transformation
- Focus on the concepts of Digital Thread and Digital Twin
- Understand how to leverage the ecosystem for successful transformation
- Get to grips with various case studies spanning industries in both private and public sectors
- Discover how to execute transformation at a global scale
- Find out how AI delivers value in the transformation journey

Packt is searching for authors like you

If you're interested in becoming an author for Packt, please visit `authors.packtpub.com` and apply today. We have worked with thousands of developers and tech professionals, just like you, to help them share their insight with the global tech community. You can make a general application, apply for a specific hot topic that we are recruiting an author for, or submit your own idea.

Share Your Thoughts

Now you've finished *Hands-on Azure Digital Twins*, we'd love to hear your thoughts! Scan the QR code below to go straight to the Amazon review page for this book and share your feedback or leave a review on the site that you purchased it from.

`https://packt.link/r/1801071381`

Your review is important to us and the tech community and will help us make sure we're delivering excellent quality content.

www.ingramcontent.com/pod-product-compliance
Lightning Source LLC
Chambersburg PA
CBHW080135220326
41598CB00032B/5075